Model Driven Architecture with Executable UML

This book offers a unique insight into a revolution in software development that allows model specifications to be fully and efficiently translated into code. Using the most widely adopted, industry-standard software modelling language, Unified Modeling Language (UML), the reader will learn how to build robust specifications based on the Object Management Group™'s (OMG™) Model Driven Architecture™ (MDA™). From there, the authors describe the steps needed to translate the executable UML, (xUML) models to any platform-specific implementation. The benefits of this approach go well beyond simply reducing or eliminating the coding stage – it also ensures platform independence, avoids obsolescence (programming languages may change, the model doesn't) and allows full verification of the models by executing them in a test-and-debug xUML environment. This is an excellent reference for anyone embarking on what is surely the future of software development for medium and large scale projects.

The authors are all experienced practitioners of the techniques and processes described in the book. They have worked in industry sectors such as defence, aerospace, automotive, telecommunications, government, health, insurance and process control, and have seen the wide-ranging applicability of object-orientation, executable modelling and Model Driven Architectures. They have worked with clients to apply executable modelling and code generation in systems ranging from small real-time embedded systems to massive distributed systems containing hundreds of thousands of lines of generated code. They know from first hand experience over several years that executable modelling and code generation works. They have also learnt how to make it work most effectively ad where caution is required. The aim of this book is to back up the technical details with this practical experience.

Model Driven Architecture with Executable UML

Chris Raistrick
Paul Francis
John Wright
Colin Carter
Ian Wilkie

PUBLISHED BY THE PRESS SYNDICATE OF THE UNIVERSITY OF CAMBRIDGE
The Pitt Building, Trumpington Street, Cambridge, United Kingdom

CAMBRIDGE UNIVERSITY PRESS
The Edinburgh Building, Cambridge CB2 2RU, UK
40 West 20th Street, New York, NY 10011–4211, USA
477 Williamstown Road, Port Melbourne, VIC 3207, Australia
Ruiz de Alarcón 13, 28014 Madrid, Spain
Dock House, The Waterfront, Cape Town 8001, South Africa

http://www.cambridge.org

© Cambridge University Press 2004

This book is in copyright. Subject to statutory exception
and to the provisions of relevant collective licensing agreements,
no reproduction of any part may take place without
the written permission of Cambridge University Press.

First published 2004

Printed in the United Kingdom at the University Press, Cambridge

Typefaces Times 10.5/13 pt. and Helvetica   *System* LaTeX 2_ε [TB]

A cataloge record for this book is available from the British Library

ISBN 0 521 53771 1 paperback

The term "UML" refers to the Unified Modelling Language and is used with the consent of the Object Management Group. The UML Cube logo, Model Driven Architecture™ and MDA™ are either registered trademarks or trademarks of Object Management Group, Inc. in the United States and/or other countries.

The publisher has used its best endeavours to ensure that the URLs for external websites referred to in this book are correct and active at the time of going to press. However, the publisher has no responsibility for the websites and can make no guarantee that a site will remain live or that the content is or will remain appropriate.

To Adam and Hannah
>CHR

To Jacky, Ethan and Callum, and to the memory of my father Peter
>PDF

To Jayne for all her love and support and to Andrew and Katharine for their love and inspiration
>PJW

To Pauline and Christopher for their unstinting support
>CBC

To Rhona, Laura and Neil for all their love and support
>ITW

In memory of David Walker, close friend and colleague, greatly missed

Contents

The authors	*page viii*
Foreword by Richard Mark Soley, Ph.D.	*xi*
Acknowledgements	*xiii*
Glossary	*xv*
List of abbreviations	*xviii*

1	Introduction	1
2	Executable model driven Architecture	19
3	Using MDA in a typical project	53
4	Use case modelling	82
5	Platform-independent modelling with domains	99
6	Class modelling in a domain	120
7	Class behaviour and interactions	162
8	Operation modelling	179
9	Dynamic modelling	188
10	Action specification	212
11	Modelling patterns	249
12	Integrating domains	293
13	System generation	326
14	Case study	364
	CD installation procedure	380

References	382
Index	385

The authors

Chris Raistrick
Chris Raistrick is the Professional Services Director at Kennedy Carter Limited [www.kc.com]. Since 1989, Chris has spent the majority of his time applying object-oriented methods on strategically important system developments. Consulting support has been provided for clients in the telecommunications, automotive, distributed control, embedded systems and healthcare sectors. Chris has five years practical experience of applying the UML to a number of successful projects. He has authored ground-breaking papers on executable modelling, component integration and code generation techniques, and has presented technical papers for all the mainstream UML, OMG and Embedded Systems Conferences.

Paul Francis
Paul Francis is a principal consultant at Kennedy Carter Limited [www.kc.com]. After spending 15 years developing hardware and software systems within the avionics industry Paul joined Kennedy Carter in 1993. Since that time he has applied executable modelling techniques across a diverse range of applications including underwater weapons, healthcare, telecommunications, aerospace, and insurance. Additionally, Paul regularly presents training courses in executable modelling techniques and has authored and presented papers at a number of conferences. He is also responsible for the ongoing development of Kennedy Carter's configurable code generator – a meta-model based application for developing translators for executable UML models.

John Wright
John Wright is a director of Aurora Consulting Ltd [www.auroraconsulting.co.uk]. After gaining a Ph.D. from University College London, investigating real-time parallel algorithms, John spent some four years at the Defence Research Agency (DRA) and pioneered the introduction of the then new Object Oriented Analysis into that organisation, a precursor to MDA and xUML. Having left the DRA in 1995 he has spent his time with a wide variety of clients in sectors as diverse as automotive, aerospace, government and consumer electronics; helping these clients with process definition and improvement as well as mentoring and teaching xUML. John has a particular interest in software architectures and code generation techniques. He is currently actively looking at ways of exploiting J2EE in UML and MDA for enterprise architectures.

The authors

Colin Carter

Colin Carter is the founder of Aurora Consulting Ltd [www.auroraconsulting.co.uk], and is a UML consultant, mentor and methodologist with over 20 years experience of the practical application of software engineering methods. He has worked with object-oriented methods since 1983 has been instrumental in introducing object-oriented development to many organisations. Since 1998 Colin has been supporting projects in their adoption and exploitation of UML ranging from real-time embedded to web-based applications. He has been involved in the development of executable modelling since its earliest days and has extensive experience of applying it to system development and business modelling.

Ian Wilkie

Ian Wilkie is Technical Director at Kennedy Carter Limited [www.kc.com] and has over 25 years experience of software development for a wide variety of target environments ranging from computation intensive systems for High Energy Physics data analysis to Real Time embedded control systems. Since joining Kennedy Carter, Ian has provided consultancy of a wide range of projects and has had overall responsibility for the development of the iUML Toolset to support MDA and xUML. Ian was also closely involved the the collaboration that developed the successful submission on "Precise Action Semantics for the UML", recently approved by the OMG, and which is a cornerstone of the xUML approach.

Foreword

The history of the computer software industry is a story of the layering of abstractions. From the early days in Manchester and Philadelphia, as the invention of software tools naturally followed the first stored-program computers, it was clear that (as in any engineering discipline) the only way to control system complexity was to create layers of abstraction, each a simplified or clarified description of the layer beneath it.

The first descriptions of stored programs – what we now call assembly language programs – were essentially one-to-one mappings to the underlying implementation infrastructure (i.e. the instruction set architecture of EDSAC at Manchester). It didn't take long to realize that reduction in programming complexity would be critical to building systems whose complexity demanded abstraction. The first of what we would later term 'macro assembler' and the first link/loader that simplified programming were to be developed even before EDSAC was up and running.

From this earliest work of Wheeler and Wilkes, through the invention of high-level programming languages (FORTRAN and Lisp in 1959, which as a side effect also introduced the concept of software portability), to the development of clarified data abstractions in Algol, the industry has sought 'orthogonality'. In essence, by abstracting away some of the complexities of software development, we have tried for decades to reduce the software development task to the specification of the 'business process', minimizing the coding complexity that is not specifically called for in specifying a process.

The 1990s brought into wide use 'middleware' systems that hid the complexity of various difficult IT development tasks (transactional integrity, persistent storage, naming and directory services and the like) while offering guarantees of portability and interoperability. I have been quoted in the past as defining middleware as 'software that nobody wants to pay for', since developers must assume it is ubiquitous on their platform choices and users don't like to pay for infrastructure. Nevertheless these packagings of critical (and difficult-to-design) implementation services have been very popular as the next step in software abstraction.

Of course, at all levels of abstraction for a given system design, each description is equivalent; but (we assume) higher-level descriptions are easier to understand and expose more starkly the business function that the programmer desires. How do we move from abstraction layer to layer? By the use of transformations; these transformations between system designs (we might call them 'models' of systems) are generally called 'compilation' and 'interpretation'.

This book is about another abstraction layer for describing systems, and one that is powerful and proven. In essence, it's just another compilation stage, but one that finally lets us bridge the gap between system design and system implementation that gives us, once and for all, the software blueprint holy grail: the ability to define system architectures divorced from system implementation, choice of programming language, instruction set architecture, networking topology and the like. OMG's Model-driven Architecture (MDA) has been proven over and over again in real IT and embedded system implementation since the set of specifications appeared.

Why, then, has MDA not taken the world by storm? In one sense it has; OMG's Unified Modeling Language (UML) and related model transformation and textual description languages MOF and XMI have become the lingua franca of the software modelling world in a very rapid time frame. In the years from the 1997 adoption of UML as an OMG standard to the end of the millennium, UML essentially displaced all of the extant object-oriented modelling languages (such as OMT, Booch and OOSE). What we need is the next step: the recognition that modelling must be an integrated part of the software development process, relating not only to requirements analysis and software design, but also supporting software implementation, unit test, system test, long-term system maintenance and integration.

This will require a change in mindset, which itself will require good sources of information and training in the modelling-based approach to building systems. You hold in your hand the right first step, a clear statement of the goals and aspirations of Model-driven Architecture, but more importantly a step-by-step guide to understanding MDA and putting MDA to work. As you read this book, don't look for the hidden revolution; though the productivity gains (and support for reuse, portability and interoperability) will make MDA a key part of your development arsenal, think of MDA as just another level of abstraction, and MDA tools as compilers.

It's just the next natural step in the progression of software development tools. But one you'll enjoy!

Richard Mark Soley, Ph.D.
Object Management Group, Inc.
Wuhan, Hubei, China

Acknowledgements

The authors gratefully acknowledge the contributions, influence and support of the following individuals:
- Glenn Webby, for his thorough and highly valued review of the whole text;
- Allan Kennedy, for his review comments and constructive guidance;
- Jeff Terrell, for his helpful review comments and suggestions, in particular regarding code generation;
- Andy Land, for the insights he has implicitly contributed over the years as a result of his dedicated work on the development of iUML;
- Bobby Doherty, for his help in putting together the Case Study;
- Chas Weaver, for his many contributions both methodological and humorous;
- Steve Lewis, for his insights on project management and organization that could easily fill another book;
- Terry Ruthruff and Bary Hogan of Lockheed Martin, for sharing their insights and experiences in applying MDA to a real project, and for producing one of the most widely used MDA presentations of all time;
- Mark Jeffard of Alenia Marconi, for his pioneering work on industrial strength Ada code generation;
- Don Stewart of Marconi Communications, for his many ideas about the development process;
- Charlie Darkins of CAP Gemini;
- Mike Clarke, for his work on the action language and code generation.

This book represents a synthesis of many years' experiences and insights, passed on down the ages through stories and anecdotes from people like Dave Fletcher, Adrian King, and the multitude of individuals with whom we have been privileged to work over the years.

We have had many enjoyable debates, conversations and beers with luminaries such as Leon Starr, Mike Lee, James Rumbaugh, Steve Mellor and Sally Shlaer. Sally sadly died on 12 November 1998 before she could see the ideas that she and Steve pioneered in the late 1980s become mainstream in the form of MDA and xUML. In the words of Mike Budd of Ovum Evaluates: 'It is a tragedy that the Shlaer–Mellor ideas have not achieved full recognition before Sally's death, but perhaps a predictable one. Shlaer–Mellor's abstract austerity and radical efficiency challenges intellectual reputations and the culture of the programmer. Wider recognition will come, but as with most challenging ideas, time will be needed before the wider software development community is willing to acknowledge Shlaer–Mellor's true importance.' This book is a testimony to that wider recognition.

Acknowledgements

The executable UML case study included on the enclosed CD-ROM, was lovingly crafted by Paul Francis. Cynics might wish to start by running this model using the included free software, to dispel their suspicion that this is just another bookload of UML hype. For the more open-minded, it provides the reader with many concrete examples of the ideas presented in this book.

To those whose names have been inadvertently omitted from these acknowledgments, our sincere apologies. Please let us know who you are, and we shall include you in the next edition!

The term *UML* and *iUML* refer to the Unified Modeling Language and Kennedy Carter's MDA product suite, respectively, and are used with the consent of the Object Management Group.

Glossary

abstract class	A class that can only be instantiated in connection with an instance of one of its descendant subclasses.
abstraction	The suppression of irrelevant detail.
actor	An entity external to the system being modelled that interacts with the system during its operation.
archetype	A rule, or rules, detailing how to transform a general xUML model element into text.
association class	A class that models an association.
association	The abstraction of a relationship that holds systematically between objects.
attribute	An abstraction of a single characteristic possessed by all entities that were themselves abstracted as a class.
bridge	A class containing a set of bridge operations for a single required interface.
bridge operation	A realization of a required operation that uses one or more provided operations of one or more domains.
business model	A model of all aspects of the system, irrespective of whether or not we intend to automate them.
class collaboration model	A model summarizing the communication between classes within a domain.
class	An abstraction of a set of things in the domain under study such that (a) all of the things in the set – the objects – have the same characteristics, and (b) all objects are subject to and conform to the same rules and policies.
class diagram	A declarative diagram showing classes, their attributes, operations and relationships.
collaboration diagram	An interaction diagram emphasizing patterns of interaction and closeness of collaboration.
domain	A separate real, hypothetical or abstract world inhabited by a distinct set of classes that behave according to the rules and policies characteristic of that domain.
domain chart	Actually a UML class diagram, showing domains, as UML packages, and their dependencies.

entry action	The sequence of processing which takes place on entry to a given state.
event	The specification of a significant occurrence that can be ascribed a time.
final state vertex	The final pseudostate in which an object is deleted.
identifier	A set of one or more attributes whose values uniquely distinguish each object of a class.
initial state vertex	Indicates the point that an object of a class that has a state machine comes into being and starts operating.
link	An instance of an association.
mechanism	Code elements that are not subject to translation rules at code generation time.
metaclass	A class whose instances are classes in a different model.
metamodel	A model of the elements of a language, e.g. a UML model of UML.
method	An implementation of an operation.
model	A formal representation of the function, behaviour and structure of the system we are considering, expressed in an unambiguous language.
multiplicity	The multiplicity on an association end indicates how many objects participate in each association.
operation	An action that can be invoked via a parameterized interface.
pattern	A generalized solution to a commonly occurring problem.
platform	A technology, or set of related technologies, e.g. CORBA, J2EE. XML, C++.
polymorphic signal	A signal that is directed to exactly one class, and is available to all subclasses of the class to which it is directed.
primary scenario	The most usual path that captures the normal behaviour associated with the use case.
provided interface	The set of services a domain offers on behalf of a single terminator. A domain may have many provided interfaces – zero or one per terminator.
required interface	The set of services associated with a single terminator that a domain will not realize itself, but which the domain requires to be realized by some other domain. A domain may have many required interfaces – zero or one per terminator.
scenario	One particular path through a use case.
secondary scenario	This captures paths through a use case that deal with less usual behaviour such as exception handling and error behaviour.
sequence diagram	An interaction diagram showing the interaction between objects, emphasizing the time ordering of the interactions.
signal	This represents an incident that causes a state transition; it is asynchronous in nature.

Glossary

state	This represents a condition of the class subject to a defined set of rules, policies, regulations or physical laws.
statechart	This provides a graphical representation of the way in which each class responds to the signals.
stereotype	An extension mechanism that allows new UML elements to be created from existing ones.
terminator	An abstraction of an entity external to a domain with which a domain interacts.
transition	Specifies the state that an object will enter from a given state on receipt of a specific signal.
use case	A specific type of behaviour required of the system, typically described as a sequence of transactions between one or more actors and the system in dialogue.
visibility	The permissions associated with how a name may be seen and used by other model elements.

Abbreviations

API	Application Programming Interface
ASL	Action Specification Language
BNF	Backus-Naur Form
CASE	Computer-aided Software Engineering
CCD	Class Collaboration Diagram
CCM	Class Collaboration Model
CM	Configuration Management
COCOMO	Constructive Cost Model
CORBA	Common Object Request Broker Architecture
COTS	Commercial Off-the-shelf
CWM	Common Warehouse Metamodel
DFD	Data Flow Diagram
EJB	Enterpise JavaBeans
GCHQ	Government Communications Headquarters
GUI	Graphical User Interface
iCCG	Configurable Code Generation
IDL	Interface Description Language
I-OOA	Intelligent Object-oriented Analysis
IP	Intellectual Property
J2EE	Java 2 Enterprise Edition
MDA	Model-driven Architecture
MISRA	Motor Industry Software Reliability Association
MoDAL	Model-driven Action Language
MOF	Meta-object Facility
OCL	Object Constraint Language
OMG	Object Management Group
OMT	Object Modelling Technique
OO	Object-oriented
OOA	Object-oriented Analysis
OOD	Object-oriented Design
ORB	Object Request Broker
PIM	Platform-independent Model
PSI	Platform-specific Implementation
PSL	Process Specification Language

PSM	Platform-specific Model
RAM	Random Access Memory
RDBMS	Relational Database Management System
ROM	Read-only Memory
ROOM	Real-time Object-oriented Modeling
RPC	Remote Procedure Call
RTOS	Real-time Operating System
RUP	Rational Unified Process
SDH	Synchronous Digital Hierarchy
SOAP	Simple Object Access Protocol
STL	C++ Standard Template Library
STT	State Transition Table
UML	Unified Modeling Language
XMI	XML Metadata Interchange
XML	Extensible Mark-up Language
xUML	Executable Unified Modeling Language
YACC	Yet Another Compiler Compiler

1 Introduction

1.1 Why should I read this book?

This is not just another Unified Modeling Language (UML) book. It is not about drawing pictures of your code. It is about realizing value from your intellectual assets. It is about formalizing expertise in any subject matter, and making it accessible, both to your colleagues and to your code generators. It represents a long overdue realignment of attitudes to software engineering, with an emphasis on the kind of process and notational rigour one typically finds in the more mature engineering disciplines. It relegates programming languages and middleware products to their rightful place in the scheme of things. They are seen as technologies, typically with transient fashionability, that underpin high-grade specifications in the form of models. It is the models that matter. They are built to last and will outlive any currently in-vogue middleware product or programming language.

This book will show you how to build specifications with longevity. They will be expressed with a simple subset of the global standard UML. They will be organized using the principles of the Object Management Group's (OMG) Model Driven Architecture® (MDA). We will show how you can translate such models onto any desired underlying technology. The book is not based upon theories and academic research, although the principles expounded have a compelling conceptual coherence that will appeal to any computer scientist. The approach developed in this book is based upon the authors' many years of practical experience applying these ideas. The weight of support that is now being thrown behind the OMG's UML notation and its MDA process means that they will almost certainly achieve global prominence. The true potential of MDA, when combined with a precise form of the UML, has yet to be fully appreciated by the global software community.

This book endeavours to provide practical insights into what the authors believe is an impending revolution in the way software is built. Those organizations that have already used this approach, on strategic or critical projects, have realized significant benefits even on their first project, despite the inherent non-recurring learning and tooling overheads associated with adopting a novel engineering process.

1.2 What you will learn

In this book you will learn:
- Model Driven Architecture (MDA);
- enhancing MDA with executable modelling;
- the benefits of executable modelling;
- how to make UML executable;
- what Executable UML (xUML) is;
- the role and application of use cases;
- how to partition a system;
- how to specify a Platform-independent Model (PIM) using xUML;
- how xUML class diagrams are developed;
- the specification of dynamic behaviour using statecharts;
- the specification of class operations;
- the benefits of the OMG's Action Semantics addition to UML;
- the practical application of an Action Specification Language (ASL);
- a range of xUML modelling patterns;
- how to integrate multiple models in order to specify whole systems;
- how to specify translation rules so that any target language can be generated from xUML models;
- how to build systems specified using xUML entirely automatically.

The OMG's MDA offers many benefits. Executable modelling also offers a number of exciting advantages. The combination of the two together, xUML models used within the MDA framework, creates a software development process which will bring you the following valuable assets:
- **Separation of concerns** – the ability to partition a problem into distinct subject matter areas and maintain that separation throughout the development life cycle;
- **Capturing intellectual property** – valuable subject matter expertise is captured in a standardized, reusable, readily shared and easily maintainable manner;
- **Executable and testable models** – early verification that requirements are understood and early demonstration of required system behaviour lowers through-life costs, reduces risk and increases confidence;
- **Clear and unambiguous models** – facilitate communication between development staff and with users and customers;
- **Early integration** – avoids big-bang integration problems and supports iterative development;
- **Component based** – strong definition of provided and required interfaces for a component;
- **Reuse** – of intellectual property at the application level, with generic services and at the platform and technology level;
- **Accelerated development life cycle** – understanding requirements first time and exploiting reuse at all levels;

- **Manageable lightweight development process** – with strong completion criteria, minimal documentation overhead and clear work breakdown structure;
- **Scalable, proven, industrial-strength process** – successfully applied to large and small projects in a wide variety of industry sectors;
- **Resilience to change** – changes in functional requirement are isolated from changes in technology and platform, therefore giving the ability to migrate to the 'next generation' platform without model rework;
- **100 per cent code generation from models** – so models are maintained and code is a derived product, models never become obsolete and are guaranteed always to be up to date;
- **No redundancy** – minimal maintenance overhead and minimal documentation overhead, supports the philosophy of 'one fact in one place';
- **Fully configurable code generation** – so valuable platform and technology expertise is captured and changes can be accommodated easily;
- **Higher quality generated software** – reduced defect injection and the ability to perform systematic system-wide tuning;
- **Reduced life cycle costs** – through early defect identification, maintained separation of concerns and minimized impact of change;
- **Traceability** – from requirements to generated code via executable PIMs;
- **Applicable to a wide variety of application areas** – real-time embedded to large distributed information systems;
- **Applicable to a wide variety of technologies** – distributed enterprise technologies, through embedded systems to 'Systems on a Chip';
- **Commercial-off-the-shelf market** – the opportunity to integrate the 'best tool for the job' in order to create a design and tool chain that is optimized for your project and organization, when taken with other OMG interoperability standards.

1.3 What right do we have to write about MDA and UML?

You may be wondering who the authors of this book are and what credibility they have to write about UML, executable modelling and MDAs.

We are experienced practitioners of the techniques and processes described in this book and have experience of object-oriented development going back to the early 1980s. Our experience of defined software development processes goes back even further than that – to about the time the first software engineer hauled himself/herself out of the primeval sea onto the land. In the 1980s some of us were involved in both structured and object-oriented styles of software development (the progress of Structured Analysis techniques with DeMarco (1978), Yourdon (1978), Page-Jones (1988), McMenemin and Palmer (1984), Ward and Mellor (1985) and Hatley and Pirbhai (1988)), and also in the birth of object-oriented design with Booch (1986; 1991), Buhr (1984), Meyer (1988), and Nielsen and Shumate (1988). At the time we were working in companies that were early adopters of these methods and between us we saw dozens of projects attempt to improve the software development processes with varying degrees of success. By the late 1980s several of us were working together at Kennedy Carter to provide consultancy and training in the use of Structured Analysis, Structured Design and, for companies moving to Ada and C++, Object Oriented Design (OOD). We started applying OOD principles to improving Structured Analysis models – in essence we did Structured Analysis in an object-oriented way.

A meeting with Steve Mellor in 1989 brought to our attention that he and Sally Shlaer had formalized an Object Oriented Analysis (OOA) approach (Shlaer and Mellor 1988; 1992). We were not only attracted to the method but also realized that our experience showed us how Structured Analysis users could move to OOA. We realized that the migration from a functional viewpoint to an object viewpoint required a fundamental change in thinking. We witnessed approaches that delivered little more than 'functions dressed as objects' and cautioned against this style of working. By presenting good training courses and providing follow-up consultancy and mentoring we succeeded in getting many organizations to 'think objects'.

One of the characteristics of Shlaer–Mellor OOA models is that they have well-defined semantics. The process models within an OOA model were made up of simple canonical processes. In principle, these models were executable although there were no tools around to execute them. In 1993 an IBM team defined a Process Specification Language (PSL) and built a Smalltalk-based toolset that made Shlaer–Mellor models executable. We used this approach and the prototype tools and recognized the potential of executable modelling with appropriate tool support.

In 1994 the initial version of the Action Specification Language (ASL) was defined and supported in Kennedy Carter's OOA modelling toolset called Intelligent Object Oriented Analysis (I-OOA). Not only could models be captured, edited, reviewed, etc., as with the other Computer-aided Software Engineering (CASE) tools of the time, but also they could be executed and debugged. Furthermore, an executable model has the information necessary to support code generation and we delivered a bespoke project with over 90 per cent of the target code automatically generated from executable models in 1994.

In parallel with developing our own process and tools for executable modelling, we kept track of developments in object-oriented methods. The Object Modelling Technique (OMT) by Rumbaugh *et al.* (1991), Martin and Odell's Object Oriented Analysis and Design (1992), Jacobson's use cases (1992), design patterns by Gamma *et al.* (1994) and Fowler's introduction to UML (1997) were all influential.

We work in industry sectors such as defence, aerospace, automotive, telecommunications, government, health, insurance and process control. We have therefore seen the wide-ranging applicability of object-orientation, executable modelling and MDAs. We have worked with clients to apply executable modelling and code generation in systems ranging from small real-time embedded systems to massive distributed systems containing millions of lines of generated code. We know from many first-hand experiences over several years that executable modelling and code generation works. We have also learnt where it works most effectively and where caution is required. In this book we aim to back up the technical details with this practical experience.

We realized the benefits and impact that UML was going to have well before it was a standard. As a result, we embarked on work with Steve Mellor to adapt the Shlaer–Mellor process and formalism to use the UML (Wilkie & Mellor, 1999). We also saw, along with Steve Mellor, Jim Rumbaugh and others, that in order to complete this work, some additions and constraints would need to be applied to UML models in order to make them executable. This led to our involvement in the OMG Action Semantics consortium which worked from 1998 to 2001 to deliver Action Semantics as part of the UML standard (OMG, 2002).

Developing code generators for executable models raises interesting issues. Potentially many different code generators are required for each target language with additional variants to cope with target platform differences. In the past, if code generation was attempted, then bespoke generators were created for each language/platform combination, making platform migration a costly business, since every platform change may require major rework of the code generator. In order to provide a framework that allows these generators to be produced and maintained easily, we formalized an approach to building these generators based on an xUML model of xUML itself. This means that the same formalism is used to specify both the code generator and the models to which the code generator is applied. This allows the subject matter of code generation to be formalized in the way we would take with any other domain. Therefore the code generator itself is fully documented (since it is expressed in xUML) and testable as any other executable model. Finally, it has the advantage that a well-structured code generator, expressed in xUML, is far easier to tailor, tune and evolve than a baroque collection of YACC, LEX and other scripts common in bespoke generation strategies.

This innovative approach required a thorough understanding of the xUML formalism and some high-powered abstract modelling skills. It is now available as a commercial product called iCCG.

Since 1992 all of the authors have provided training, consultancy and mentoring in object-oriented analysis and design, and more recently UML. Typically, this has been in projects using executable modelling and code generation. With the insights gained in this wide range of practical project support, a CASE tool has been specified and developed that provides best-practice support for xUML. The effort needed to specify such a product has resulted in an in-depth understanding of the method, which naturally benefits our consulting work.

In November 2000 the OMG published a paper on MDA (Soley, 2000). We immediately recognized the synergy between our approach to code generation and MDA, and continue to be actively involved in the OMG's work to establish MDA standards (www.omg.org/mda).

1.4 What is Model Driven Architecture (MDA)?

First, a verbatim summary from www.omg.org/mda:

MDA development focuses first on the functionality and behavior of a distributed application or system, undistorted by idiosyncrasies of the technology or technologies in which it will be implemented. MDA divorces implementation details from business functions. Thus, it is not necessary to repeat the process of modeling an application or system's functionality and behavior each time a new technology (e.g., XML/SOAP) comes along. Other architectures are generally tied to a particular technology. With MDA, functionality and behavior are modeled once and only once. Mapping from a PIM (Platform Independent Model) through a PSM (Platform Specific Model) to the supported MDA platforms will be implemented by tools, easing the task of supporting new or different technologies.

The most valuable asset in any software organization is its accumulated expertise. It follows that this expertise should be formalized and made accessible within the organization. It seems logical that we should capture the expertise that will eventually be manifest in a software system in a form that lends itself to systematic, potentially automatic mapping to implementation. This rules out informal approaches such as narrative text and suggests we embody our intellectual property in the form of a precise UML model.

But if we are to build models that will retain their value over decades, then we must desist from the long-standing tradition of using languages that come with a guarantee of impending obsolescence. If we express our business rules and system behaviour using a traditional programming language such as Java, our model will diminish in value quickly, as new technologies supplant the language in which it is expressed.

The key premise behind MDA is that if we are to preserve our investment in system specification, then the system must be specified in a platform-independent way, in a 'platform-independent model'. This immediately raises the question, 'What platform are we to be independent of?' After all, even assembly language can be seen as being independent of the hardware platform, and Java is often promoted as a 'platform-independent language'. So is it OK to specify our system behaviour using Java? We must decide whether we think Java will be the preferred programming formalism for the next few decades. Historical precedent suggests otherwise. It shows that the fashion in programming languages changes every two or three years, although all retain some vestige of a cult following.

So we must aim for a higher level of platform independence. A level at which programming language changes do not impact PIMs. If our models are to survive for decades, we must be independent of:
- the hardware;
- the operating system;
- the programming language.

But we cannot build models that are truly platform-independent. The wily reader will have already surmised that we are about to propose the use of the UML for construction of PIMs. So we will. UML is the key enabling technology for the MDA; every application built using MDA is based on a platform-independent UML model. By leveraging OMG's universally accepted modelling standard, the MDA allows creation of applications that are portable across, and will interoperate easily between, a broad range of systems types, from embedded, to desktop, to server, to mainframe and across the Internet.

However, these models are still technically platform-dependent. For example, they assume that something in our execution platform provides the ability to do such things as:
- create objects and links;
- send signals;
- make state transitions.

In other words, we are dependent upon a platform that supports the capabilities inherent in the UML itself. We call this platform the 'UML Virtual Machine', and it is this upon which our 'platform-independent models' depend. It is this virtual machine that provides a standard set of capabilities upon which the modeller can depend and maps them to the capabilities of the specific underlying platform.

So is it a problem that we depend upon the UML virtual machine? It would certainly become an issue if the UML ceased to be the language of choice for building PIMs. Of course, the UML will continue to evolve as it has for years, but as yet it shows no signs of becoming unfashionable. If it were to be replaced with another formalism, then the enormous global investment in UML models would certainly stimulate the development of products that would map existing UML PIMs onto PIMs expressed in the new formalism. Therefore, this level of platform independence appears to be low risk. MDA changes the emphasis for software development in two ways:

- Models are more valuable than code – the essential intellectual property of a business is formalized in models that allow the organization to exploit different technology platforms as these become appropriate;
- Models are precise rather than 'high level' – all the essential intellectual property of the business is captured in the models, no arbitrary analysis/design split is admitted allowing 'difficult' issues to be deferred to design time, or even worse, coding time.

'So what is the role of middleware platforms in the MDA?' you may ask. Every middleware vendor would have you believe that adoption of their technology will address all your portability problems. Again, we quote from the OMG website:

'In the MDA, a specification's PIM is used to define one or more PSMs and interface definition sets, each defining how the base model is implemented on a different middleware platform. Because the PIM, PSMs, and interface definitions will all be part of an MDA specification, OMG will adopt specifications in multiple middleware platforms under the MDA. While CORBA's platform and language independence and proven, deployed transactional and secure nature continue to make it the best choice for systems from embedded to desktop to Internet, MDA makes porting to, and interoperating with, other middleware platforms easier and cheaper.'

But what about the code? Surely we have to model that somewhere? Well, how often do you look at the machine code generated from your Java? Do you really care what it looks like? If we have a validated, executable PIM, do we care what the generated code looks like? Of course we care about its performance characteristics, but we can address that issue by specifying rules in our code generator. So where does the code generator come from? Imagine you were to take all the software implementation expertise in your organization and formalize it in a UML model. Now imagine you could generate code from any UML model. What will you get if you generate code from your 'software implementation' model? You will get a code generator, which embodies the expertise that your organization has accumulated about how to build platform-specific implementation for your type of application on your type of platform. This is our preferred approach to implementation. We formalize expertise about the subject matter of code generation in exactly the same way as we formalize any other expertise – using UML.

1.5 Introduction to the OMG

Frequent mention has been made of the Object Management Group (OMG). It is worth digressing a little to explain the composition of this organization, its history, background credentials and goals. This supports our assertion that the MDA is an initiative that has broad industry support and comes from an organization that has *gravitas*!

The OMG was founded in April 1989 by eleven companies and has now grown to incorporate about 800 organizations (www.omg.org). Its mission statement starts:

> The OMG was formed to create a component-based software marketplace by accelerating the introduction of standardized object software. The organization's charter includes the establishment of industry guidelines and detailed object management specifications to provide a common framework for application development.

Its members work together, in this not-for-profit organization, to produce and develop standards that are both vendor-neutral and commercially viable specifications for the software industry. Its members represent the broadest interest groups from the software development community, with representatives from all market sectors. It has produced the well-known CORBA standard (www.omg.org/corba), which is the seminal component-distributed architecture definition. Another major success is the standardization of UML (www.omg.org/uml). It has other standards which include the MOF, XMI and CWM (www.omg.org/).

In 1990 the OMG produced the Object Management Architecture, which is a framework to achieve ease of integration and interoperability, based upon the CORBA standard. This work has been subsumed into what the OMG style their 'flagship specification', the MDA, that makes use of their UML standard and frees the architectural vision from a single middleware technology, CORBA, allowing integration of diverse systems based upon a myriad of changing middleware technologies (or none at all!).

The best way to learn more about the OMG is to visit their website at www.omg.org. We shall be exploring MDA in some depth in this book and showing you how to exploit it with xUML.

Suffice to say that the OMG has emerged as one of the most significant standards bodies in the software industry and one that influences the way that we shall all build systems in the future!

1.6 History of software methods; the road to MDA

It all began at the end of the life cycle. In the old days, when memory and processor resources were measured in kilobytes and kilohertz rather than gigabytes and gigahertz, programmers necessarily worked at the machine level and scorned anything more abstract than a one or a zero. They worked alone, and got away with it, because they mostly built systems that were small enough for one person to understand.

However, the computers started to get bigger and more powerful and the systems started to get more complicated. Newfangled, more abstract notations, such as assembler, came along. Then came the third-generation programming languages, in wondrous varieties such as C, C++ and C#. Sometimes, programmers had to look at other programmers' code. They did not like it. Each programmer quickly realized that all the other programmers needed psychological help. Their code was perverse and clearly the result of a diseased mind.

1.6.1 Structured programming

Of course, they weren't really mad, just anarchic. They needed rules. Programs needed to be more structured and the imaginatively named 'Structured Programming' movement was instigated in the early 1970s by the likes of Michael Jackson (no – the other one!) (Jackson, 1983).

Strangely, this did not seem to solve all software development problems. Even programmers in the most fascist development regime, where everything was done using sequence, selection and iteration, with definitely no 'goto's, still produced systems that did not work properly.

1.6.2 Structured design

Perhaps the problem was in the earlier life cycle stages. Perhaps the designs needed to be more structured. Edward Yourdon and Larry Constantine thought so and led the 'Structured Design' movement. They proposed two key quality criteria against which any design should be assessed. They were:

1. **Coupling** – a measure of the degree to which modules are dependent upon one another. The more loosely coupled, the better. The loosest possible coupling is a key facet of our approach to MDA. We shall show you how to specify and integrate systems composed of multiple platform-independent modules, each independent of one another – the modules (or domains) are 'anonymously coupled';
2. **Cohesion** – a measure of relatedness of the contents of a module. High cohesion is desirable. Again, this principle pervades our style of MDA. We shall show you how to specify and integrate systems composed of multiple platform-independent modules, each of which encapsulates an entire and distinct subject matter.

Yourdon and Constantine (1978) also offered a design formalism, 'Structure Charts', which allowed designers to visualize their code, and subjectively assess its quality in terms of coupling and cohesion, before they wrote it. Meilir Page-Jones, who proposed a number of mapping strategies that would reliably generate structured designs from structured analysis models, reinforced these ideas (Page-Jones, 1988). This principle is firmly embedded in MDA, in the form of PIM-to-PSM mapping rules.

1.6.3 Structured analysis

The focus on earlier parts of the life cycle continued with the advent of 'Structured Analysis', popularized by Tom DeMarco (1978) and Edward Yourdon (1989). The key principle here was to separate the specification of the problem from that of the solution. Yourdon proposed construction of two primary models. The first, known as the 'essential model', was an implementation-free representation of the system. The second, known as the 'implementation model', was a functional, behaviour-specific realization of the essential model, highlighting the organization of the hardware, software and code. It is here that the MDA principle of platform-independent and platform-specific modelling has its roots.

These analysis and design models were expressed using notations that were more abstracted from the implementation than Structure Charts. Data Flow Diagrams (DFDs) were used to show dependency between functions, State Diagrams were used to specify state-dependent behaviour, and Entity Relationship Diagrams were used to show the information structure within the system. All of these notations, with the curious exception of DFDs, have been assimilated in some form into the UML.

Ward and Mellor (1985), with their event-driven approach to modelling, proposed that a system did not have to be partitioned 'top-down', but could be done 'outside-in'. Their approach was based upon understanding the environment in which the system lived, in terms of terminators and events, and then deriving a specification of the behaviour that must exist in the context of that environment. Some might see some correspondences here between Ward and Mellor's 'terminators' and 'event-response', and Ivar Jacobsen's actors and use cases.

Hatley and Pirbhai (1988), with their Process Activation Tables, and David Harel (Harel and Politi, 1998) with his Hierarchical State Models, all made significant contributions in the area of modelling state-dependent behaviour. State modelling has been widely perceived as being applicable only to real-time systems. As we shall demonstrate later, this is a false belief.

1.6.4 Object-oriented methods

While the structured methods were evolving, there was a growing realization that there might be an alternative partitioning principle to the functional approach. It was realized that the 'things', or objects, in a system are far more stable than the functions performed upon them. So the object-oriented development movement followed a similar path to the structured movement. As with the structured movement (DeMarco, 1978), the development of object-oriented methods began at the point where all the issues manifest themselves – the code. Object-oriented programming books by Booch (1986), Meyer (1988) and others were followed by books about OOD, and then OOA. This reflects the realization that it is at the front end of the life cycle, where the emphasis is upon user requirements, desired behaviour and business rules, that the most important decisions are made. MDA embodies this view in the emphasis it places on the PIM.

Bertrand Meyer (1988), inventor of the Eiffel object-oriented programming language, advocated the principles of 'programming by contract'. This is achieved by specifying interfaces using pre- and post-conditions and invariants. These ideas, which were also present in the formal languages such as Z (Potter, Sinclair & Till, 1996) and VDM (Jones, 1991), are central to the UML's Object Constraint Language (OCL) (Warmer & Kleppe, 1999).

1.6.5 Patterns

Grady Booch popularized the notion of patterns to allow systematic implementation of Ada systems, with his book *Software Components with Ada* (1987). He was to publish *Object-oriented Analysis and Design with Applications* in 1993, detailing the Booch method of object-oriented development. The principle of using patterns, or archetypes, as the basis for a platform-specific implementation has been widely advocated, by such luminaries as Kent Beck (1996), Sally Shlaer and Stephen Mellor (1992), Buschmann, Meunier and Rohnert (1996), and Peter Coad, who described a pattern-based approach to system design in *Object Models: Strategies, Patterns and Applications* (1997). Martin Fowler, in *Analysis Patterns: Reusable Object Models* (1997) provides example patterns, drawn from various vertical markets. The so-called gang of four, comprising Gamma, Helm, Johnson and Vlissides, published one of the best-known pattern books in *Design Patterns* (1994).

Ivar Jacobson, with his use-case driven approach (1992), raised the level of awareness about the importance of organizing requirements in a way that makes them easier to model and trace through the life cycle.

Dr James Rumbaugh *et al.*, developed the Object Management Technique (OMT) (1991) and associated set of notations to support building analysis and design models. This was more formal than the Booch method, and if any method can be said to have achieved dominance in the 1990s, OMT would be the one.

Steve Cook and John Daniels, in their 'Syntropy' method (1994), emphasized the value of modelling rigour based on OMT style diagrams.

1.6.6 Executable modelling

Shlaer and Mellor, with their 'recursive design' approach, emphasized the value of building precise PIMs that could then be systematically (and if desired, automatically) translated to generate the target system. Further, they postulated that the most appropriate partitioning principle was by subject matter, or domain. This resulted in platform-independent modules that were highly cohesive and loosely coupled – exactly the properties that Yourdon and Constantine had sought in software modules 20 years earlier.

Bran Selic *et al.*, working on similar principles to Shlaer and Mellor, developed the Real-time Object-oriented Modeling (ROOM) method (1994), which focussed upon real-time system development, with particular emphasis on the importance of defining interfaces and protocols, clearly a key component of the MDA approach.

All of these method gurus developed different formalisms to allow graphical depiction of their models. Thankfully, they have now been synthesized into the UML.

Leading lights such as Ed Colbert, Anthony Wasserman and Rebecca Wirfs-Brock (Wirfs-Brock, 1990), and James Martin and James Odell (1997) also made major contributions in this field.

The current state of the art, as manifest in our approach to MDA, can be seen as the culmination of this history. It embodies a number of key principles that have been tried and tested over the years:

1. Partitioning based upon subject matters (domains), resulting in large, cohesive and loosely coupled 'virtual' components;
2. Separation of platform-independent behaviour from platform-specific behaviour, as achieved to some extent by the 'essential model versus implementation model' split;

3. Definition of pattern-based mappings to allow systematic creation of any number of platform-specific models from a PIM. These mappings can be applied manually, or, as this book advocates, they can be automated;
4. Use of an abstract but semantically tight formalism, xUML, which allows models to be expressed so fully and exactly that they are executable. This property of executability allows modellers to assess the correctness of their formalization of the system in an objective way.

1.7 What is executable UML (xUML)?

UML, as its name suggests, provides a Unified notation for the representation of various aspects of object-oriented systems. It is not the purpose of this book to introduce the whole range of notations and semantics that the UML represents. Rather, it is our intention to use a carefully selected subset of the UML notation to build models with an important property – they can be executed.

UML is a widely adopted standard notation for the representation of aspects of object-oriented systems (www.omg.org/uml). We shall use UML extensively in the MDA process, however whilst the core UML specification is necessary, it is not sufficient for executable modelling. In recognition of this, the UML specification has been enhanced to incorporate the Action Semantics (OMG, 2002). This serves to resolve many of the ambiguities in UML and adds definitions of execution behaviour for appropriate UML model elements. Core UML plus the Action Semantics, termed the xUML, is sufficient to build executable PIMs, see Figure 1.1.

Let us examine the contrast between UML and xUML:

UML specifies a diagrammatic language that allows systems to be defined using a number of diagram types, however it is quite informal about how the different diagrams are to be used. For example, state models can be used to describe use cases, subsystems and objects. This means that a reader must first establish the context for the diagram they are reading before being able to understand it. In xUML, notations are used for a specific purpose: a statechart is always associated with a class so a state model always describes the behaviour of an object.

Not all elements of UML have execution semantics. For instance, the component and deployment diagrams do not have run-time behaviour and the UML action semantics have nothing to say about them. Component and deployment diagrams can be used, if appropriate, with xUML models but in exactly the same way as they would with a conventional UML model. This book, therefore, has little further to say about the component and deployment views.

Fig. 1.1 The basis of xUML

Use cases are part of an xUML model although they are an informal notation and no specific execution semantics are required of them. They are included because an executable model will realize the behaviour specified by each use case scenario and are therefore valuable in tracing from requirements through to executable behaviour. We discuss use cases and how they are best exploited within the context of MDA in Chapter 4.

Activity diagrams could potentially be used to specify actions in part. However, activity diagrams alone would not be enough. It is still necessary to define the required behaviour inside each action state. As a result, xUML does not use activity diagrams to specify the executable behaviour of operations or state actions. Activity diagrams may prove useful for informal descriptions, such as business workflows, for example. The use of activity diagrams for these purposes is the same as for standard UML and not covered further by this text.

The statecharts defined in UML are most certainly used in xUML and the action semantics submission clarified much of the behaviour definition for state machines. The statechart notation defined in UML is very rich, but, particularly for novices, potentially very confusing. For example, state actions can be defined:

- on entry to a state;
- on exit from a state;
- on a transition between states;
- on an internal transition;
- as an ongoing activity in a state.

So the UML offers us a great number of tempting ingredients to support a wide range of modelling styles, but hear what James Rumbaugh (1999), a master chef of the UML says about its rich set of constituents:

It's like being a cook. You have to know how to use all the possible ingredients. Just not all at the same time....'

MDA recognizes the challenge of using UML in a way that enhances correctness, clarity and maintainability of our models.

xUML also adds a model view that is absent in UML. For all the different styles of specifying state behaviour in UML, there is no notation that is well suited to defining the completeness of a state model. The concept of state transition tables has been around for some time (Hatley and Pirbhai, 1988) and whilst they are not effective at providing a visual description of dynamic behaviour they are an effective adjunct to a state transition diagram that allows the completeness of the state model to be checked. For executable models this is a valuable addition.

There is broad agreement that one of the fundamental purposes of analysis is to **manage the complexity** of the problems, so that humans can understand and visualize them clearly. Unfortunately, simply using the full myriad of UML notations can easily produce models that are incomprehensible to normal people (i.e. those who are not software engineers and many who are). Is this a problem? Well, since another stated purpose of analysis is **to capture our understanding of the customers' requirements**, it can be rather demoralizing when a customer, presented with a painstakingly constructed analysis model, reacts with incomprehension at the sight of the models.

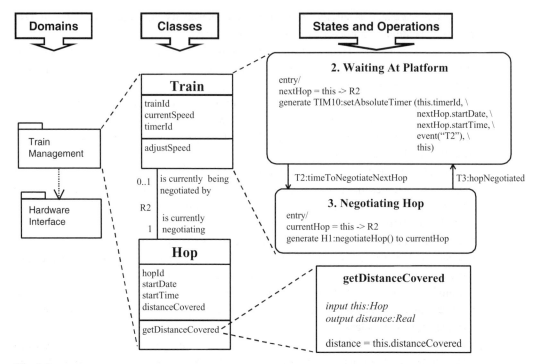

Fig. 1.2 xUML Model layers

It is critical to a successful analysis phase that the models can be understood not just by the analyst but also by subject-matter experts and customers. To this end, xUML prescribes a simple, coherent subset of the UML notations that allow us to formalize our understanding of the distinct subject matters, the **domains**, present in any system. The resulting models are based upon the **structure of the real world**, not the individual constructs that make up a software system. Not only are the notations themselves simple, but the way in which they are organized and integrated conforms to strict rules that preserve the overall clarity of the system specification. There is no scope for individual analysts using their own structural style, or building hierarchies of indeterminate depth. The structure of an xUML model is based upon layers of abstraction (information, behaviour, operation) rather than layers of successive detail, where each layer is effectively made redundant by the next layer down in the hierarchy. This means that the development process can be more effectively monitored and controlled, since a project manager knows exactly how many layers are to be constructed. It eliminates the idiosyncrasies that characterize models built using large, complex notations. Figure 1.2 gives the layers of a model expressed in xUML.

Experience shows that people with no software background whatsoever can quickly understand and provide feedback on the xUML models. Indeed, some years ago, during the development of a hospital patient administration system, it was only a matter of days before the medical staff who were allocated by the customer to act as subject-matter experts were making significant refinements to the models constructed by the analysts. They were drawn into the analysis activity and were eager to interact with the analysts and the analysis models. The deployed system met their requirements!

Another key benefit of using a UML subset is that it leads to a more consistent modelling style across the development team. If there is only one notational device to express a particular construct, then the developers are spared the agony of choice!

UML provides many different ways of specifying the same or similar concepts, such as the many ways of specifying state behaviour. xUML uses a necessary and sufficient subset of these various specification techniques. This allows clear, precise semantics to be defined for each xUML model element.

In xUML we shall show you how to use:
- packages to represent the partitioning of the system into platform-independent domains;
- the package diagram to give a layered system overview showing the contract-based dependencies between the domains;
- use cases to capture the functional requirements for the system;
- sequence diagrams at the domain level (a lifeline corresponds to a domain) to show how the domains collaborate to achieve each use case scenario;
- a single class diagram to show the classes and associations in each domain or subsystem;
- operations with typed input and output parameters to specify state-independent behaviour of the objects of a class;
- a single collaboration diagram for each domain to show the patterns of interaction between the classes and interfaces of a domain – operation calls are synchronous interactions and signals are asynchronous interactions;
- a statechart for each class that has dynamic state-dependent behaviour;
- each statechart stimulated by signal events with typed parameters;
- a state transition table for a complete specification of the dynamic behaviour of each class;
- Action Specification Language (ASL) to describe each state entry action;
- ASL to describe each operation;
- ASL to describe how each domain makes use of the services provided by other domains;
- ASL to establish the executable model simulation environment.

In summary, xUML is an executable version of the UML, with:
- clearly defined simple model structure;
- a precise semantics for actions;
- an action specification language;
- an accompanying process, MDA.

1.8 Structure of the book

As you can see from this brief introduction, there is a lot to cover. The book is structured as follows:
- In Chapter 2 you will learn more about the **MDA process**. It introduces the principle elements of xUML models and shows how they are related. It is a road map for Chapters 4 to 10 in which xUML is covered in detail;

- In Chapter 3 you will learn how **MDA fits into a typical project life cycle**. It uses the process framework from the Rational Unified Process (RUP) (Kruchten, 1999) as an example development life cycle because it is well documented. MDA is a technical development process that will fit into a whole range of software project life cycles from an ultra lightweight process to a heavy, formal process with full traceability. The purpose of Chapter 3 is to illustrate how the technical activities map onto a typical project life cycle;
- In Chapter 4 you will learn how **use cases** are used as a technique for extracting and organizing functional requirements. Use cases are the most informal part of the models we build and we make no attempt to make the use cases executable. Rather, the use cases capture the user's perspective on how the system being analysed should behave. The executable models we build later should be able to demonstrate that they can satisfy each of the use cases. The use cases are therefore valuable in identifying the behaviour the models should have and in specifying the test cases to be applied to the models;
- In Chapter 5 you will learn how **system partitioning (into multiple platform-independent models or domains)** is used in MDA. Any system will consist of a set of domains and finding the right ones is important. Domains are not just functional chunks of the system; a well-partitioned system will deliver domains that can be reused. Although this chapter visualizes a system as a set of domains it does not look at the detail of how we build a system by integrating sets of domains – that comes later (Chapter 12) after we have introduced the details of the modelling formalism;
- In Chapter 6 you will learn how to use the primary UML notation, the **class diagram** to reveal the detail within a domain. In xUML models we are not using the class diagram to represent the elements of an object-oriented programming language but to represent abstractions of the subject matter being modelled. We emphasize these points and try to highlight the often encountered pitfalls. The chapter introduces the various elements of the notation with many examples and looks at practical techniques for building class diagrams;
- In Chapter 7 you will learn how **behaviour can be modelled**. The differences between state-dependent and state-independent behaviour are described and UML's forms of interaction diagram are introduced. Recommended ways of using sequence diagrams and collaboration diagrams are described;
- In Chapter 8 you will learn how to **model state-independent behaviour as operations**. The different types of operation are described along with their execution rules;
- In Chapter 9 you will learn how to **model state-dependent behaviour by state modelling**, using UML's statechart notation. The UML standard describes a very rich set of statechart notation. There are many ways of modelling the same state-dependent behaviour in UML. Our approach is to use a subset of the full notation that is sufficient to describe any state-dependent behaviour but is simple enough that it can be easily understood. The resulting state models have behaviour that is deterministic and fully specified;
- In Chapter 10 you will learn how to **specify operations and state actions by using the Action Specification Language (ASL)**. ASL is a language for manipulating model

elements, it complies with the UML Action Semantics and it is the part the makes the models executable. It is a simple language that makes no assumptions about underlying implementation languages and allows executable PIMs to be built. ASL is used for additional purposes including, to define test methods for domains and to integrate sets of domains together. Chapter 10 completes the coverage of the model elements required to build executable domain models;

- In Chapter 11 you will learn from the authors' experience of building xUML models, expressed in terms of common **modelling patterns**. Patterns range in size from the very small scale – frequently observed ASL structures – right up to common groups of domains for particular types of application. In addition to providing patterns which readers may find directly applicable, the patterns also act as examples of good modelling practice;
- In Chapter 12 you will learn how to **integrate a set of PIMs**. The key principle here is to maximize the cohesion of each PIM (domain) and minimize the coupling between them. Each domain makes services available for other domains to use through its provided interface and each domain states which services it requires from others through its required interface. Integration is achieved through the matching of required interfaces to provided interfaces using bridges. Each domain is executable in its own right and as we build the bridges between domains we get sets of domains executing together. Eventually, we have an executable model of the system as a whole. Performing this level of integration with models greatly mitigates conventional integration risks. The chapter also introduces some of the more powerful techniques that can be used to specify bridges;
- In Chapter 13 you will learn how to achieve **code generation by applying a mapping** from the PIMs and how to express such mappings in xUML. Since the PIMs are executable they have sufficient information in them to generate **all of** the code and not just structural code such as headers. The authors have considerable experience of this and the techniques described in this chapter are in use in real projects. In MDA terms code generation is a realization of the mapping from PIM to PSM. The power of this technique comes with the ability to specify the form that the mapping takes and therefore determine the type of code that is generated. The target language can be selected and so can the way the target language is used in the target environment. Different properties will be required in the code for a small, real-time embedded system compared with a large distributed information system;
- In Chapter 14 you will learn how all elements in xUML and MDA come together in a **case study** that includes a complete executable model together with description and walkthrough. This model is also available on the accompanying CD-ROM. The case study model covers all of the key points of xUML introduced in the book and provides a fairly simple, but still representative, example of a complete model.

The emphasis in this book is to cover all of the aspects of MDA with xUML so that the big picture is complete. However, there are detailed aspects relating to some topics that cannot be addressed in a single-volume book. We therefore invite the reader to use the resources found on the OMG's website (www.omg.org) and on the Kennedy Carter website (www.kc.com) where detailed technical papers can be found.

1.9 How you should read this book

You certainly can read the book from front to back. The later chapters make use of ideas introduced in the earlier ones. However, you can be selective in the order you read the chapters if you have a particular motivation for reading this book.

We would recommend that you read Chapters 1 and 2 – you've presumably read Chapter 1 by this stage! These set the scene and the context for all that follows.

If you have a particular interest in the software development process then Chapter 3 should be read early. However, if you would prefer to get stuck into the technical details of making models execute then leave Chapter 3 until much later. You could read this chapter last if you so desire.

Use cases are dealt with in Chapter 4 so if you are familiar with use cases you can merely scan this chapter to find any differences in style.

Chapters 5 to 10 introduce the technical details of xUML. We recommend you read these chapters in order. You may like to keep a bookmark in Chapter 14 on the case study as you read them since it will help build up the big picture of how it all hangs together.

Chapter 11 can be read any time after Chapters 5 to 10. However, it is not a prerequisite for the chapters that follow it.

Chapter 12 closes the loop on the technical description of xUML since it uses details introduced in Chapter 10 in the context of the concepts from Chapter 5. Therefore like Chapter 11 it can be read any time after Chapters 5 to 10 and is not a prerequisite for any others.

Chapter 13 describes how some of the key concepts of MDA, the mappings from PIMs to PSMs can be achieved in practice, and can be read any time after Chapter 10.

Chapter 14 illustrates a complete executable model specified using xUML. You may like to refer to this chapter each time a new concept is introduced in the earlier chapters. Keep a bookmark in Chapter 14 for easy cross-reference.

2 Executable Model Driven Architecture

2.1 Introduction

The aim of this chapter is to provide an outline of the process that will introduce many of the key topics in the book and serve as a 'road map' for the chapters that follow. This chapter is not intended to explain the entire notation or how to build a model but to provide the framework on which we can hang the details as we subsequently introduce them. Model fragments are shown to illustrate the type of model elements that are developed at each stage of the process. Don't worry if there are aspects of the model that are not clear, we shall explain all of the notation and syntax in the subsequent chapters.

At this point, it is necessary to emphasize an important, and frequently confused, distinction between process and notation. We often encounter questions about the 'UML method'. UML is a notation, a language, for expressing abstract models. It provides a way of representing a set of concepts and the interactions between them. It says nothing about how to produce models. UML is not a method. A process or method is required to define how to use UML to build sets of models with particular properties. An analogy may be found in writing. Language is made up of characters, words, clauses, sentences and paragraphs. Grammar provides the rules that dictate the acceptable ways different types of words can be put together to make clauses and sentences. However, expert knowledge of grammar isn't enough to make you a novelist. There is a method to writing a novel, as well as considerable creativity. The plot needs to be built and interwoven with sub-plots and characters need to be developed. You can go on a training course to learn the method for writing a novel. The distinction between the process of creating a novel and the language used to represent it is clear. The same distinction applies to modelling and its language, UML.

We therefore take care to distinguish between language and method. We use the term Executable UML (xUML) to refer to the style of using UML to represent **executable** models. Extending this idea we shall consider the use of xUML within an executable form of the Model Driven Architecture (MDA). We shall show how such a technical development process will deliver the extensive list of benefits described in Chapter 1.

2.2 Background to MDA – software engineering and process

The concepts that underpin MDA have a long history. To give a full and proper historical perspective of how we got here would take a book in its own right, so this section is intended to give a flavour of the highlights of the evolution of software engineering and software development processes which lead to the MDA approach.

The term 'software engineering' was probably first coined in a NATO workshop in 1968 (Naur and Randell, 1969). It was at about this time that the start of the metamorphoses of the production of software from a craft to a professional engineering discipline was started (a transformation that many argue is not yet complete).

Finch (1951) gives a generic roadmap of the evolution of an engineering discipline, reproduced in Figure 2.1.

Finch argues that the exploitation of a technology starts with a limited number of skilled 'craftsmen', who know how to exploit the technology effectively, but on a limited scale. Production techniques are then incorporated to allow procedures to be put in place that permit the production to scale in a limited way. Often the techniques employed are ad hoc and not reproducible across organizations. It is also apparent that processes derived at this level of maturity are pragmatic rather than theoretic and so largely based upon subjective argument and heuristics. The next step is to incorporate scientific theory and process to the practical production process to give the method the necessary conceptual foundation to allow it to become a proper engineering discipline. This is what MDA with xUML provides.

The software process is, at its heart, the means by which we characterize and structure the practice of software construction. Since the very early days a process, of sorts, was followed. This could be loosely termed hack-it-and-patch-it (a less pejorative term is an 'ad hoc' or 'heuristic' approach). This 'process', which is code-centric and model-free is viable for programming-in-the-small, but scales very badly. Often there is a reliance upon hard working 'heroes', who are able by Herculean efforts to get something working. Typically, when the heroes move on the code is found to decay over time and successive maintenance and upgrades are performed, until eventually we must start afresh. It was realized early on that a more structured approach to processes was required to allow software production to scale, both in size and in the time over which the product is maintained and evolved.

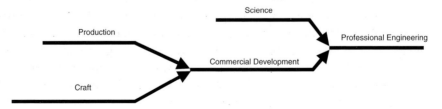

Fig. 2.1 Evolution of engineering disciplines (Finch, 1951)

Sebastián Tyrrell (2001) gives seven goals for any software engineering process, paraphrased below:
- **Effectiveness** – the process must help us produce the right product, assisting us in determining what the customer needs, producing what the customer needs, and, crucially, allowing us to demonstrate that what we have produced is what the customer needs;
- **Maintainability** – one of the goals of a good process is to expose the designers' and programmers' thought processes in such a way that their intention is clear. Then we can quickly and easily find and remedy faults or work out where to make changes;
- **Predictability** – a good process will help us to plan and estimate the work. The process helps lay out the steps of development. Furthermore, consistency of process allows us to learn from the designs of other projects;
- **Repeatability** – if a process is discovered to work, it should be replicated in future projects. Ad hoc processes are rarely replicable unless the same team is working on the new project. Even with the same team, it is difficult to keep things exactly the same. A closely related issue is that of process reuse. It is a huge waste and overhead for each project to produce a process from scratch. It is much faster and easier to adapt an existing process model;
- **Quality** – quality in this case may be defined as the product's fitness for its purpose. We can bear in mind the definition of quality from the British Standards Institute, 'The totality of features and characteristics of a product or service that bear on its ability to satisfy a given need' (BSI 2002). One goal of a defined process is to enable software engineers to ensure a high quality product. The process should provide a clear link between a customer's desires and a developer's product;
- **Improvement** – no one would expect their process to reach perfection and need no further improvement itself. Even if we were as good as we could be now, both development environments and requested products are changing so quickly that our processes will always be running to catch up. A goal of our defined process must then be to identify and prototype possibilities for improvement in the process itself;
- **Tracking** – a defined process should allow the management, developers and customer to follow the status of a project. Tracking is the flip side of predictability. It keeps track of how good our predictions are, and hence how to improve them.

These seven process goals are unsurprisingly closely allied to McCall's quality factors for software (McCall, Richards and Walters, 1977), implying that it is practicable to apply the same quality criteria to our software process as to the software we wish to produce by applying it.

In order to meet these seven aspirations of a software process, it is necessary to document the process. The first such attempt at a formalization of a development process was made by Royce (1970), in the now famous **waterfall life cycle**. The process is depicted in Figure 2.2. Each stage, from software requirements through to system maintenance, is serially dependent upon the completion of its predecessor. This simple interpretation of the waterfall model is actually a distortion of Royce's original intent but is sadly the way most legacy development processes view it. Each stage builds on its predecessors, adding detail as we get closer to the software implementation, leading to inevitable redundancy and

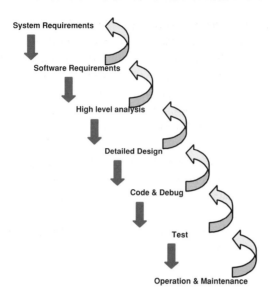

Fig. 2.2 The waterfall life cycle

so to maintenance headaches for the products produced. The shortcomings of the simple model are now well understood, nevertheless it remains the basis for many of the standard 'elaborative' approaches to software construction. We shall illustrate, in some detail, the deficiencies of these in Chapter 13 and contrast this approach with MDA with xUML.

There have been many subsequent life cycles defined, such as the spiral model by Boehm (1988), the rapid prototyping, incremental model (Sommerville, 2001) (originally from 'Evolutionary delivery' (Gilb, 1988)), synchronize and stabilize (Cusamano and Selby, 1997), the evolutionary development model for software (May and Zimmer, 1996), and others. Many industry observers contend that such approaches have failed to deliver the promise of putting software production firmly into the arena of an engineering discipline, given well-known evidence of massive budget and time overruns, etc., so prevalent in modern software production.

The scene set, we shall now look at how the OMG's MDA with xUML attempts to address these shortcomings in the software development process.

2.3 Model driven architecture

It is widely accepted that the rate of change of technology is increasing in every sphere of software development. Once the choices for the developer were well characterized and relatively limited. They included operating system and language alternatives. Nowadays the range of options is bewildering in all aspects of technology, ranging from the myriad of middleware options (CORBA, EJB/J2EE, NET, XML/SOAP, etc.), architecture frameworks (client–server, 3/N-tier, peer-to-peer, etc.), persistence and transactional services, quality of service specifications, security infrastructures and languages. It has also become somewhat of a truism to say that companies are expecting decreased time to market whilst business requirements are changing as fast as, or faster than, ever. The heady mix of rapid technology advancement and business change has lead to an effort to decouple the models of business requirements from the technologies that realize them. This allows development teams to exploit the 'next big thing' in technology whilst preserving the essential intellectual property (IP) of their business models. In realizing these goals the MDA provides an open, vendor-neutral approach to building robust scalable systems, based upon well-documented standards such as UML and avoids the 'deadly embrace' of intertwining business requirements with a particular technology. This is why the OMG style the MDA as 'The Architecture of Choice for a Changing World^{TM}'.

In order to discuss MDA we must first examine a few concepts that are key to understanding how MDA delivers its benefits.

It should be noted that although MDA is based upon well-understood engineering disciplines its formal definition is still evolving.

2.3.1 The model

Models, in MDA, are formal representation of the function, behaviour and structure of the system we are building. The rather intimidating term 'formal' is used here to say that the language that we use to describe the model must have-well defined syntax and semantics. The UML, newly enhanced by the Action Semantics work, fits this bill perfectly, providing the necessary apparatus to fully specify complete and rigorous models in MDA. It should be noted that this definition rules out the use of block diagrams and other forms of 'bubble-ology' as proper models (though on occasion we may resort to such techniques in order to help communicate our intent to non-UML literate stakeholders).

There are two basic forms of model that are used in MDA, the platform-independent model (PIM) and the platform-specific model (PSM). Before we discuss these we require a definition of platform.

2.3.2 The platform

In the MDA, the term **platform** is used to refer to technological and engineering details that are irrelevant to the fundamental functionality of the software. That is not to say that the platform is unimportant, quite the contrary in fact, but it is not germane to specifying the essential nature of the system we intend to build. It is important when we come to construct our system because the platform is a specification of the technologies that we shall employ. Examples of platforms include CORBA, J2EE, C, NET, and a Real Time Operating System (RTOS), etc.

2.3.3 The platform-independent model (PIM)

The PIM is a model that captures all the business requirements of the system we are building. It includes everything we wish to specify about the system free from implementation (platform) bias. In the approach described in this book, it is therefore not a vague and woolly specification but rather it is expressed as a precise executable (xUML) model. Each individual PIM captures an aspect (or viewpoint) of the system. In MDA we term these **domains** (see Chapter 5), which are combined to give a complete, executable description of the system we wish to construct.

Since we express PIMs as executable models, using xUML, the testing of such models is much easier at the PIM level, since they are uncluttered by platform concerns.

This book will cover all aspects of building PIMs and their specification using xUML.

2.3.4 The platform-specific model (PSM)

The PSM comprises all the functionality expressed in the PIM with the added design concerns of the realizing platform. The PSM can be expressed in two primary ways: as a design model in UML and as a language implementation (in CORBA or C++, for instance). The latter is sometimes termed the Platform-specific Implementation (PSI).

There is, of course, a relationship between the PIM and its PSM. The PSM is derived from the PIM by applying a set of systematic transformation rules, called a mapping. Figure 2.3 shows how conceptually a PIM of an air traffic control domain may be transformed into its corresponding PSM by apply systematic mapping rules.

The mapping from PIM to PSM is itself captured as an UML model. In the approach taken in this book the mapping is actually an xUML model. Because the PSM is derived from the PIM by applying a mapping, it is not a maintained deliverable, but is **always generated** from the PIM. If business requirements change, these are reflected and tested in the revised PIM and then we apply the mapping rules again to derive the new PSM. In this way the PSM can never get out-of-step with the PIM and we have no redundancy headaches. We can accommodate changes in the platform or in the system's non-functional requirements

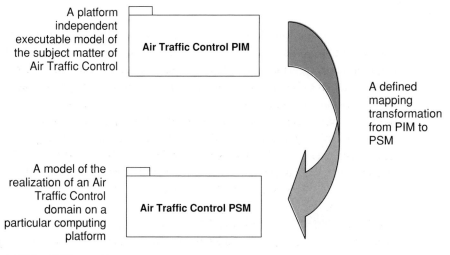

Fig. 2.3 A PIM to PSM mapping

Executable Model Driven Architecture

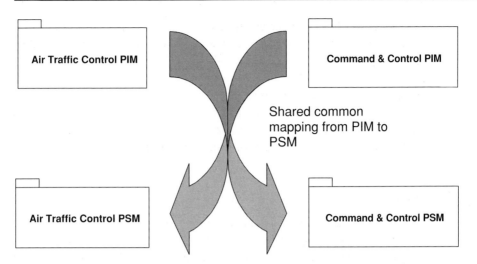

Fig. 2.4 Shared mappings from PIM to PSM in MDA

by reformulating the mapping rules and regenerating the system. In this way, the PIMs never become obsolete due to changes in the technology and are therefore enduring and reusable.

The conceptual coherence of the MDA approach gives us another advantage: the mapping from PIM to PSM is itself a highly reusable entity (in fact such mappings and commercial automated tools to perform the mappings are readily available). This is shown in Figure 2.4.

Now we have laid the foundations of the MDA but before we can go on to discuss it further in an executable context, we must first explore what is meant by xUML.

2.4 Executable UML (xUML)

UML, as its full name suggests, provides a unified notation for the representation of various aspects of systems. It is not the purpose of this book to introduce the whole range of notations and semantics that the UML represents. Rather, it is the intention to use some of the UML notation to build models with specific and exciting properties, namely that the models themselves are executable.

To specify executable models in UML requires the execution semantics of UML to be fully defined. Although the original UML specification defines some execution semantics it does not define enough. This has been recognized and the OMG has extended the UML standard to allow the models to be executable. This is a fundamental part of xUML.

In summary, xUML is an executable version of the UML, with:
- a clearly defined simple model structure;
- a precise semantics for actions, which has been incorporated into the UML standard;
- a compliant Action Specification Language;
- an accompanying process, which is oriented towards:

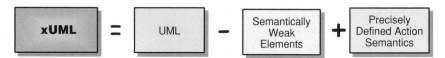

Fig. 2.5 The basis of xUML

- executable modelling;
- large-scale reuse;
- pattern-based design.

Informally, Figure 2.5 shows how we can consider xUML.

2.5 The need for process improvement

So the standards infrastructure is in place to be able to define executable models in UML. But why bother? Well, we have already hinted at the problems that afflict the traditional 'elaborative' software development processes, so what is the problem with building models the way we have for years?

> You cannot solve a problem using
> the same kind of thinking that caused it.
> Albert Einstein.

For the past two decades many of the software systems produced:
- do not meet the needs of the users because insufficient emphasis was placed upon understanding the users' requirements;
- achieve minimal levels of component reuse because any candidate reusable components are so implementation-specific and often fine-grained, that they are suitable for reuse only within a very narrow range of target environments;
- have obsolete analysis and design documentation because later requirements changes were only reflected in the code;
- have excessive maintenance costs and eventually reach a stage where, due to maintenance changes, the software has so little integrity that further maintenance becomes infeasible and it has to be re-implemented from scratch or abandoned.

Why do these major problems occur? Over a similar period, software engineering has been dominated by a development process that can be summarized as:

1. Build an informal analysis model of what the system should do in terms of functions;
2. Use the analysis model as the starting point for a design model, adding to it and restructuring it to reflect how the system will be built. Add functions as required to address issues not included in the analysis model;
3. Write code to implement the functions and fix the code as problems materialize during testing.

In the last ten years, a plethora of new object-oriented methods has emerged, a few of which have achieved mainstream status. These new methods propose a change to the above process and can be summarized as 'Follow steps 1 to 3 above, but change the word 'function' to 'object' or 'class''.

This may not be a big enough change. It is based on the same kind of thinking that caused the problem.

A system development process based upon the principle of building executable models is a **fundamental change** in the way we perceive software development. MDA with xUML seeks to remedy the shortcomings in the 'legacy' elaborative approaches.

One of the difficulties of describing an abstract notation such as a software development process is that we have to use words that may already have a particular meaning for the reader: a meaning which is different from the one we intend. Even the word 'object' has had several meanings thrust upon it over the years. Two words that inevitably occur in any description of a software development process are 'analysis' and 'design'. The activities which occur during 'analysis' and 'design' in the MDA process are different to most other processes. We have tried to use these words with care and will explain them as the book progresses.

2.6 The principles of MDA with executable models

To build high quality executable PIMs, it is important to understand the fundamental principles upon which MDA with xUML is founded. In particular, it is necessary to appreciate the reasons that the MDA process differs so significantly from the other mainstream methods. The MDA process embodies these distinctive characteristics:
- Precise, complete analysis models, captured as PIMs, that can be subjected to rigorous testing by simulation;
- Scalable partitioning that organizes models according to distinct subject matter areas;
- Unambiguous standard notations that can be understood by human readers;
- A conceptually coherent process that provides a small but sufficient set of techniques to address all subject matters, including 'design', in a uniform way;
- Implementation by translation, in which the entire xUML system is automatically generated from the PIMs, using a set of rigorously specified rules that deliver a system with the required performance characteristics;
- Large-scale reuse in which entire sets of classes are reused as a single component.

Let us examine each of these features in turn.

2.6.1 Precise, complete analysis models

The terms 'precise' and 'analysis' are rarely found together as they are here. There is a prevailing view that analysis is in some sense an 'overview' activity, in which the most salient aspects of the problem are described and the detail is left for a subsequent phase. This attitude is revealed in the document titles associated with an analysis phase. How many times have you been confronted with a document named something like 'High Level System Specification'? What does the phrase 'High Level' mean here? Upon studying the document, the answer often becomes immediately apparent, it means 'Vague'. However, the report's author was wily enough to realize that a document entitled 'Vague System Specification' would possess limited reader appeal and appears to contain a terminology conflict; after all, if it is vague, how can it be a specification?

Many analysis approaches offer the analyst a very convenient escape route. They require the analyst to construct a model that is **implementation-free**. The rationale behind this is very sound. By constructing an implementation-free statement of a problem, the model is:
- simple;
- free from premature implementation decisions;
- reusable for systems that are to run in different target environments.

Whilst these ideals have noble intent, sadly the term 'implementation-free' is often interpreted to mean 'detail-free', or even 'free from all the difficult parts'. In the old days, when we used to be involved in reviewing Yourdon analysis models, we were often faced with this phenomenon. While reviewing a model of a telecommunications network management system, we would ask a question such as 'How does the system allocate network resources when it receives a request to set up a telephone call?' This would be met with a gleeful 'Aha! You asked a question beginning with the word 'How', this is **obviously** implementation detail, and therefore not an analysis issue! The coders will sort that one'. This is **obviously** an unhelpful view of analysis. The question clearly pertains to the subject matter of telecommunications network management and one would not expect different answers depending upon the software implementation decisions that are made subsequently.

One helpful way to distinguish detailed analysis questions from implementation questions is to ask, 'What type of expert would I consult to obtain an answer to this question?' In the above example, one would consult a network management expert and not a software designer or programmer (although, of course, one person might be capable of performing both roles). This is, therefore, an analysis question. But what if the question was 'How do you store the data about two party calls in progress?' In this case, the intention is to solicit information about the most appropriate data structure to use for data of this type (i.e. low volume, high access rate). Clearly, this is an issue that requires access to a software engineering expert. It is implementation detail and would have no place in an analysis model for network management.

So how much detail should be included in an analysis document? In MDA, where the PIM fulfils the role of analysis documentation, the answer is straightforward:

> **Executable MDA principle**
>
> A PIM contains all the information pertaining to the subject matter in question – including the hard parts.

This principle of MDA relieves the analyst of the dilemma created by the simplistic and ambiguous 'Analysis is **what**, design is **how**' distinction.

How do you know when to stop analysing? With traditional methods, the technique used to determine whether the analysis is complete is typically to call a review. In such a review, attendees are equipped with:
- copies of the analysis models, which are to be reviewed for compliance;
- copies of the requirements against which the analysis models are to be reviewed (although this seems to be regarded as an optional extra in many cases);
- very little time.

Our experience of such reviews is that, because the reviewers have been given such short timescales, they arrive having reviewed only the first few pages of the analysis model document. They will have found many defects, such as non-conformance to the document numbering scheme, typographical errors, split infinitives and other errors that will be very unlikely to compromise the quality of the delivered software. During the review meeting, the attendees will rummage through the endless diagrams in the model, trying to find errors. Some diagrams are simple and so errors are easy to find. Some diagrams, however, are rather complicated and hard to understand. Each reviewer assumes that another reviewer has understood and checked those. Once everybody has had a really good look at all the diagrams, they will form a collective opinion on whether the model is finished. Of course, all the serious defects are usually hidden in the 'hard parts', but, because nobody found them, they are assumed to be absent. The model is therefore deemed to be complete.

Not all reviews follow this pattern, some are run on a very formal basis, and there is genuine effort to find and rectify defects based on tracing model elements back to requirements, but the principle is the same. It is a **matter of opinion** that the model is complete.

The completion criterion used for the implementation phase is rather different: imagine a customer's reaction if, upon project completion, the developers handed over the system source code, declaring 'Behold! Your system is ready. We've not actually compiled or executed any of this code, but don't worry – we've had a **really good look** at it and nobody found anything wrong. Good luck!'

Clearly, this is an unacceptable completion criterion for the implementation phase, so why is it accepted for analysis? The answer is, of course, because we **have no choice** with traditional analysis methods, since they yield informal diagrams that lend themselves only to visual inspection by humans.

In stark contrast to the lamentable situation with traditional approaches, MDA provides a strong analysis completion criterion:

> **Executable MDA principle**
>
> The PIM is complete when it delivers the expected results for all specified test cases.

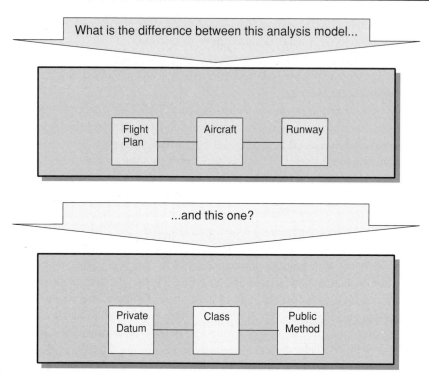

Fig. 2.6 Distinct subject matters in a system

This criterion is a reflection of the fact that PIMs are both **precise** and **complete** and as a result they can be executed. PIMs are therefore subject to the same rigorous testing regime and completion criteria as the delivered system, that is they deliver results rather than just documentation.

2.6.2 Scalable partitioning

Try this exercise: spot the difference between these two outline UML models shown in Figure 2.6.

Clearly, they are topographically identical – they both show three classes and two associations (we shall meet these concepts soon). The difference is in the type of expertise that would be required to complete the models. To complete the top model, we would need access to an expert in air traffic control. Completion of the lower model would require access to an object-oriented programming expert. The difference, therefore, is the type of expertise that each model formalizes (Figure 2.7). It is unhelpful to refer to the upper model as 'analysis' and the lower model as 'design' – this simply reflects the type of subject matter being formalized: real world or technology. What matters is that both subject matters are inherently complex and that complexity needs to be addressed by building xUML models to formalize the expertise in that subject matter.

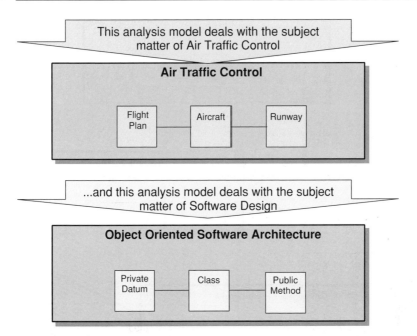

Fig. 2.7 Domains are subject matter related

> **Executable MDA principle**
>
> Each subject matter area is called a domain and we use a separate PIM to capture each domain.

For those readers who enjoy metathinking, it should be noted that while the result of executing the upper model would be a working air traffic control system, the lower executable model would be a code generator (about which there will be more in Chapter 13)!

A good domain model encapsulates a subject matter, which may be 'problem-oriented' or 'solution-oriented' (Figure 2.8). Such aspect engineering promotes a divide and conquer philosophy without slipping back into the quagmire that is functional decomposition.

With this view on the world, we can see that the way we wish to use an implementation language can be regarded as a subject matter amenable to analysis. Contentiously, we might state this as; **design is just another subject matter for analysis**. This injects a level of conceptual coherence and simplicity into system development that the 'Analyse–Design–Implement' processes overlook.

To recap, each domain captures valuable intellectual property (IP) relating to a single subject matter area. A domain is captured as a PIM. Each PIM is executable in its own right and can be tested. Of course, real systems are made up of many domains, some of which will be end user and problem related, others may provide some generic capability. We therefore build systems by putting domains together. MDA allows us to put domains together without the domains having any explicit knowledge of each other, which ensures

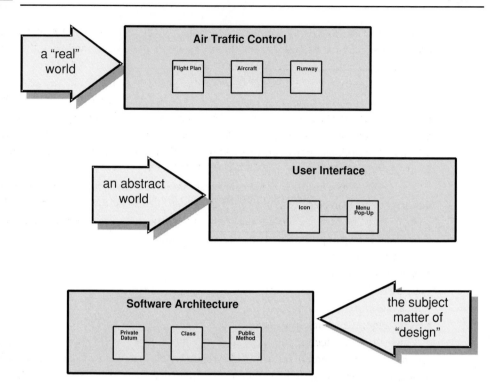

Fig. 2.8 Domains covering different abstraction levels

a loosely coupled system that is easier to maintain. Such assemblies of domains are also executable 'in-concert' and so can be tested to provide overall system level assurance.

2.6.3 Unambiguous standard notations

UML is a widely adopted standard notation for the representation of object-oriented systems (www.omg.org/uml). We shall use UML extensively in the MDA process, however, whilst the original core UML specification is necessary, it is not sufficient for executable modelling. In recognition of this the UML specification has been enhanced to incorporate the Action Semantics. This serves to resolve many of the ambiguities in UML and adds definitions of execution behaviour for appropriate UML model elements. Core UML plus the supplementary action semantics, termed the xUML, is sufficient to build executable PIMs.

Not all elements of UML have execution semantics. For instance, the component and deployment diagrams do not have run-time behaviour and the UML action semantics have nothing to say about them. Component and deployment diagrams can be used, if appropriate, with xUML models but in exactly the same way as they would with a conventional UML model. This book, therefore, has little further to say about the component and deployment views.

Use cases are utilized in the MDA process, they are an informal notation and no specific execution semantics are required of them. We discuss use cases and how they are best exploited within the context of MDA in Chapter 4.

Activity diagrams could potentially be used to specify actions in part. However, activity diagrams alone would not be enough. It is still necessary to define the required behaviour inside each action state. As a result, xUML does not use activity diagrams to specify the executable behaviour of operations or state actions. Activity diagrams may prove useful for informal descriptions, such as business workflows for example. The use of activity diagrams for these purposes is the same as for standard UML and not covered further by this text.

The state charts defined in UML are most certainly used in xUML and the action semantics submission clarified much of the behaviour definition for state machines. The state chart notation defined in UML is very rich, but, particularly for novices, potentially very confusing. For example state actions can be defined:

- on entry to a state;
- on exit from a state;
- on a transition between states;
- on an internal transition;
- as an ongoing activity in a state.

Using all of these techniques is very powerful but any state-based problem can be specified just by using entry actions as a so called Moore state machine or by transition actions as a Mealy state machine. Now a Moore state machine may have a few extra states and a Mealy state machine may have a few extra transitions compared to the minimalist state machine possible using all of the above. However, MDA recognizes that they will be easier to get right, easier to understand and easier to maintain. Correctness, understandability and maintainability are considerably more important criteria than minimalism.

Our experience is that many users of UML tend to avoid the state machine parts because they are too complex. It is also our experience that the effective use of state machines to specify systems is very important. As a result we have chosen to use a subset of UML state machine notation that is sufficient to describe any state-based problem, but that is simple enough to be unintimidating. This subset uses entry actions only and avoids hierarchic and concurrent states. This subset also avoids any situations where non-computability may arise.

There is broad agreement that one of the fundamental purposes of analysis is to **manage the complexity** of the problems so that humans can understand and visualize them clearly. Unfortunately, simply using the full myriad of UML notations can easily produce models that are incomprehensible to normal people (i.e. those who are not software engineers and many who are). Is this a problem? Well, since another stated purpose of analysis is **to capture our understanding of the customers' requirements**, it can be rather demoralizing when a customer, presented with a painstakingly constructed analysis model, reacts with incomprehension at the sight of it.

It is critical to a successful analysis phase that the models can be understood not just by the analyst but also by subject matter experts and customers. To this end, xUML prescribes a simple, coherent subset of the UML notations. The notations are based upon the **structure**

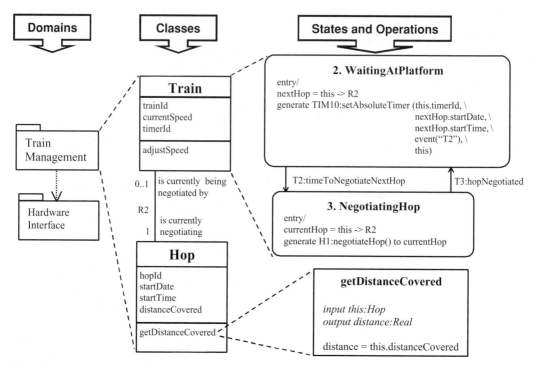

Fig. 2.9 xUML model layers

of the real world not the individual constructs that make up a software system. Not only are the notations themselves simple but the way in which they are organized and integrated conforms to strict rules that preserve the overall clarity of the system specification. There is no scope for individual analysts using their own structural style or building hierarchies of indeterminate depth. The structure of an xUML model is based upon layers of abstraction (information, behaviour, operation) rather than layers of successive detail, each layer effectively being made redundant by the next layer down in the hierarchy. This means that the development process can be more effectively monitored and controlled because a project manager knows exactly how many layers are to be constructed. It eliminates the idiosyncrasies that characterize models built using large, complex notations. Figure 2.9 gives the layers of a model expressed in xUML.

Another key benefit of using a UML subset is that it leads to more consistent modelling style across the development team. If there is only one notational device to express a particular construct then the developers are spared the agony of choice!

We have already seen that one of the cornerstones of the MDA approach is the idea of translating the PIM to PSM by applying a well-defined set of mapping rules. To fully comprehend why this is a valuable idea we revisit the legacy elaborative approach and highlight its particular shortcomings when a non-translational approach to software construction is taken.

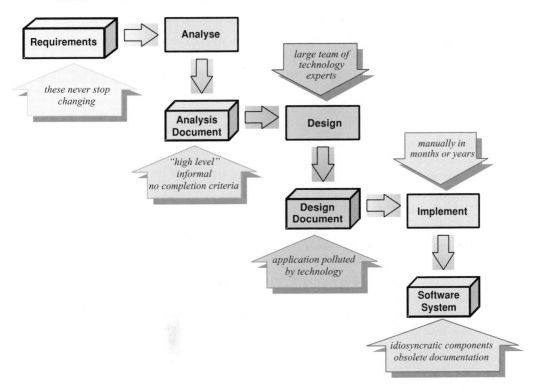

Fig. 2.10 An elaborative process

2.6.4 A conceptually coherent process

In order to appreciate the concepts underpinning MDA we first revisit the 'traditional' elaborative approach.

The fundamental aspects of the elaborative development approach are summarized in Figure 2.10. Of course, there is iteration within each phase and across all phases. We shall come to this.

Let us focus on the 'Design' phase. What is design? In the traditional, so-called elaborative life cycle, it is the process of taking an analysis model – a statement of the **problem** – and elaborating it with additional design detail describing the proposed **solution**. This would include decisions about tasking and concurrency, data distribution and replication, data structures and message passing, but it also includes addition of detail about the problem. This is a dilemma: the resulting model becomes schizophrenic. It is a mixture of statements about both the problem and one particular solution to it. These two completely separate issues soon become irretrievably intertwined and will never again be separable. The analysis investment is steadily diminished as the design progresses. We can liken this to software entropy.

Anybody who has been involved in software development knows that the requirements never stop changing. Of course, we can make them appear to do so by freezing a requirement specification for a particular system release but they are still changing and, if we do not deal with the changes at some stage, the system will not be fit for purpose.

It is a feature of elaborative approaches that a project spends a considerable amount of time in the implementation phase, in which each design component is manually translated into code. In these later stages of development, when a requirement changes, where is that change first reflected? In the analysis model, in the design, or in the code? Of course, many organizations have a strict process by which the requirement changes are 'rippled down through the analysis, design and finally into code'. Some even enforce this process and some of these are even still in business, but they operate at a short-term competitive disadvantage. The more common approach, driven by the realities of hard customer deadlines and limited resources, is to reflect the changed requirements in the code, test it and deliver it. If the developers ever had any spare time, they would, of course, update the analysis and design to reflect the changes they have made to the code. But they never have any spare time and the inevitable happens: the analysis and design models fall into obsolescence, software entropy tends to increase!

The developers in this process are not incompetent or negligent. They each know that they should be keeping the analysis and design in step and that the maintenance task in the future will be made difficult by the fact that the only way to know what the system does is to look at the code. Anyone who has ever been given the task of 'maintaining' a piece of software will realize that the amount of time expended in trying to understand why the software looks the way it does typically exceeds that spent in making the required changes. This phenomenon is simply a consequence of the decision to iterate the design process over all software components, effectively making every component a 'special case', implemented using the programmer's personal coding rules. Such rules, although obvious to the originator, will form the basis of endless intrigue for those sent later to maintain the code. To these engineers, with their own personal programming styles, the original code can often look perverse or arcane and they will probably conclude that the only responsible approach is to recode the entire unit, using a different personal style. Hundreds of lines of code are rewritten and retested, instead of a few dozen lines.

We can therefore conclude that **the elaborative development process is faulty**. It has a number of facets that act as a barrier to success:

- The manually maintained deliverables of the process are 300 per cent redundant. Every fact in the analysis model appears again in the design model and every fact in the design model appears again in the code. With constantly changing requirements, this level of redundancy is untenable. There is barely time to specify the system once, let alone three times. For this reason, the analysis and design phases are often perceived as a means for getting to the first code release. Thereafter, code is the primary commodity;
- Each component of the design is subjected to its own, personal, design phase. This results in a plethora of different design and coding strategies, which impedes maintenance. It also results in code units whose quality is proportional to the quality of the coder. In most projects, there will inevitably be a number of coders who are climbing the learning curve and consequently deliver code units of a relatively low quality. This manifests itself at the system level as compromised performance and reliability.

Although elaborative methods do have a reasonably good track record of delivering working systems, thanks mainly to heroic programmers, they have a much more convincing track record of delivering:

- **Obsolete, reusable analysis models** – these are implementation-free and therefore highly reusable components. Unfortunately, they are incomplete and out of date;
- **Up-to-date, un-reusable code units** – these are, of course, current and meet the needs of the system. Unfortunately, they tend to be heavily contaminated with knowledge of the specific environment in which they reside. This limits their reuse potential for other target environments;
- **Idiosyncratic, expensive-to-maintain code** – because elaborative methods do not require the developers to formalize the universal design policies, there tends to be inconsistent use of the available technologies, resulting in software units that are incomprehensible to other engineers.

The use of such methods has given rise to an increasingly common activity, often referred to as 'reverse engineering'. This is the process by which an implemented system is studied and the design documentation is derived from it. It becomes necessary when, for the reasons discussed earlier, the code has become so incomprehensible, with patches to patches, that nobody can understand how the system works in order to make further changes.

The term 'reverse engineering' like the term 'prototyping' is often (but not always) used to lend legitimacy to an activity that might be more accurately named 'producing the documentation too late'. Reverse engineering sounds easy but when faced with the realities of colossal amounts of poorly structured code, reality soon intervenes. It is hard to reverse engineer something that was never engineered in the first place.

We now examine the contrasting MDA approach of **translating** models from PIM to PSM and outline how this exhibits a uniformity of approach that is both coherent and rigorous.

2.6.5   Benefits of the translation-based process

The fundamental aspects of the translation-based development approach (producing the PSM by applying mapping rules to the PIM – that is **translation**), as used in MDA, are summarized in Figure 2.11. Of course, iteration occurs in this process too. It's entirely reasonable to develop some primary use cases and then develop some key threads through the model right through to prototype implementation. Additional threads can then be developed and the functionality of the system expanded.

The MDA process overcomes the problems of the elaborative methods by refusing to consider each component of the analysis model as a special case, worthy of individual design attention, and regarding the subject matter of computer science ('design') to be the same as any other subject matter, that it formalized in the mapping from PIM to PSM.

Imagine a mature business, which is building another air traffic control system based on those it has built before. Unlike the systems it has built before, this one is to be implemented on new target hardware, using a new target language, sophisticated but unproved database technologies and a new distributed RTOS. Where is the technical risk? In any such system development, there will be a number of subject matters in which expertise needs

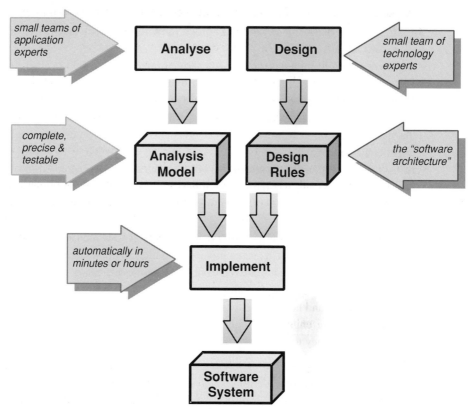

Fig. 2.11 MDA is a translation-based process

to be acquired and formalized. For example, in this case, there is the subject matter of 'air traffic control', in which this organization has a wealth of experience, and there is the subject matter of 'real-time persistent distributed software architectures', with which the organization has minimal experience and expertise. Clearly, it is this latter subject matter that presents the most significant technical risk to the project. This is typically the case in mature organizations whose core business is the supply of a particular type of system. The technology changes faster than the application.

Again we have to consider the terminology issues here. The term architecture has numerous definitions. For an exposition of the profuse definitions of architecture see the Carnegie Mellon SEI web site dedicated to software architecture and its definition (www.sei.cmu.edu/architecture). The term architecture here is used in a particular way to mean the application-independent view of the system design. It spans a number of topics from language features used to implement model elements through persistent storage requirements to support for distributed implementations. Architectures vary enormously with the type of target environment in which they operate. The architecture to support an automotive engine management system will be characterized by being highly constrained in terms of memory and available runtime and it will have simple (or no) persistence and distribution requirements. In contrast the architecture to support a typical information system will be transaction-based with large-scale persistence and distribution requirements. MDA recognizes the importance of these classes of architecture.

Use of an elaborative method relegates the issue of the software architecture to the 'design phase', wherein it is implicitly assumed that the designers understand how best to deploy this complex technology to address the problem of air traffic control. No explicit study of the 'problem' of the technology is required. The technical risk is addressed **informally and late**.

MDA regards analysis as the process of gaining understanding **in any subject matter**. The results are formalized in an xUML model. In just the same way that the study of the subject matter of 'air traffic control' requires formalization of the policies for classes like AIRCRAFT, RUNWAY, and FLIGHT, study of the **subject matter of design** requires formalization of the policies for classes like PROCESS, SHARED DATA REPOSITORY and INTER-PROCESS MESSAGE. This brings us to one of the most important insights within xUML:

> **Executable MDA principle**
>
> Design is a subject matter for analysis.

Having taken this view, we now have a process that requires that we formalize our expertise in how to construct software systems with the required levels of performance and robustness based on particular types of technology. This expertise is never formalized in the elaborative approach and yet it represents an enormous corporate asset. It is a fundamental aspect of MDA that we capture, and make accessible, expertise that resides in the heads of the subject matter experts, whether they are experts in, for instance, air traffic control, or in distributed real-time software architectures.

This process of acquiring expertise, capturing it in a form that is uncontaminated by other subject matters, and making it accessible to others, is the essence of the MDA approach. It is fundamental that if a component is to be reusable it does not intertwine separate subject matters, such as network management, user interface and C++ structures, because it has historically proven impossible to untangle these to reuse any one of them. This is why MDA is primarily based upon **reuse of expertise** expressed as domains, rather than reuse of implemented code, although code reuse can be achieved using implementation domains if desired, as described later. We consider how we capture and construct the models that allow us to realize a translation-based approach in Chapter 13.

MDA is promoting the model from being a **means to an end** to being an end in itself. Traditionally modelling has been a tool for representing, understanding and documenting a design, whilst the code is considered to represent the **real** system. Where a good process has been followed faithfully then the models correspond closely to the code, although, as discussed, this can only be achieved at some cost. More frequently, the models fade into obscurity and become 'shelfware'.

With MDA the focus remains firmly on the models. They represent the **real** system. This means that developers are working at an appropriate level of abstraction on development artefacts that are guaranteed to remain current. Even after the code has been generated the work continues on the models; the generated code is now merely a means to an end – it is the means of getting the xUML models to run in the target environment.

This change of approach is an important aspect of MDA. Developing executable models and then proceeding with traditional design and implementation phases is possible and will deliver the benefit of increased confidence in the models. However, this does not exploit the full potential that executable models can bring.

2.6.6   Large-scale reuse

Hardware engineers have always had an edge on software engineers.

In the 1970s, whilst hardware engineers were busy solving problems by picking up and assembling systems based on existing transistors and logic gates, software engineers were busy rewriting the access code for linked list data structures. Why were they not reusing the linked list code they wrote on the last project? Because the components were expressed at the level of a target programming language and had no reuse potential in systems based on a different language.

In the 1990s, whilst hardware engineers were busy building systems from existing integrated circuits, software engineers were flying into fits of ecstasy because they have deployed the immensely powerful technique of reusing a binary tree class.

Of course, this is a step forward. It's like reusing a transistor. But they could really push the boat out and set their sights on integrated circuits.

In MDA, the integrated circuit is the **domain**. A domain does not necessarily correspond to a software unit, because the choice of software units is affected by the degree of concurrency and distribution needed in a particular system, although the concept of domain may map to a software unit in due course.

A domain **represents a subject matter**. Or, if that does not sound object-oriented enough, it **encapsulates** a subject matter or **aspect** of the system. This is a straightforward idea, really. Consider an air traffic control system with a graphical user interface (GUI). In such a system there will be classes such as RUNWAY, AIRCRAFT, ICON and POP-UP MENU. We could build an AIRCRAFT class that knows how to represent itself as an icon and how to present a pop-up menu on request but, although a popular approach, this would be a transgression of the MDA paradigm. We have committed the ultimate transgression of mixing up unrelated subject matters. It is this that compromises reuse potential because we now have an AIRCRAFT class that cannot be reused in a system that is not based on a GUI. We have **polluted** the subject matter of air traffic control with the unrelated subject matter of user interfaces.

We consider domains in detail in Chapter 5, however, even with this brief discussion, we can start to reason about whether two classes are in the same domain by asking questions like:

In the air traffic control domain, does it make sense to have a AIRCRAFT class without an ICON class?

The answer is clearly 'Yes.' Aircraft can exist without icons. Icons can exist without aircraft. These classes exist in separate subject matters and can be analysed independently and concurrently.

In the air traffic control domain, does it make sense to have a RUNWAY class without an AIRCRAFT class?

The answer is 'No.' The only purpose of a runway is to support arriving and departing aircraft. They are inseparable ideas. It is for this reason that MDA does not explicitly encourage reuse of individual classes. They are too tightly coupled to the other classes in the same domain.

2.7 Mapping of models

MDA offers the potential automatically to transform PIMs to PSMs (Figure 2.12). Modellers will often tag model elements in their PIMs with information to control the translation (e.g. to determine if a class has a dynamic population). They will specify and tailor the translation rules to achieve a system with their desired performance characteristics.

Figure 2.13 shows how metamodels may be exploited in MDA to formalize the mapping from PIM to PSM. We cover this in depth in Chapter 12.

Once built, the mapping from the PIM metamodel (essentially the xUML formalism) to the PSM metamodel, itself expressed using xUML, is a highly reusable commodity. It

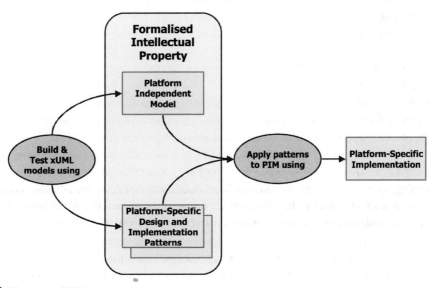

Fig. 2.12 The basis of MDA

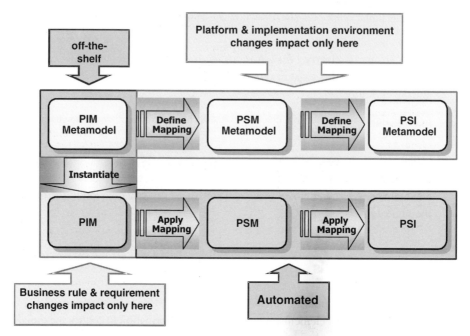

Fig. 2.13 The Role of metamodels in MDA

can be used to generate a PSM and PSI from any PIM, in much the same way as a C++ compiler will generate code for any C++ program. With this scheme, all the business IP is captured, in reusable form, as xUML models. Because this IP is held in PIMs, its longevity is guaranteed, as it will not become obsolete as today's technologies become obsolete.

2.8 Summary of the MDA process

It is the case then that MDA represents a change in the way we have traditionally thought about system development. On the other hand, MDA represents a coherent synthesis of a number of well-established and proven software engineering techniques.

The MDA process delivers a level of sophistication that puts software engineering on par with other, more mature, engineering disciplines, such as hardware engineering, aeronautical engineering and civil engineering. Such engineering disciplines are characterized by the fact that the engineers:

1. Build **precise, predictive models**;
2. Subject the models to **rigorous testing prior to implementation**;
3. Establish a **well-defined** and **automated construction process**;
4. Construct the product from **large reusable components**.

Executable Model Driven Architecture

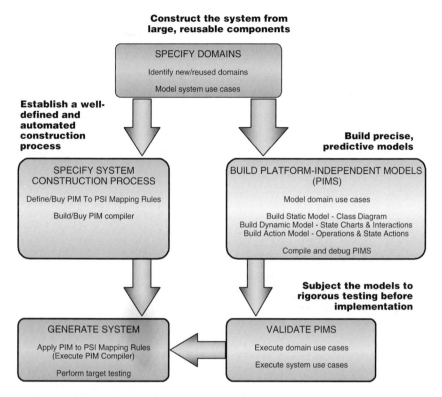

Fig. 2.14 The MDA process and work products

The MDA process and primary deliverables are summarized in Figure 2.14.

The following sections give a whirlwind tour of the MDA process steps, each of which will be considered in further depth in subsequent chapters.

2.9 Specify domains

2.9.1 Identify new/reused domains

The MDA process is based upon the assumption that the development of any system requires the development team to acquire and formalize expertise in a number of diverse, and largely unrelated, subject areas or **domains**. These might be **application domains**, such as 'air traffic control', 'insurance' or 'network management', or **service domains** that support the application, such as 'user interface', 'hardware interface' or 'alarms'.

The expertise for each domain is captured in a PIM, expressed in xUML, which addresses that single subject matter only and is uncontaminated by knowledge of other system aspects. This results in large components that are exceptionally reusable. The set of domains that are

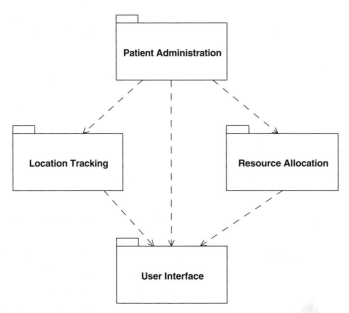

Fig. 2.15 An example domain chart for a hospital admissions system

to be used in the context of a particular system development is shown on a **domain chart**, with **dependencies** showing the client–server associations between them (Figure 2.15). This will include any pre-existing software that is either bought for the purpose or being reused. Each piece of pre-existing software can be treated as a domain and incorporated into the domain chart. This approach provides projects with a migration strategy to MDA since it is not necessary to have a 'green field' project.

We consider domain modelling in Chapter 5.

2.9.2 Define system use cases

A set of use cases is defined, representing the various capabilities that the system must deliver from the viewpoint of the users, human and non-human, of the system. The role that each of the domains plays in realizing each use case is captured using domain-level sequence diagrams (Figure 2.16). These describe the interactions between the domains. The domain chart, use cases and the domain-level sequence diagrams are system-wide views that define the partitioning of the system into domains; the functionality the system must support and the allocation of that functionality to each of the domains.

A domain-level sequence diagram describes the set of interactions between domains required to achieve a single use case scenario. If we look across the whole set of domain-level sequence diagrams from the perspective of a single domain we can see all of the interactions that domain must support. This list of interactions will identify

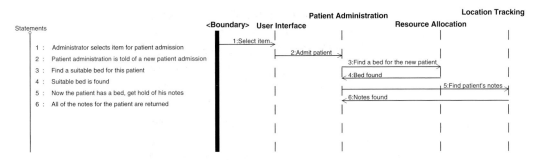

Fig. 2.16 Sequence diagram for primary scenario in the use case **admit a patient**

the scope and responsibility that the domain will have together with an outline of its interfaces.

Each domain will have its own PIM based on a class diagram for that domain. The scope, responsibility and interfaces to the domain, identified from the use cases, will serve to drive the incremental development of the PIMs, which can be built on a use case by use case basis. They will also form the bedrock of the domain use cases and the simulation scenarios used to test the PIMs.

We consider use cases in Chapter 4.

2.9.3 Define domain contracts

A realistic system is always composed from many domain models. Each platform-independent domain model, by the very nature of platform independence, assumes that other domains will provide certain services or deal with other aspects of the system requirement. Thus each domain model may publish a number of operations that define contractually the services it provides. Similarly each domain model may publish a number of contracts that define the services that it requires. The contract specification includes detail about the synchronous and asynchronous behaviour as well as expected returns from stimuli (Figure 2.17).

2.10 Integration of PIMs

At the point where two or more platform-independent domain models are to be integrated, a bridge specifies the PIM-to-PIM mapping between a required operation and a provided operation.

The PIM-to-PIM mapping between two domain models is often based on the idea that a class in one domain has an associated **counterpart** class in the other domain. For example, an instance of 'patient' in the patient administration domain, representing a person receiving

> **PA1::admitPatient**
>
> **Description**
>
> *Perform the activities necessary to admit a patient (whether in- or out-patient).*
>
> **Contract**
>
> *Closed Non-blocking*
>
> **Contract Description**
>
> *The operation will reliably perform all the activities necessary to admit a patient. This includes ensuring that all the resources are obtained for the admission and treatment of the patient.*
>
> *If no such resources are available then the caller is informed and no resources are allocated.*
>
> **Input Parameters**
>
> *newPatientNumber Type: Integer*
>
> **Closure Description**
>
> *On successful patient admission or on failure due to resource allocation failure.*
>
> **Closure Notifications**
>
> *Outgoing Synchronous Call A1:patientAdmitted*
>
> *Outgoing Synchronous Call A2:noBedsAvailable*

Fig. 2.17 Contractual definition of the 'Admit Patient' operation

treatment, may have a counterpart in the location tracking domain called 'item', representing a set of medical case notes located in a specific place.

The domains on the domain chart are assembled by specifying bridge mappings that identify the counterpart class pairs and the way in which operation requirements of one domain are satisfied by operations provided by classes in other domains.

Domain integration is considered in Chapter 10.

2.11 Build PIMs

2.11.1 Impact of use cases

The system-wide use cases, described above, are considered to identify the impact of each use case on a single domain. The object interaction for each scenario of the use case can be captured as an object-level sequence diagram (Figure 2.18). The impact of each use case on a domain will serve to drive the incremental development of the state and process models and also direct the simulation and testing of that domain.

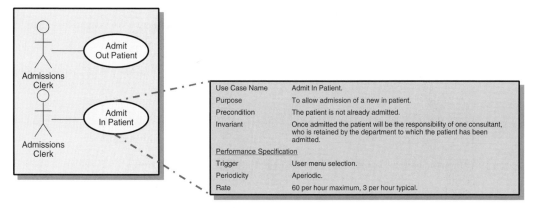

Fig. 2.18 Domain level use case for `Patient Administration::admitInPatient`

2.11.2 Build the class model

The PIM for a single domain is constructed in three layers. The first layer is the **Class Model**, describing the classes in the domain in terms of their characteristics and associations with other classes. This is a crucial model as it is the foundation for the domain PIM as all other model elements are based on it (Figure 2.19).

Class modelling is considered in Chapter 6.

2.11.3 Specify state machines and operations

In the second stage the **dynamic behaviour** of the domain is defined. State models are used to specify the way in which an instance of a class responds to asynchronous stimuli (**signals**). Operations are used to specify the response to synchronous invocations.

A subset of the UML state machine formalism is used for its unambiguous semantics. The chosen subset is powerful enough to model any situation. The state models for classes are constructed by considering the **life cycles** of the objects in the class.

The state models are represented as both **statecharts** (Figure 2.20) and **state tables**. Having two automatically maintained views makes it easy for the analyst to specify **all** of the state-dependent behaviour.

Synchronous operations are represented on the class diagram in the operations compartment of each class.

In considering these life cycles and synchronous operations, there is a strong emphasis on collaboration and delegation of responsibility. The models define how the classes collaborate with one another to achieve the behaviour required from the domain.

An important aspect of the xUML formalism is that all the signals not only have a defined destination (appearing as the events on statecharts), but must also have a defined source. This enables testing and review to confirm that models are complete.

Another important feature is a set of rules defining what cannot be assumed about the ordering and timing of signals exchanged between classes. These rules have been carefully

48 Model Driven Architecture with Executable UML

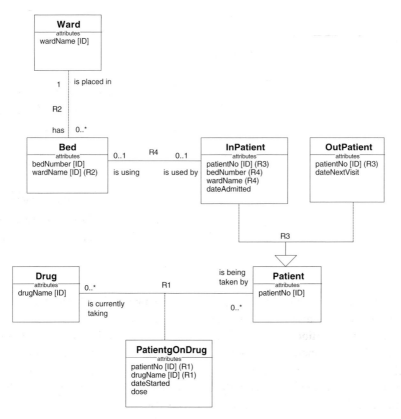

Fig. 2.19 Class diagram for the patient administration domain

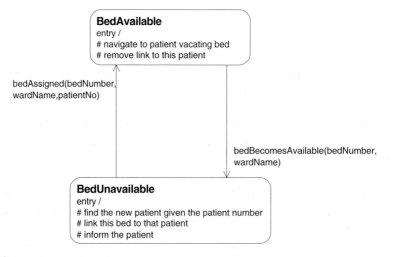

Fig. 2.20 Statechart for BED class

Executable Model Driven Architecture

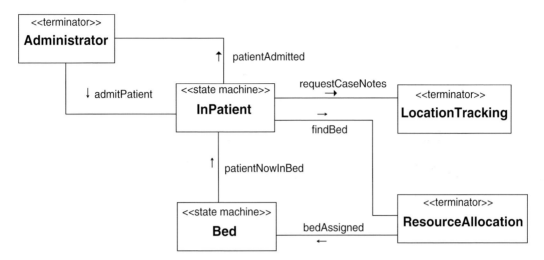

Fig. 2.21 Class Collaboration Diagram for Patient Administration domain

constructed so as to permit optimal mappings to a wide range of platforms, including those with characteristics such as distribution and multi-threading.

The **class collaboration diagram** (CCD) (Figure 2.21) provides a summary of the synchronous and asynchronous interactions between the classes in the domain model.
- Synchronous interactions, or **invocations**, are depicted as full-headed arrows;
- Asynchronous interactions, or **signals,** are depicted as half-headed arrows. These are associated with the state transitions in the statecharts.

A key use of this view of a domain model is to provide an understanding of the layering of responsibility between the classes. The **class collaboration view** shows the interface classes, called terminators, which model connections to other domains.

2.11.4 Define action behaviour

During the third stage the actions are defined in detail. Actions are defined in two places: in statecharts in response to signals and in the methods that implement operations (Figure 2.22).

The actions are expressed using the **Action Specification Language** (ASL) (Wilkie et al., 2002), which is a UML action language compliant with the OMG's Precise Action Semantics for the UML.

A key feature is that ASL provides facilities, formulated at the UML level of abstraction, to manipulate the elements of the PIM. This means that the actions are specified in a platform-independent manner. No assumptions are made about middleware, implementation language or software design policy.

ASL is also used to specify the initial conditions for model execution as well as external test stimuli. In addition, features are provided to allow models to access pre-existing legacy code and other platform-specific domains models.

Dynamic modelling is covered in Chapters 7, 8 and 9.

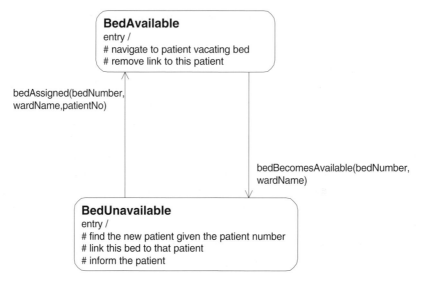

Fig. 2.22 Entry actions for Bed.BedUnavailable and Bed.BedAvailable states

2.12 Validate PIMs

2.12.1 Execute use cases on each domain

The PIMs for each domain are tested individually using a simulator to verify that they exhibit the required behaviour. When errors are detected, the simulator will be used in 'debug' mode to step through the modelled behaviour and isolate the errors. This is the xUML equivalent of 'unit testing'. Domains can be built and tested, one use case scenario at a time if required, providing full support for iterative and incremental development.

2.12.2 Execute system use cases

Once domain level testing has been successful, sets of domains can be tested together to verify the bridge mappings. This is the xUML equivalent of 'integration testing'. The ability to do this early, even with only partially completed domain models, is a major benefit in reducing the risk of late integration problems and thus avoiding the 'big bang'.

2.13 Specify system construction process

2.13.1 Specify generic design patterns

The xUML domain models specify the intended behaviour of the system in a platform-independent manner. In order to produce production quality target code, a route must be found to define the mapping rules from PIM to PSM taking into account issues such as:

- use of **implementation domains** – databases, languages, operating systems and so on;
- distribution;
- persistence;
- task and code structure.

Instead of elaborating such detail into the analysis models (with all the problems that that involves), MDA uses abstract design patterns that show how any particular element of the xUML formalism is to be mapped to a platform-specific form.

Such abstract specification is possible because the models to which they are applied have well-defined execution semantics. MDA has the concept of specifying the mapping from PIM to PSM and PSI. The formal capture of these abstract design patterns and how to use them is the MDA mapping. Therefore, the 'design' is not captured as an elaboration of the analysis but, in the form of a mapping from PIM to PSM and PSI, that is independent of any analysis concepts.

With the (tested) domain models and the design patterns, programmers have a solid base upon which to create efficient and correct run time code.

2.13.2 Evaluate and test the software design

The software design specified using abstract patterns must, of course, meet the necessary space, performance and reliability constraints placed on the system. It may therefore be necessary to carry out tests to ensure that the chosen design is adequate. Since the design can be specified in a domain model independent way, such tests can be carried out on 'test' models while the domain models are being developed. As work proceeds, key timing requirements can be combined with metrics from the domain models to feed into the design process.

2.13.3 Automatic code generation

Since the PIMs are built with an unambiguous and executable formalism and the mappings from PIM to PSM and PSI can be fully specified, it is feasible to generate 100 per cent of the code required to execute the system (Figure 2.23). The automation of the mappings from PIM to PSM and PSI results in a code generator. Code generators may be bought off-the-shelf or can (with suitable tooling) be constructed or tailored for a particular project or organization.

Code generated in this way will be functionally correct, but may not meet all performance and time constraints. In this case there are a number of strategies for optimization.

First, a sophisticated code generator can be made to recognize constructs within the xUML models that are amenable to optimization.

Secondly, directives can be planted within the model to guide the code generator in its work.

Finally, a mixture of automatic and manual code generation can be used with the code generator performing the bulk of the mundane and repetitive work. This leaves the skilled designer with the job of performing tricky optimizations.

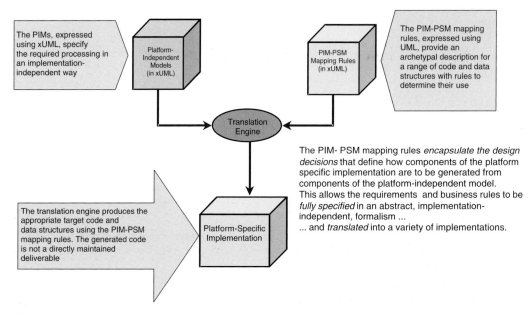

Fig. 2.23 Generating the target system

With the code generation approach, any defects found during system testing will be a manifestation of defects in either the PIMs or the mappings. The target code need not be maintained directly by hand. This eliminates the redundancy that characterizes traditional development processes, in which the target code, analysis and design have to be kept in step manually.

2.14 Summary

In this chapter we have provided an overview of the key elements of an xUML model and looked in outline at the MDA process.

In the next chapter we are going to focus on how MDA might be mapped to a typical project process. If you feel you would like to get stuck into the modelling first then skip the next chapter and return to it later. On the other hand if you feel a good understanding of how MDA works in a real project is needed first then continue to read in the sequential order of the chapters.

3 Using MDA in a typical project

3.1 Introduction

The OMG's Model Driven Architecture (MDA) (www.omg.org/mda) defines a technical development process that can be incorporated into a project life cycle, however it does not itself constrain the choice of which type of life cycle is selected. The choice of project life cycle will be driven by many factors:
- Existing standards within an organization;
- The type of application (e.g. embedded real-time versus database management system);
- The customer's requirements;
- The operation life cycle of the delivered system (e.g. long-term maintained versus unmaintained).

MDA is particularly well suited to iterative and incremental development and so to illustrate how MDA might be applied in a typical project the project management framework as described in the Rational Unified Process (RUP) (Jacobson, Booch and Rumbaugh, 1999; Krutchen, 1999) will be used. The RUP is selected because it is a widely documented, iterative and incremental development process. This chapter will use the project phases defined by the RUP but change the technical process, in particular the analysis and design workflows.

The RUP is based upon the iterative nature of real projects at various levels. Software development occurs as a series of cycles where each cycle delivers a major release of functionality. A cycle may last from many months to a small number of years. Some projects will have only one cycle that goes from requirements to delivery. However, it is more common that the first cycle is followed by one or more subsequent cycles that add major functionality or develop a major product revision. For example, the development of Microsoft's Windows 3.1, Windows 95, Windows 98, Windows Me, Windows XP, would each correspond to a cycle.

Within each cycle the RUP defines four phases:
1. Inception;
2. Elaboration;
3. Construction;
4. Transition.

Iteration also occurs within these phases. The RUP is described as use case driven and use cases can be used to drive the iterations through the phases. Do bear in mind throughout this chapter that the RUP is being used as an example development process. If the number of deliverables (which RUP calls artefacts) and the distinction between the phases doesn't appear to suit your type of application or organization, then consider how MDA would fit into your preferred project life cycle. MDA may be and is used with other software development life cycle models including ones that are very lightweight and streamlined.

In this chapter, and indeed throughout the book, we refer to MDA as meaning the application of the OMG's MDA principles in a context that fully exploits the strengths of xUML.

3.2 The inception phase

> The goal of the inception phase is to understand the system to be developed well enough that there is confidence that the project is feasible and that the estimates for cost and timescales are reasonable.

The inception phase comprises a number of activities (Figure 3.1), which follow in more or less sequential fashion:

1. Produce project vision document (informal);
2. Identify stakeholders;
3. Draw up candidate list of use cases (informal);
4. Agree primary use case set;
5. Interview appropriate stakeholders to start documentation of use cases;
6. Document use cases with primary scenarios;
7. Add representative alternate scenarios;
8. Review use cases, with particular attention to non-functional requirements;
9. Identify primary partitions;
10. Perform object blitz;
11. Produce initial cost estimates, plans and risk assessment;
12. Check completion criteria are met.

In addition, a requirements engineering process must be established. This is done once only and so is not repeated for each inception phase performed within the organization (although it may be modified to reflect a better organizational understanding gained from the first iteration of the process).

3.2.1 Project vision document

With any project there must be a definition of the high level goals that will be addressed by the proposed system. Such a document provides a starting point to eliciting further, detailed requirements. The **vision document** should be relatively short (a few pages at most) and not attempt to provide detailed requirements. It is one of the functions of this document to define the scope of the project, defining the approximate boundaries of development effort.

Using MDA in a typical project

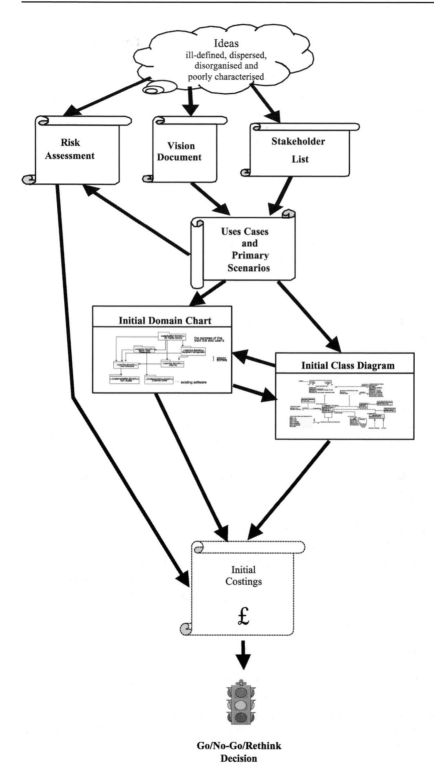

Fig. 3.1 Overview of the project inception process

It is important that all development team members have access to this document and that it is updated as required reflecting the current shared understanding of the aims of the project.

3.2.2   List of stakeholders

A **stakeholder** is defined as anyone that has a vested interest in the outcome of a project. It includes customers, end users, subject matter experts, analysts, developers and managers. The **stakeholder list** is a way of capturing a catalogue of the interested parties and their relationships with the project. This list should contain the name, contact details, role (e.g. customer, user, subject matter expert) and authority. It is important that this document is maintained as, inevitably, the stakeholder list expands.

It is important to identify one individual who is nominated as the **requirements authority**. This role is used to resolve contention over requirements' acceptance, requirements' change or in resolution of queries. The requirements authority may call upon the services of individuals with specialist knowledge of some aspect of the requirements for the system under development. These specialists are referred to as **viewpoint authorities**. It is permissible for the requirements authority to delegate decision responsibilities to designated viewpoint authorities (who should be named in the stakeholder list).

3.2.3   Overview of the initial analysis artefacts

Given the list of stakeholders and vision document, it is now appropriate to refine our understanding of the proposed system by applying techniques that will elicit and clarify requirements. A set of use cases must be developed that reflects the essential functionality of the system. The first task in this part of the phase is to gather from the stakeholders a set of use cases that represent, in their opinions, the core functionality of the system. It is important to elicit these use cases from a wide variety of stakeholders (not just from, say, the customer). Using differing stakeholders will ensure that multiple **viewpoints** are considered when drawing up the candidate list of use cases. Considering many viewpoints can have its downside. It may be very hard to converge on a coherent set of requirements. This is where the requirements authority plays an important role by being prepared to 'draw the line' and resolve differences expressed from different viewpoints.

The next step is to refine the candidate list of use cases to form the set, known as the **key use case** set, which will be considered in greater depth. This list, comprising the set of primary requirements for the project, should be finalized in agreement with participating stakeholders. In order to document and refine this set of use cases, it is necessary to bear in mind the definition of a requirement.

> **Definition**
>
> A requirement is a statement with which the project must in due course demonstrate compliance.

Use cases are not suitable for capturing all requirements. Some projects choose to capture and engineer requirements in the form of textual requirements statements. This approach can be used for requirements that can't be captured as use cases or for all necessities. Caution is needed if requirements are captured in textual form and use case form since their maintenance can then become onerous.

It is important to capture and understand requirements that have an impact on the system architecture, and not all of these will be evident through use cases. Requirements that constrain the choice of implementation solution have an impact on distribution and persistence and demand provably high quality (e.g. safety or security related system) will all have a major influence on the plans and costs.

Each of the use cases identified in the primary set should be documented. In order to do this the relevant stakeholders must be identified and interviewed.

The principal subject matters present within the system may be identified by domain partitioning, leading to an initial **domain model** in the form of a UML package diagram. The domains are represented as a layered set of packages with the package dependencies identified. The initial domain partitioning activity is not driven by the use cases since that would lead to a functional decomposition rather than being based on subject matter. Instead it is an orthogonal activity. Once domains have been identified then the next stage is to see if each domain can be bought or reused. Domains that are not ideally suited but could provide a starting point should be considered. Typically, some domains will have to be developed. These choices could of course have a big impact on costs and plans.

The final step is to perform an **object blitz** in the domains of most interest and highest risk. The object blitz produces a first-cut platform-independent model (PIM) for each domain. This should not be thought of as a complete model, but rather as an aid to documenting the principal classes within each domain. The aim is to produce class diagrams for each of the chosen (typically highest risk) domains that contain well-named classes with associations and attributes captured and named. The purpose is to capture the abstractions and vocabulary of the domain. A good blitz will reveal approximately 80 percent of the classes and associations and probably 60 percent of the attributes. Operations on the classes are not critical at this stage and will emerge from the detailed consideration of how the use cases will be realized by the collaboration of objects in each of the domains.

The UML sequence diagram is typically used to show the time-ordered sequence of interactions between objects. These will be referred to as **object level sequence diagrams**. The sequence diagram also plays a valuable role in describing the interactions between domains. The lifelines on the sequence diagram correspond to domains rather than objects. This gives an abstract view of key interactions. This style of sequence diagram is called a **domain level sequence diagram**. If time permits, domain level sequence diagrams can be produced to bring together the use case view and domain partitions. This will capture the required interaction between domains and determine the types of operation one domain will require of another. Object level sequence diagrams are left until the elaboration phase.

3.2.3.1 Initial costings

Having produced a number of analysis artefacts in the steps above, it is possible to apply costing heuristics to the use cases, domains and class diagrams to arrive at an initial cost model. There are a number of techniques that may be applied to arrive at the initial costings, these include:

- Complexity estimates (COCOMO[1], function point analysis (Boehm, 1981; Symons, 1991));
- Class count;
- Thread count;
- Use case based estimates (Schneider and Winters, 1998);
- Staff skill analogy.
- Weighting.

Estimating techniques go beyond the scope of this text. However, the subject is well documented (Brooks, 1975; Albrecht, 1979). To illustrate the use of class count as a rule of thumb a typical project should estimate five to seven effort days per class. If the project is large, with an inexperienced team, stringent documentation and testing requirements, then the figure may rise to 12 to 15 effort days per class. An experienced team with an effective requirements authority has been known to achieve two effort days per class.

3.2.3.2 Initial risk assessment

A list of major risks to the project should be drawn up. This list should be started alongside the vision document and updated as the initial analysis artefacts are produced.

These fall into a number of categories, including:

- **Requirements**:
  - Risk that requirements change;
  - Risk of unavailability of requirements;
  - Risk that subject matter experts will not be available;
  - Risk that requirements authority is not available;
  - Risk that the requirement engineering process is poorly defined, understood or managed.
- **Technical**:
  - Introduction of new techniques;
  - Introduction of new languages/middleware;
  - Use of obsolete techniques;
  - Use of obsolete languages.
- **Personnel related**:
  - Insufficient committed resource;
  - Insufficient/inappropriate training;
  - Staff retention;
  - Staff recruiting;
  - Inadequate mentoring.

[1] COCOMO–COnstructive COst MOdel.

- **Management-related**:
  - Inappropriate organizational structure;
  - Insufficient organizational change.
- **Environmental**:
  - Risk that the business environment changes too rapidly for the development process;
  - Risk that environmental changes are not detected or appreciated.

Each risk must be acknowledged, prioritized and planned for in the project plan. This is the start of a coherent risk management strategy. Risks clearly have a bearing on cost and schedule estimation and contingency.

3.2.3.3 Inception phase completion criteria

The aims of the inception phase are to produce a well-founded basis for costing, sizing and scoping the system. Once the activities of domain partitioning and object blitzing have occurred then it is possible to proceed with the production of the costing model. It is best to impose a time constraint on the inception phase in order to force closure on the production of the analysis artefacts. There is typically only one iteration in the inception phase and it should last less than three months.

The inception phase is complete when costed plans for the system development can be produced with confidence. This is a subjective decision point since more work will reduce risk and increase confidence. As a result it is common to timebox the inception phase.

If it is decided that the project is to proceed, then the deliverables from the inception phase are used to feed into the next stage of the system development process, namely the elaboration phase.

3.3 The elaboration phase

The goal of the conventional elaboration phase is to develop the analysis and design of the system sufficiently that construction can proceed at low risk. The difficulty with the conventional elaboration phase is defining meaningful and objective criteria against which the completion of the phase can be assessed. Executable modelling permits more effective analysis and design during elaboration and provides well-defined completion criteria for the phase.

> The goals of the elaboration phase using MDA are to develop a set of executable PIMs that demonstrably support the primary scenarios of each selected use case, and to develop the mappings for transforming PIMs through PSMs to target code.

The inception phase will have produced a number of artefacts that feed into the elaboration phase. This could include the following, though you may decide that a subset of these is suitable for your project:

- Functional requirements expressed as architecturally significant use cases (usually 10–20);
- Non-functional requirements expressed as a supplementary specification;
- Scoping document and/or vision statement;
- Pamphlets covering architectural design topics;
- Layered domains and their dependencies;
- An initial PIM for each domain including a class diagram together with labelled relationships, attributes and operations;
- An interaction diagram showing the realization of the primary scenario for each use case (at the domain level and/or the class level);
- A quantitative model;
- Candidate Commercial Off The Shelf (COTS) product selection;
- Project plans and estimates;
- Risk areas;
- Component definitions;
- Agreed priorities for implementation;
- Dependencies on other domains.

And less tangibly:
- Confidence;
- Motivation.

3.3.1 Elaboration artefacts

The focus for the elaboration phase is to deliver a specification of the required behaviour of the system. Although this is the focus, it is entirely appropriate within the elaboration phase to perform early construction activities to attack key risk areas and/or deliver early limited capability versions of the system. The iterative and incremental pattern followed through this phase supports this approach.

The elaboration phase will deliver a set of executable PIMs supplemented by additional information. Specifically, it will deliver the following artefacts, although again a subset of these may be all that is required for your project:

- Additional use cases;
- A refined set of layered domains (represented using packages on the domain chart) and specified dependencies;
- A completed class diagram per domain with classes, associations, attributes and operations;
- Statecharts for classes with state-based dynamic behaviour;
- Interaction diagrams at domain and class level;
- Executable descriptions of operations and state actions;
- PIMs to be tagged with quantitative information where appropriate;
- Component diagram showing allocation of domains and classes to components;
- Deployment diagram showing allocation of components to nodes;

- Fully specified interfaces to other domains;
- Identification of domains (or parts of domains) that can be reused or purchased (if not done in inception);
- Development plan showing the use case scenarios to be included in each iteration;
- Refined quantitative model;
- Managed risk list.

3.3.2 The Benefit of executable PIMs in elaboration

3.3.2.1 Specification of the system's behaviour

A set of executable PIMs, one per domain, forms an easily measurable deliverable from the elaboration phase. The PIMs provide a complete behavioural specification of each domain and they can be tested, using simulation, to demonstrate that they meet the behavioural requirements of the use cases. The PIMs are therefore of higher quality and more useful than models that aren't executable. The quality assessment of executable PIMs is also more straightforward. They can be assessed quantitatively (did they pass or fail their tests?) and they can be assessed qualitatively (is that the best way of modelling that aspect of the system?)

3.3.2.2 Platform and implementation independence

By abstracting away from platform- and implementation-specific issues, the PIMs concentrate on the business issues. The PIMs are abstract but precise, and uncluttered by knowledge of platform and implementation technologies. This makes them easier to understand and verify, easier to maintain in the light of changes in business requirements and applicable to any platforms and implementation technologies.

3.3.2.3 A more focused deliverable

Developers are more motivated to produce executable PIMs because they do something useful. They are also more 'fun' to produce than models that are seen as little more than a documentation exercise. Therefore more effective effort goes into them, problems are identified and resolved earlier and a better quality artefact is produced.

3.3.2.4 A potentially contractual specification

Where implementation is to be carried out by an external party, an executable PIM forms an ideal specification to hand over to the implementers since it fully specifies required behaviour. Testing compliance of the final system against the specification is straightforward because the same tests used to exercise the executable PIMs can be used on the final system. Additional specifications will also have to be provided describing implementation constraints and performance requirements but many of these can be annotated to the executable PIMs.

3.3.2.5 Choice of implementation strategies in the construction phase

xUML models allow a range of implementation strategies to be adopted.

- Mappings to PSMs and to the implementation-specific environment can be defined. Manual implementation can proceed following these mappings, which define how to map UML elements to code structures. Mappings for C++, Java, etc. can be provided as necessary to ensure consistent styles of implementation;
- By automating the PIM to PSM mappings, executable PIMs can be translated automatically into the target language and environment.

Even if the translation route is not adopted initially, it remains an option for future implementations of the system.

3.3.2.6 Exploiting commercial translation technology

There are commercially available translators or code generators that use executable PIMs as their staring point. These may be suitable for the project's target architecture and should be evaluated. Even if they are unsuitable there are benefits to be gained by using commercially available code generators. They can:

- be used to prototype domain models in order to gain valuable metrics;
- act as a demonstrator;
- provide confidence in the process;
- be modified to suit the project's target architecture.

3.3.2.7 A maintainable specification

The executable PIMs should be maintained through the lifetime of the system. It forms the clearest and most precise description of system behaviour. Where translation technology is used it is the only form of information that needs to be maintained since the implemented system is updated by translation from the changed model rather than be modifying the code. If the system is manually-coded from the executable models, then the PIMs should still be maintained.

3.3.3 The role of iteration

The RUP and other modern development processes are based upon iterative development. The primary reason for this is to avoid so-called 'big bang' integration where the first time the components of the system come together is at the end of the project. Ironically, this approach ensures that this does not turn out to be the end of the project because the following painful integration activity can, typically, take 40 per cent of the total project elapsed time. Big bang integration does not work because a large number of critical risks are not addressed until late in the project. In contrast, the iterative approach makes integration progressive.

There is, perhaps, a perception in some quarters that component-based development reduces integration risks. There is every reason to expect that big bang integration in a component-based development will be just as difficult as any other. Components have clearly defined syntactic interfaces but it is incompatibility in interface semantics that will make integration so problematic. Component-based development has not substantially developed the state of the art in interface semantics.

Organizations that have worked with a waterfall style process for many years find the transition to iterative development understandably challenging. The adoption of iterative development tends to be somewhat half-hearted with iteration permitted between major project milestones. For example, having a project milestone called 'Requirement Specification Complete' which needs all of the use cases to be fully documented with primary and secondary scenarios before analysis can start is not iterative development. If the project manager claims that it is iterative because the use cases are specified one by one then you know you're in trouble!

The process proposed below does make use of iterative and incremental development.

3.3.4 What goes on in elaboration?

The inception phase will deliver a set of layered domains represented on the domain chart (see Chapter 5). The domain chart is represented using the UML package diagram (Figure 3.2), showing domains as packages and the dependencies between them.

It also delivers the primary set of use cases (see Chapter 4). Strictly speaking it delivers the primary scenario for each use case in the primary set. Remember the primary set represents

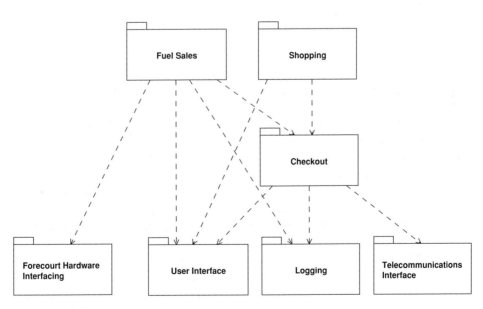

Fig. 3.2 Initial domain chart

Use Case Name	Make Fuel Delivery.
Purpose	To allow a paying customer to deliver fuel of a customer selected grade.
Preconditions	The desired fuel grade is available.
Invariants	Tank level > 4% tank capacity while pump is turned on.
Primary Scenario	1. Customer removes gun from holster;
	2. Attendant enables pump;
	3. Customer selects one fuel grade;
	4. Pump motor is started;
	5. Customer uses trigger to control fuel delivery, causing the system to engage/disengage pump clutch as trigger is depressed/released;
	6. Customer replaces gun in holster;
	7. Pump motor is stopped.
Postconditions	At least two litres of fuel have been delivered.
	The pump gun is in the holster.
	The pump motor is off.
Secondary Scenario 1	Customer delivers less than two litres of fuel.
Postconditions	Less than two litres of fuel have been delivered.
Secondary Scenario 2	Tank level falls below 4%.
Postconditions	All pumps connected to tank are disabled.
Secondary Scenario 3	Customer releases trigger but does not return gun to holster.
Postconditions	Pump switched off after timeout period.
Performance Specification	
Trigger	Customer removes gun from holster.
Periodicity	Aperiodic.
Rate	60/hour.

Fig. 3.3 A use case description

a small proportion of all of the use cases but they should be the ones of greatest significance to the business and to the design of the system architecture. In the elaboration phase the remaining use cases are captured. In all cases the primary scenario is of chief importance. Secondary scenarios can also be described but care is needed here as this could be an open-ended task and it is possible that many alternatives will be handled by generic solutions (Figure 3.3).

As a guide, do not spend longer on the secondary scenarios than on the primary one, unless the effort is justifiable. There are other places in the modelling where error conditions and alternative behaviours are considered.

These use case scenarios represent required threads of behaviour that will be realized by the behaviour of classes within the domains. A typical scenario will interact with the classes in many domains. The inception phase will deliver a domain level sequence diagram (see Figure 3.4 as an example) for each use case scenario (see Chapter 5). This goes some way to describing the required and provided interface for each domain.

At the start of each iteration select a number of use case scenarios to specify. Typically, there will be three to six related scenarios. These scenarios should be used to develop the executable models in all of the domains with which they interact. This allows for parallel working. When each domain has specified the behaviour required to realize the scenarios, the set of domains are brought together so that the behaviour across the set of domains can be

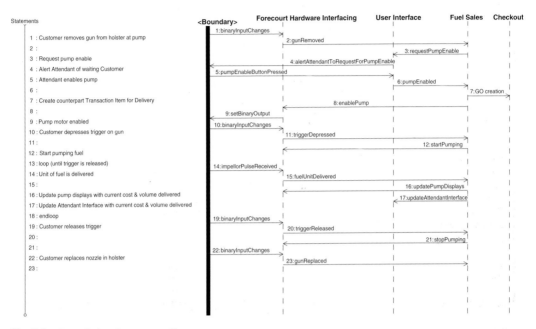

Fig. 3.4 Domain level sequence diagram

verified. Domains should be version controlled and so it is strictly a set of domain versions that are brought together. A set of domain versions is known as a **build set**. There will be at least one build set for each iteration. This ensures that the domains are progressively integrated as the elaboration phase proceeds and avoids the big bang integration of the domains at the end of the phase.

Develop and integrate the scenarios in the following order:
1. primary scenarios of the key set of the use cases;
2. primary scenarios of the other use cases;
3. secondary scenarios of key set of the use cases;
4. secondary scenarios of the other use cases.

Each domain should have a small team of one or two developers. One developer is sufficient for a domain of 20 classes or less (adjust this figure depending on the experience of the developer). The developers will take responsibility for the domain and all of its classes. They will need to consider every use case that has been selected for the current iteration to see if it affects their domain. Separate more senior roles are required:
- to have overall responsibility for the system use cases;
- to have responsibility for the domain integration.

This style of working has a number of benefits:
- it maximizes the opportunities for parallel working;

- it relates the most stable aspect of the system model (the class groupings) to work breakdown;
- individuals can be assigned on the basis of existing expertise or they can develop focused expertise;
- integration happens continuously throughout the project;
- there is considerable flexibility;
- stronger team members can be used to support weaker team members;
- communications paths within the team are efficient.

This approach to progressive integration can be extended beyond one team. If the system under development is very large and the work has been broken down to a number of subsystems that are being treated as separate projects, then progressive integration between the subsystem teams should be actively planned and individuals in each team should be nominated to be responsible for specific interfaces. This requires coordination between the teams to ensure that the system-wide use cases are realized in a suitable order across the whole project. This extra effort and discipline is worthwhile to avoid the perils of 'big bang' integration.

3.3.5 Modelling within a domain

3.3.5.1 The domain class diagram

The foundation stone of the domain PIM is its class diagram. The class diagram from the inception phase is, hopefully, a good starting point. A successful inception phase will have produced a class diagram that will be the core of the final class diagram. Extra classes, associations, attributes and operations will be added but the abstractions identified in the inception phase will still be found (although refined) in the final model. Before developing it further check that the class diagram is fully specified, and that it has role names on the associations and descriptions for the classes, associations, attributes and operations. Note that 'fully specified' doesn't mean finished, it means that all of the elements that have been included in the model so far are fully defined. A PIM class diagram that has associations with missing role phrases or missing multiplicity isn't fully specified and probably isn't useful. We have found models in projects, and regrettably in books, with most role phrases missing. As a result only the author of the model can understand it and then only for a short time. As a model is developed, classes, associations, attributes and operations are added. Each new class should have a clear name and a description. Each association should have role phrases at both ends and multiplicity at both ends. Each attribute should have a clear name, a defined type and a description. During the inception phase, finding operations is not a priority but any that are added should have a brief description. Figure 3.5 shows an example class model from a system for fuel sales.

We will use the class diagram to capture all of the classes in each domain. The UML allows class diagrams to be used in other ways too. Class diagrams can also be used to show the subset of classes and operations required to realize a given use case. This is called a use case realization. Clearly a separate use case realization diagram is required for each

Using MDA in a typical project

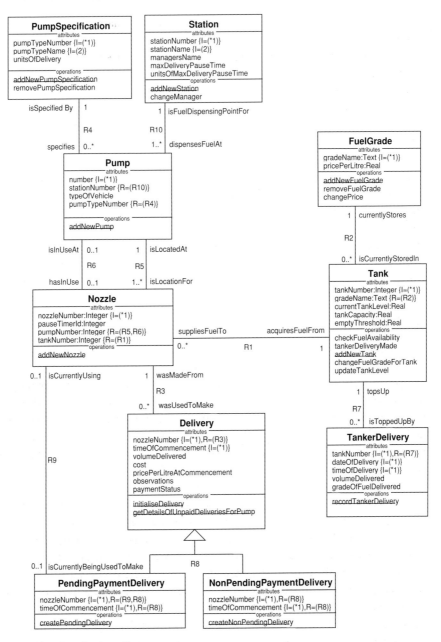

Fig. 3.5 Fuel sales domain class diagram

use case. However, maintaining a large set of use case realization diagrams can become onerous and can be omitted provided that the whole domain class diagram is maintained (see Chapter 6). The MDA process does not require use case realizations to be developed and maintained, but they may prove useful as sketches to work through a problem when the overall class diagram is being developed.

3.3.5.2 Using statecharts

Classes divide into two broad categories, sometimes called active and passive (see Chapter 7). The behaviour of an active object (an instance of an active class) changes over time as the object moves through its life cycle. The way that an active object reacts to a message it receives depends upon its current state. Active classes are represented using statecharts. Passive classes have the same behaviour at all times and so operations on the class are sufficient to define its behaviour. Active classes have operations for any behaviour that is state independent.

The UML statecharts support all forms of state modelling. The developers of UML did not intend that all forms of state modelling be used. In fact, a small subset of the UML statechart notation is entirely adequate. This has the benefit that it is simpler to learn and easier to interpret. Think of the statechart as describing the states of an instance of the respective class. Each instance will be in a given state, but each instance is independent of other instances. This is obvious when you consider one instance of an aircraft can be airborne whilst another instance of aircraft is on the ground.

There is a common belief that statecharts are only useful where classes are complex. The complexity of the class is not the issue but rather its behaviour over time. If the set of operations the instance of the class can respond to varies over time, or in response to previous events, then a statechart may well be useful. For example, consider a class called JOB, which can be started, paused, continued or stopped. An object will only respond to `pause` if it is started and it will only respond to `continue` if it is paused. The JOB class should capture this with a statechart. The number of classes that require statecharts will depend upon the nature of the domain being modelled. Some domains will not require any statecharts whilst for others, as many as 50 per cent of the classes may benefit from statecharts.

Statecharts are not only applicable to real-time systems. Information systems such as hospital administration systems have been seen to have very important state-dependent behaviour. Statecharts can effectively describe the behaviour of classes such as PATIENTREFERRAL, PATIENTAPPOINTMENT, WAITINGLISTPATIENT and INPATIENTSPELL. Some largely query-based information systems may have no use for statecharts, though this is fairly rare. Figure 3.6 gives an example, from the fuel sales domain.

The recommended form of state modelling is the Moore state model in which actions are specified on entry to states and an individual event can only cause a single transition out of any given state. To conform to this form of modelling describe all state behaviour in entry actions only. This provides clear unambiguous semantics and is capable of modelling any state-related problem. The use of additional statechart features will be discussed in Chapter 9.

3.3.5.3 Using the class collaboration diagram

The domain level sequence diagram for each use case is useful in the inception phase as it shows the extent of interaction implied by a given use case scenario. During elaboration there is a need to specify the details of the interactions between the classes and also to visualize the pattern of interaction between the classes as a whole. This latter stage is often

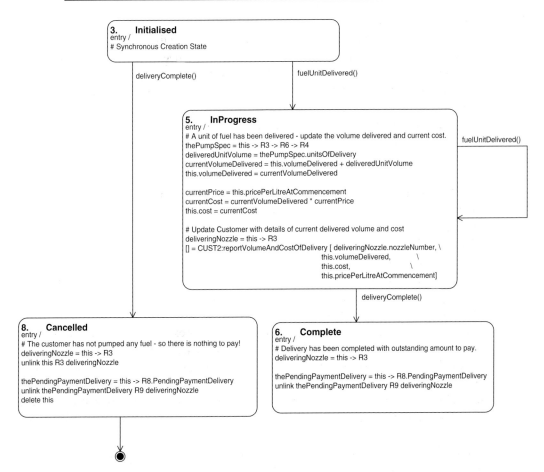

Fig. 3.6 Statechart for the DELIVERY class

overlooked but is a major contributor to the correct distribution of responsibilities across the classes in the domain. To see these patterns of interaction all interactions between the classes need to be summarized on a single diagram. Time order is immaterial and so the specification level class collaboration diagram (CCD) is the appropriate representation.

A good CCD identifies layers of responsibility. Use a left to right layout with interactions with clients of the domain on the left side of the diagram and interactions with servers to the domain on the right side of the diagram. This means that abstract requests and responses appear on the left side and detailed interactions with other domains on the right side. The classes are layered with those with most responsibility to the left of the diagram and those with limited responsibility (and potentially specialist knowledge) on the right side of the diagram. Successful use of layered responsibility in CCDs is a valuable aid to the understandability and maintainability of the model. Fuel sales again provides an example in Figure 3.7

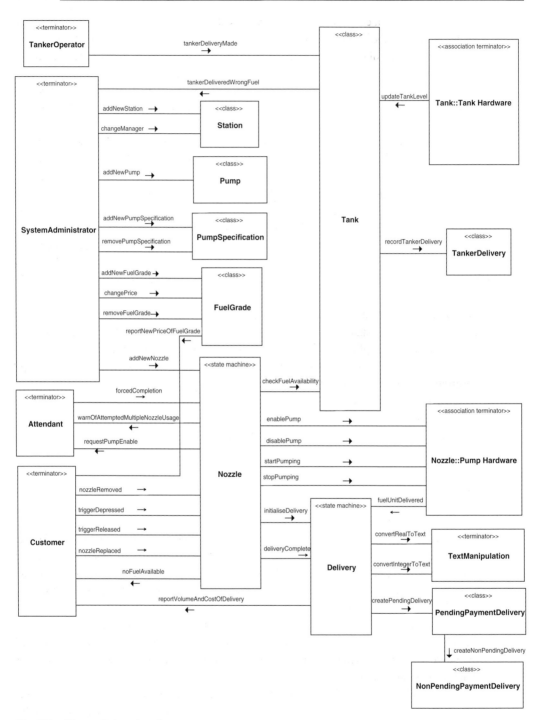

Fig. 3.7 Class collaboration diagram

Conventionally, a collaboration diagram is used like a sequence diagram at the instance level to represent the set of interactions between instances in a given use case. Using object level collaboration diagrams and sequence diagrams is an overkill since the same aspect of the model is being specified in two ways. In xUML the collaboration diagram is as being used at the class level. Each 'box' on the diagram represents a class and not an instance of a class. This allows the pattern of communication between classes to be observed across the set of use cases (see Chapter 7).

Detailed interactions including the linking and navigation of associations need to be captured. The technique for describing these is discussed in the following section. Object level sequence diagrams may be found to have limited value at this stage since the pattern of interactions is described by the CCD and the detailed interactions are specified in the operation and state action descriptions. Maintaining large numbers of object level sequence diagrams is unlikely to be worth the effort. When the models are executed it is possible to obtain a trace output that catalogues the actual interactions that take place. As the domains are integrated, these can be compared against the original domain level sequence diagrams to see if the expected behaviour of the integrated system has been achieved.

3.3.5.4 Describing the operations and state actions

Every operation and state action needs to be described (see Chapter 10). This should specify the behaviour implied by the invocation of the operation or arrival in the state. The description should be in terms of the manipulation of model elements. The types of things described in operations and actions include:

- create and delete objects;
- manipulate sets of objects;
- find objects meeting specified criteria within a class;
- link and unlink associations;
- navigate associations;
- read and write attributes;
- call operations;
- send signal events.

The descriptions can be informal and use natural language or they can be more formal and use an agreed syntax. The benefits of an agreed syntax are consistent use and easier readability. The disadvantage is that the syntax has to be learned by those who will be reading and writing it. A suitable syntax will limit itself to the types of object manipulation described above and be readily readable as a natural language.

An important review criterion for any model is the quality of its descriptions. Clear unambiguous descriptions of operations and state actions are vital to the effectiveness of the models. Simply relying on a name and a set of parameters is not sufficient to specify what the operation or action does.

One suitable form of defined syntax for describing operation and actions is the Action Specification Language (ASL) (Wilkie *et al.*, 2002). This has the added benefit that it results in models that are executable. Models built in such a way have the following characteristics:

- They are fully specified;
- Objective completion criteria (the models work or they don't);

- They can be validated using a test specification that can later be applied to the final system and used for regression testing;
- Implementers are free to make implementation choices but not behavioural choices;
- Generic models are easier to understand when executed because they can be viewed at the instance level;
- They utilize a semantically precise subset of UML and therefore remove ambiguity in interpretation of the models;
- They allow early refinement and high confidence in the quantitative model;
- They amenable to qualitative appraisal (e.g. is this the best way to model support for these requirements?).

ASL provides the ability to manipulate model elements. It is both abstract and fully specified with unambiguous semantics.

Consider the situation where the first use case primary scenario is being developed. If the operations and state actions are specified using ASL then the model can be executed. Even though it is only a partial model it is possible to test that the model performs the behaviour implied by the use case. In order to do this the only additional work that is required is to set up the initial conditions for the use case and this can be done in a number of ways (e.g. static data, user-specified data, data accessed from file). Typically a small number of scenarios will be modelled before the model is executed but this is up to the individual users. The model execution environment allows the status of the model to be inspected at the beginning and end of the execution and at any intermediate step. Therefore if the model does not behave as expected, it can be fixed and re-executed. Inspection of the model allows any value of any attribute of any object of any domain to be viewed and any association to be navigated so the developer has the tools to build a correct and verifiable model that can be demonstrated.

As an example, the following is the state action ASL for a state in the DELIVERY class state model.

```
# A unit of fuel has been delivered - update the volume delivered
# and current cost.
thePumpSpec = this -> R3 -> R6 -> R4
deliveredUnitVolume = thePumpSpec.unitsOfDelivery
currentVolumeDelivered = this.volumeDelivered + deliveredUnitVolume
this.volumeDelivered = currentVolumeDelivered

currentPrice = this.pricePerLitreAtCommencement
currentCost = currentVolumeDelivered * currentPrice
this.cost = currentCost

# Update Customer with details of current delivered volume and
# cost deliveringNozzle = this -> R3
[] = CUST2:reportVolumeAndCostOfDelivery
   [deliveringNozzle. nozzleNumber,
     this.volumeDelivered, this.cost,
     this. pricePerLitreAtCommencement]
```

The model execution environment allows the packages of work from different developers to be brought together and executed. This tests the semantics of the interfaces between the domains and not just the syntax. Again this can be done iteratively through the development process so that integration is progressive.

3.3.6 Process for each iteration in each domain

Having introduced the various model elements that are required to model a domain, here is a typical process that is used to iteratively develop the model:
1. Consider the use case scenarios selected for this iteration;
2. Add or refine classes, associations, attributes and operations to the class diagram as required;
3. Create or refine statecharts for classes with dynamic behaviour;
4. Create or refine a suitably layered CCD;
5. Create or refine the object-level sequence diagram for each use case scenario (if required);
6. Specify operations and state actions;
7. Test executable models against use cases.

Each of these modelling activities is described in detail in the subsequent chapters.

3.3.7 Domains not suited to executable modelling

There are aspects of systems for which executable modelling is not necessarily the best approach. One would expect this to be a relatively small proportion of the overall system but these areas need to be accommodated.

One of the parts of a system most frequently developed outside of a conventional or executable modelling formalism is that of graphical user interface (GUI) development. The very nature of GUI design and the range of tools available means that GUI development is usually treated as a separate domain and not developed using xUML. Good GUI design shares many of the principles of domain partitioning. The GUI itself should be free of application knowledge – it should capture the style and policies for user interaction with the system and not the business logic that needs to be performed. In an MDA project the business logic would be captured in one or more executable model domains whilst the GUI will be treated as a separate domain developed outside of the executable modelling formalism. The bridges between the GUI domain and the xUML domains will map from one world to the other. A good GUI domain model can still be platform-independent; it may refer to GUI policies such as using dialog boxes, drop-down menus, etc, but not be tied to a specific platform (e.g. Microsoft Windows). Clearly, if the GUI model were tightly coupled to a particular platform then it would be a platform-specific model.

Another area commonly developed separately is algorithmic specification. Algorithms are used in many types of system and they may capture complex mathematical operations or control laws for example. Typically the xUML models provide the framework in which the algorithms run and the algorithms are treated as separate domains that capture the specialized subject matter of the algorithms themselves. This allows the existing expertise, tools and even code relating to the algorithms to be used in the context of an xUML model. Again the bridge achieves the mapping between the xUML domains and the specialized domains.

Legacy code, reusable code and hardware-related code could be integrated with the xUML models in the same way, by treating them as separate domains. The benefit of this flexibility is that MDA can be adopted in projects which are not 'green field' but are developments or adaptations of existing systems. This allows the adoption of MDA to be progressive.

3.3.8 Low risk domains

Some of the domains identified in the inception phase may be regarded as low risk. They don't provide primary functionality of the system or they are not required in early releases to gain user feedback. Examples of these types of domains might be:
- Logging;
- Audit;
- Management information services.

If domains can be identified which are not critical to early releases of the project and therefore represent a low risk to the project plan, then their development can be deferred into the construction phase.

The preceding examples do not imply that domains of this type will always be low risk and not of critical importance to a project. For instance, in a financial system the audit domain may be a crucial element of the financial security of the system.

3.3.9 Integration

The work on a domain can proceed in parallel with work on the other domains. As each domain covers a distinct subject matter there is little overlap of the modelling work. However, as the work is phased by iterations based on sets of use case scenarios, the set of domains can be brought together at the end of each iteration to check that the set of PIMs as a whole supports the use case scenario correctly. This requires that the interfaces between the domains be fully defined.

Model execution can then be extended from single domain model execution to multi-domain model execution. This serves to test that the domains work correctly together and demonstrates the correct semantics of the interfaces between the domains.

The concept of the **bridge** plays a major role during integration. Domains in MDA have no explicit dependence on each other. This has the enormous benefit that domains are decoupled. However, to build a system involving many domains requires that the domains can be 'connected together'. The bridge provides the mechanism for connecting a pair of domains. During a single domain build, that is before any integration, the bridge operations are stubbed – they don't refer to any other domains, that is the respective bridge method is specified without making any references to services provided by other domains (for example, the invocation of an operation provided by another domain). This allows the single domain PIM to be executed and tested stand-alone. When two domains are to be integrated, the stubbed bridge operations are replaced with full bridge operations, which specify how the required service of one domain maps to the provided service of another domain. The replacement can be done one operation at a time if required. This provides fine-grained control over the integration steps. The combination of domains and the realized bridge operations between them is known as a build set. Build sets are the essential tool of integration right through the development process. Several versions of each bridge operation are used: the single domain stub and one or more multi-domain realizations. These tend to be very simple and are expressed using the action language as illustrated.

```
# Bridge operation for Motor M01:startMotor
# Invoke Pump Interface service for motor start
[] = PI3::startPumpMotor[]
```

All of the versions are kept, in the context of the build sets, so that regression testing is supported. Integration planning should be based on the build set with each iteration planned to deliver one or more build sets. Integration therefore happens incrementally at each iteration.

This principle can be extended to testing the interfaces in large projects where the work of separate subsystem development teams needs to be integrated. By agreeing system-wide use cases developed in parallel by the team members in each subsystem team, it is possible to plan for early integration. It is probably not realistic to plan for system-wide integration at every iteration so a more pragmatic approach would be to schedule system-wide integration activities every three or four iterations. Priorities in selecting which domains to integrate should be driven by risk.

Both of these approaches achieve valuable early integration and reduce the risks of major surprises occurring with a big bang approach to integration.

Integration of domains, bridges and build sets are all covered in detail in Chapter 12.

3.3.10 Software architecture and code generation

The other major development, which occurs during the elaboration phase, is progress on the definition and the realization of the mappings from PIM to PSM and the implementation-specific environment. These mappings can be fully automated by developing a software architecture and code generator (see Chapter 13). We will use the terms software architecture and code generator to mean the element of the project that automates the mappings from PIM through to target code. The project team will make the decision on the extent to which it exploits executable modelling. Three approaches can be considered here:
- Executable modelling for PIM verification only;
- Executable modelling and code generation for prototype purposes;
- Executable modelling and target code generation.

Let us consider the type of development work that needs to be undertaken during the elaboration phase for each of these.

3.3.10.1 Executable modelling for PIM verification only

This approach exploits executable models to verify that the models exhibit the intended behaviour. The models are run and tested in a simulation environment. These are available as part of commercial xUML toolkits. The project does not have to plan any software architecture development activity but it will have to define its design and implementation strategy, which will be very different from the code generation route.

This approach is a 'half-way house' to full MDA. It retains the benefit of high quality verified analysis models but it introduces the problems of separately maintaining PIMs, PSMs and code in step as the project life cycle progresses. That is because this approach is still basically elaborative and therefore retains the problems common to elaborative approaches.

3.3.10.2 Executable modelling and code generation for prototype purposes

This approach goes further. It uses executable modelling to develop the high quality PIMs as above, and also uses code generation to produce a demonstrable system or prototype. This early deliverable would have many potential uses from soliciting user feedback and refining requirements to quantitative measurements of the executing system.

The software architecture required to support this type of project profile would, typically, be commercially available. At most some simple tailoring of a commercially available software architecture would be sufficient to meet the project's needs.

This approach might be utilized as a stepping-stone to full target code generation. It would enable early demonstrable systems to be produced that would serve to increase confidence and reduce risk, whilst the software architecture for full target code generation is being developed.

3.3.10.3 Executable modelling and target code generation

To fully exploit MDA and gain all of the potential benefits, automatic code generation to the target environment needs to be part of the process. This will often, although not always, require a software architecture and code generator to be developed as part of the development project. Much of that work is performed during the elaboration phase.

Software architectures and code generators can be and have been developed from scratch. This requires a small team of highly skilled developers. The role of software architect demands a thorough understanding of:
- The xUML formalism;
- The target programming language;
- The chosen implementation technologies (e.g. database, ORB, operating system);
- The proposed system architecture.

A team of two or three is typical for a software architecture development.

A more commonly used development strategy is to derive a software architecture and code generator from an existing one. There are commercially available products that support the reconfiguration and tailoring of code generators. This approach tends to be much quicker and less risky, since the starting point is a software architecture and code generator that works.

The development of the software architecture and code generator proceeds in parallel with the modelling activities during the elaboration phase. Again an iterative approach is beneficial. The first iterations will develop support for the core architectural concepts:
- Representation of classes;
- Storage of instances;
- Iteration of instance sets;
- Realization of associations;
- Realization of statecharts;
- Signal queuing and delivery.

These core features will not provide sufficient capability to support the models emerging from the modelling team but they will allow simple test models to be implemented. The properties of these can then be characterized and adjustments to the architecture made as necessary.

Further iterations will add support for the remaining elements of the formalism including support for the action language. Once all of the required features of xUML are supported by the software architecture, the architecture development iterations can continue with the focus on performance tuning the software architecture and developing alternative implementations guided by tagging of the PIMs. These later iterations will typically continue into the construction phase.

Good communication between the modelling team and the architects is vital. The optimum selection of architectural structures will depend upon the characteristics of the system being developed. For example, a system with a small static population of class instances will have very different requirements on its architecture from one with vast numbers of dynamic instances. This type of difference is obvious but it is the modellers who are likely to discover properties of the required system, through their modelling work, that the architects can exploit in their development. Architects can also inform the modellers about the architecture and code generator as it develops. Modellers will be less likely to be lazy in their use of xUML features if they understand the consequences in architectural terms. A lazy modeller may use a `find` over a large instance population when a `navigate` followed by a `find` over a smaller population would be better modelling and result in better performance in the generated system.

The software architecture under development may need to support a complex system architecture with many of these features:
- Distribution;
- Transactions;
- Mixed persistence;
- Load balancing;
- Online upgrades;
- Redundancy.

In this case the iterative development of the architecture is also driven by risk. The highest risk items should be tackled first. Software architecture and code generation issues are discussed thoroughly in Chapter 13.

3.3.11 Elaboration phase completion criteria

The plans for the elaboration phase will have nominated the use case scenarios to be modelled in each iteration. A modelling iteration is complete when the selected use case scenarios execute correctly and have been subject to a style review. Tests are specified for each use case and form objective criteria against which the iteration can be assessed. The style review is a subjective assessment and serves the purpose of checking that good modelling style has been adopted.

The developing software architecture and code generator can also have objective criteria specified for them. A test model can be defined which can be successfully built and executed on the software architecture. This could include use of appropriate implementation technologies. A review of the software architecture design will provide a subjective assessment of its suitability as the basis for the final architecture required.

Combining these two elements together, it may be possible to identify elements of the models that can be built on the early software architecture to provide a capability demonstrator. This has the benefit of providing a very tangible deliverable from the elaboration phase promoting team and customer confidence.

3.4 The construction phase

The construction phase in the standard RUP focuses on the coding effort. Some analysis and design continues but coding is the main activity during this phase. An MDA process using full code generation changes that emphasis considerably. The coding activity disappears since the code is automatically generated from the models. Instead, the activities of the elaboration phase continue but with the focus on completion.

Iterations continue through the construction phase. The remaining use case scenarios need to be modelled. These will be secondary scenarios for many of the use cases plus low priority primary use case scenarios that have yet to be considered. Many of these scenarios will deal with required behaviour in error conditions. Each iteration should cover a small number of scenarios and teams should be organized around the domains as in the elaboration phase.

Domains that were classified as low risk during the elaboration phase may well have been deferred. These domains should be fully developed and integrated during the construction phase.

Regular integration of the domains to prove end-to-end execution of the use cases remains important. The use of build sets, which were introduced during the elaboration phase, continues as the primary means of controlling the integration activity. A build set defines the set of domains and the realized bridge operations between them. The primary deliverable from each iteration during construction is therefore a completed build set.

The software architecture and code generator will also be further developed during the construction phase. Complete coverage of all elements of the xUML formalism will be a target. The provision of alternative design patterns to be exploited by tagging the PIM will allow the implementation to be refined. Performance testing and tuning of the software architecture and the generated code will also be activities performed by the architects during the construction phase.

The other major activity during the construction phase is testing. The test cases devised for the models can be reused to test the target code. Additional tests may be required to test aspects of the target architecture.

The inception and elaboration phases tackled the perceived high-risk aspects of the project. At the start of the construction phase risk should be under control. However, taking your eye off the risks can be risky! Risks can change: for example, the planned use of a commercial off-the-shelf (COTS) product proves impractical because it proves to be unsuitable or unavailable.

At the end of the construction phase the developed system is at the stage of a beta release. It is suitable for distribution to a controlled set of beta customers or users. It should support all of its primary functionality and most of its secondary functionality. Supporting information such as user guides and training tutorials should be available in a beta state.

3.5 The transition phase

The nature of the transition phase varies from project to project. Software that is being developed for a single system and a single client will have a different transition phase to a software consumer product. However, in all cases the purpose of the transition phase is to prepare the delivered software for full operational use.

The essence of a successful transition phase is to roll out the software into a realistic operational environment and to monitor its behaviour and performance and observe user reaction to it. Errors and failures should be fixed. Some problems related to scalability and performance, which were not foreseen in the earlier phases, may come to light during realistic operational deployment. These will need to be addressed. Perhaps the most difficult issues are those of subjective user feedback about ease of use where there is no consistent view from a pool of users. Training may be a more effective solution than any changes to the product.

The key benefit that MDA brings to the transition phase is that any required changes to the application or business logic are reflected in the models and the relevant parts of the code are regenerated. The code is not maintained directly. This ensures that all project deliverables remain current and avoids the conventional problems of trying to keep PIMs and PSMs in step with the code. Alternatively, during the transition to operation, improvements to the software architecture and code generator might be identified, for example to improve performance. In a traditional project this may have serious consequences because the impact of change may be substantial. With the MDA approach, the enhancements can be made to the software architecture and code generator whilst the domain models, which capture the application and business logic, are totally unaffected. This allows systematic improvements to be introduced in a manner impossible with hand-coded systems.

For software delivered as part of a contract, it is likely that part of the transition phase will include a formal acceptance test. Formal acceptance tests are typically very thorough and may well take weeks to run. It is extremely desirable to be sure you are going to pass a formal acceptance test and so a number of 'dry runs' may be necessary. This can be a very time consuming activity and should not be underestimated in the planning of the transition phase.

The transition phase is the final phase of the project cycle but this does not necessarily correspond to the end of the project. During the transition phase consideration should be given to the requirement for subsequent cycles of the project. Software developed for various purposes, whether it is a large capital project or some consumer software, will typically go through a series of major releases. These releases will correspond to cycles of the project, each with its own inception, elaboration, construction and transition phases. Therefore during the transition phase, planning for the subsequent cycle, if there is one, should be carried out.

3.6 Impact of changes in requirement

The inception and elaboration phases have focussed on delivering a small volume, high value set of model artefacts. One of the reasons for this is to ensure that the models produced are **maintainable**. Models of distinct subject matter areas organized as a set of domains represent important intellectual property for the project and the organization. There is considerable potential for reusing these models (whether or not their corresponding implementations are reused). These models should not be seen as a means to an end and simply thrown away once the implementation has been achieved. They should be a living part of the project deliverables and should be maintained throughout the life of the project. Requirements changes must be assessed in the context of the models.

A major benefit of this approach is that it is economically and organizationally feasible to maintain the models since they are the source from which the system is generated. Changes in requirement will result in the models being updated and the relevant parts of the system being regenerated. The target language source code, even if it is a high level language, is not maintained. Any attempt to maintain at the code level only renders the models out of date, and hence useless. This will result in degradation of code quality over time. The model is a much more suitable medium for assessing the impact of any change on the 'big picture'.

This approach is very different to round-trip engineering where the code is updated and pulled back into the models. For round-trip engineering to work the models must be at a lower level of abstraction than they are with executable UML; they are, in effect, pictures of the code.

3.7 Impact of changes in design decisions

Not all changes that affect a project will relate to changes in required behaviour. Many will affect the design decisions made. For example, the selected database is changed; an alternative class library is chosen; a different ORB is selected; or a new multi-threading policy is adopted. None of these types of change have any impact on the required behaviour of the system. The xUML models will be unaffected by these changes. It is the software architecture and code generator that must be updated to reflect these changes. Once these updates are complete, the system can be regenerated from the models and retested.

The principle of separation of concerns is very strong in software engineering. The MDA process allows separation of concerns to be maintained right through to the maintenance phase of the project.

3.8   MDA and other life cycle processes

The purpose of this chapter has been to illustrate how MDA might be used in a defined project management process. The RUP was chosen as a representative example. The technical process of MDA could be tailored to fit into a variety of life cycle processes including waterfall and V life cycle models if required.

There are some fundamental characteristics of MDA that need to be considered when adapting a life cycle model to be able to use MDA:
- The PIMs are executable and can therefore be verified. This brings testing forward in the life cycle;
- The design activity is captured by the software architecture and code generator. This does not need to follow on from the platform-independent modelling but can proceed in parallel;
- Automatic code generation eliminates the coding phase;
- Testing of the target code is still necessary but can be reduced in duration because some of the testing was moved forward to the platform-independent modelling activity;
- Integration can be performed throughout the project: there is no need to wait until everything is complete;
- The models are the source so the code stays in step with the maintained models rather then the models trying (and usually failing) to stay in step with the maintained code.

4 Use case modelling

4.1 Introduction to use cases

A **use case** describes a 'usage of the system' from a black box, external perspective (Jacobson, 1994, Schneider and Winters, 1998). They are a useful tool in reasoning about what the system should do in response to requests or stimuli. We can think of use cases as a bridge between the informal and very often unstructured world of requirements and the formal models that we shall build using xUML.

A full set of use cases, along with their related **scenarios**, completely characterizes the required functional behaviour of the system, allowing most of the system's requirements to be captured.

In this chapter we shall explore ways that use case modelling is useful within the MDA process. We shall also see that use cases can be a powerful tool right through the system development life cycle, not just when we perform requirements' capture.

4.2 Objectives

Use cases are produced with the following goals in mind:
- To clarify and gain an understanding of the requirements of the system;
- To help define the boundary of the system to be developed;
- To provide input into production of the initial domain and class models and so provide initial costing estimates;
- To gain insights into how the system will be used operationally and how these aspects influence the requirements;
- To establish a starting point for developing the dynamic models of the system;
- To establish an initial definition of the interfaces (both required and provided) for the domains in the system;
- To provide customers, users and system testing staff with a functional 'way in' to the documentation of the system.

In other words they provide a way of bridging the highly informal world of implicit or badly formulated requirements and the formal world of the software development process. In that sense they may be thought of as a 'semi-formal' technique.

It should be noted that the MDA approach does not preclude additional or alternative ways of capturing, expressing and engineering requirements, such as stylized textual requirement documents.

4.3 Identifying actors and use cases

There are two primary concepts we must consider in use case modelling, **Actors** and the **Use Cases** themselves.

An **actor** is an entity external to the system being modelled that interacts with the system during its operation.

Actors often represent the roles played by the users of the system; it should be noted that a single actual user might play different roles over a period of time. It is the role or 'type of user' that is significant not the actual individual involved. Actors can be subdivided into two categories: **primary actors** are those who typically instigate behaviour of the system and are involved in the normal operation and **secondary actors** usually have a more passive role and support rather than instigate behaviour.

Actors are not limited to human operators of the system but may include external equipment (e.g. a printer) or other computer systems outside the scope of the development but which interact with the system. Time is often considered to be an actor; this concept is useful in systems where periodic or time-based tasks take place.

Many actors are easy to identify. Less obvious ones can be located by considering system interfaces, documented requirements and the work carried out by users of existing systems.

A **use case** is a specific type of behaviour required of the system, typically described as a sequence of transactions between one or more actors and the system in dialogue. The set of use cases for a system would completely describe the required behaviour from the perspective of the system's users.

A sensible place to start when looking for use cases is to consider all the 'things' one would expect to be able to do with the system. As an example consider an automated teller machine (ATM) of a bank.

Initial use cases might include: 'Make Withdrawal', 'Make Deposit', 'Balance Enquiry' and 'Statement Request'. Note that each proposed use case is complete in its own right and provides a valuable service to the user. An easy mistake would be to identify operations such as 'insert card' and 'enter PIN' as use cases. These fail to qualify on the grounds that they are neither stand-alone nor offer value in themselves.

This system would clearly have a 'customer' actor; but it might also have a 'maintainer' actor. By considering what use the maintainer might make of the system we might come up with the following additional use cases: 'Request Diagnostics' and 'Load Supplies'.

In practice the use case analyst should ask the question 'what things can this actor do to the system or require of the system?' to each actor in turn.

An alternative approach is to identify all the external events that can occur in the system's environment for which the system is expected to respond and then associate these with actors and use cases. This approach may yield use cases that are associated with the passage of time, rather than an unsolicited event from a tangible actor. As mentioned above it is sometimes useful to consider 'time' as an actor.

The best sources of information for use case modelling are any existing requirements documents, 'concepts of operation' documents and experts in the application subject matter. In situations where the system replaces an existing automated or manual process, the users of the existing process will be able to provide valuable input.

In some respects use case modelling is a restructuring of information that may already be captured in the requirements specification (if this exists). Even if this is so, the use case adds value by linking together sets of requirements and by adding the operational perspective to the picture.

Use cases, when used in the development of a new system, should be a classic 'black-box' view (exposing just the 'why' of the system, shunning the 'how'); however, they are also a useful tool when considering extant computer or business systems, in which case we may choose to formulate 'white-box' use cases that expose both the how and the why of the system under consideration. Care should be taken, of course, if these use cases are then fed into the requirements process for a new system, not to confuse genuine requirements with previous solutions to those requirements.

4.4 The use case diagram

Use cases and actors can be depicted graphically on a use case diagram, an example of which is shown in Figure 4.1.

The use case diagram is a little unusual compared to the other UML diagrams that we shall meet in subsequent chapters. The actors are depicted as 'stick-men' (in a Lowry-esque manner), whilst the lines that link them to the ellipses that represent the use cases are associations. For those of you that have met associations (or are just about to read Chapter 6) then you would (or will) expect an association to be named, have a pair of role phrases and have a cardinality expression at either end; in the use case diagram none of these standard UML artefacts appear on the use case diagram and we simply infer that the actor communicates with the use case. The direction of communication, or an indication of the actor that initiates the use case (many actors can communicate with the same use case ellipse) is simply not indicated on this diagram. The actors themselves are, in fact, stereotyped classes, that do not have any attributes or operations (again see Chapter 6 for an explanation of a class); they may participate in only one type of relationship with other actors, that of specialization/generalization (see Section 4.6.5). The boundary of the system is depicted, optionally, by the enclosing named rectangle. Finally, UML permits use cases to have attributes, operations and statecharts; these are not used in the approach taken in xUML since such adornment of use cases has no part to play in UML Action Semantics.

In practice it is found that use case diagrams have limited usefulness. They act as a visual index to the use cases and system boundary, but should not be considered a major deliverable

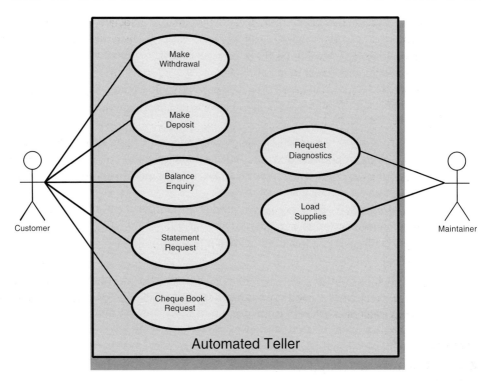

Fig. 4.1 ATM use case diagram

of the process. In fact, many projects omit the use of this diagram type. We shall see that the real value of use cases is in their description.

4.5 Documenting use cases

The process starts with identifying and documenting as many of the actors that can be thought of at this stage. This will not be the final list but provides a starting point.

4.5.1 Describing actors

Each actor that is identified during construction of the use case model should have the following information recorded:
- **Actor ID** – a number that uniquely identifies the actor;
- **Actor name** – a name for the actor that accurately and meaningfully captures its role in the model;
- **Description** – a short description of the actor including an outline of its involvement with the system;

4.5.2 Describing use cases

Each use case that is identified needs to be named and described. The following information is essential:
- **Use case ID** – each use case should be provided with a unique identifying number, letter or mnemonic. This provides a shorthand way to refer to the use case in other documents, etc;
- **Name** – the use case should have a meaningful name that captures its essential nature, for example 'Make Withdrawal', 'Balance Enquiry';
- **Purpose** – state the purpose of the use case, for example 'To allow a customer of the bank to withdraw a specified amount of cash';
- **Initiating actor** – the name of the actor(s) that initiates the use case;
- **Precondition** – state any conditions that must exist before the use case can take place;
- **Invariants** – any conditions that must not be violated throughout any scenario of the use case;
- **Minimal guarantee** – state any significant conditions that hold after completion of all scenarios of the use case (including failure scenarios);
- [**Requirements satisfied** – list the identifiers (e.g. paragraph numbers) of any documented user requirements that are satisfied by the use case. This will not be done if requirements are purely gathered using use cases.]

Each use case can typically exhibit a number of different patterns of behaviour. Picking one particular path through a use case yields a **scenario**. There will always be a **primary scenario** that captures the normal behaviour associated with the use case. Sometimes there will be a number of **secondary scenarios** that capture less usual behaviour such as exception handling and error behaviour.

For the primary scenario and each of the secondary scenarios the following information must be documented:
- **Scenario name** – give each scenario a meaningful name that captures the essential differentiator between it and the other scenarios. Prefix the primary scenario with the word 'Primary' (e.g. Primary – Successful withdrawal, Invalid PIN, Insufficient funds, etc.);
- **Scenario description** – describe the sequence of transactions between the use case and the actors. Scenarios should be described in terms of what is required to be done rather than how it will be implemented, from the viewpoint of the initiating actor. This should be documented as a numbered sequence of short, declarative statements. Exact details of interface syntax should be avoided. Emphasis should be placed on capturing the essential business processes independent of technology considerations;
- **Post conditions** – state any significant conditions that hold after completion of the scenario.

We insist that use cases and all their related scenarios terminate, so we do not admit use cases or scenarios that never reach their post conditions.

An alternative style is to omit the post conditions in each scenario description and introduce a **success guarantee** for the use case as a whole. This states the set of conditions that must hold if the use case is to be deemed successful.

During the working of the use case it is common practice to concentrate initially on describing the primary scenario. At each decision point the scenario description should follow the most likely course of events. Where significant departures points are envisaged during the primary scenario description definition, these should be informally noted, to be revisited when the secondary scenarios are investigated. Secondary scenarios may reference common steps from the primary scenario; we should avoid repeating common steps from the primary scenario and focus on the differences. It is recommended that although their detailed description may be postponed until later, the secondary scenarios are at least listed at an early stage. Although common sense should be applied when identifying secondary scenarios, we should not attempt to document every path, no matter how banal, through the system at this stage.

4.5.3 ATM use case and scenarios example

Table 4.1 ATM use case with primary and secondary scenarios

Use Case ID	ATM1
Name	Make Withdrawal
Purpose	To allow a bank customer to withdraw a specified amount of cash
Initiating Actor	Customer
Precondition	The ATM is 'In Service' AND the Cash Supply Balance > Min Withdrawal Amount AND Customer Proximity Detected
Invariants	Customer account shall never fall below agreed limit
Minimal guarantee	Customer activity is logged. Cash amount in the machine, after the transaction, equals the amount in the machine beforehand less any dispensed to the customer.
Requirements	Banking Systems plc.: ATM2000 SRS V1.0 : 2.3.1-2.3.4; 3.1; 5
Scenario	Primary
Name	Successful withdrawal
Description	1 customer approaches ATM & the system raises the 'shield'
	2 the customer inserts their bank card & the card details are verified
	3 the customer is prompted to enter a PIN to authenticate the customer
	4 the system displays a list of options
	5 customer selects the 'Withdraw Cash' option
	6 the customer selects an amount of cash to withdraw
	7 the machine delivers the requested cash
	8 the customer selects quit from the menu
	9 customer activity is logged
	10 ATM ejects the card and lowers shield
Post condition	The customer's account balance, daily withdrawal balance and the ATM Cash Supply Balance are reduced by the withdrawal amount.
Notes	
Scenario	Secondary 1
Name	Daily Limit Exceeded, customer then enters a lesser amount successfully

(*cont.*)

Table 4.1 (*Cont.*)

Description	The use case proceeds as for the primary scenario up to the point when the customer enters the requested withdrawal amount (step 6).
	7 the system checks the sum of the withdrawal amount and the daily withdrawal balance against the daily withdrawal limit and finds that the limit will be exceeded.
	8 the ATM informs the user that the withdrawal request can not be honoured & the customer is offered the chance to enter a lesser amount or to exit.
	9 the use case then proceeds as in the primary scenario (step 8).
Post condition	The customers account balance, daily withdrawal balance and the ATM cash supply balance are unchanged.
Notes	
Scenarios TBD	Stolen Card
	Invalid PIN
	Insufficient funds in account
	Insufficient funds in ATM

4.6 Managing large or complex use case models

For some systems, the use cases can become large, complicated or both. Sometimes this complexity is increased by the apparent need for repetition within the model. A number of techniques are available to add structure and to allow for reuse in the use case documentation.

4.6.1 Extend relationship

It is possible for one use case to **extend** the behaviour of another use case. This allows the capture of specialization relationships within the use case model. Consider the ATM example from above:

'Make Withdrawal', 'Make Deposit', etc. all require the customer to insert a card, enter PIN numbers and select the appropriate option from a menu before the main content of the use case gets going. As things stand, this behaviour and the associated secondary scenarios for 'invalid PIN', etc. would have to be repeated in many of the use cases.

An alternative approach is to consider a new use case that generalizes the behaviour of the other use cases. A 'Perform Customer Transaction' use case would capture the common behaviour of the customer inserting a card, etc. but would not be specific about the particular transaction type that the customer will perform. The original use cases will be used to capture the particular specialized behaviour and will no longer include the common behaviour in their descriptions.

This is an example of an extend. The 'Make Withdrawal' use case and its peers are said to extend the behaviour of the 'Perform Customer Transaction' use case. This is depicted diagrammatically in Figure 4.2. Where the extending use case is linked to the extended use case with a **dependency** arrow, labelled with the keyword «extend».

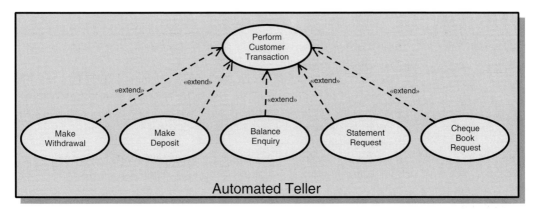

Fig. 4.2 ATM example with «extend»

The specialized behaviour of 'Make Withdrawal', 'Make Deposit', 'Balance Enquiry', etc, is inserted into that of 'Perform Customer Transaction' at the point where the customer makes a menu selection.

Extends are typically used in the following situations:
- The system performs one of several possible functions depending on some user-oriented selection mechanism (e.g. user chooses a function from a list);
- A use case has non-trivial secondary scenarios;
- It is desirable to model a use case which we believe will need to be extended in the future with minimal impact on the system.

Note that the extended use case must be complete in its own right and have no knowledge of what the extension does. The extended use case names the points at which it *may* be extended.

The next example illustrates the distribution of behaviour between extended and extending use cases.

4.6.2 ATM use case example with an extend

Table 4.2 Use case with «extend»

Use Case ID	ATM0
Name	Perform Customer Transaction
Purpose	To allow a customer to perform a number of banking transactions.
Initiating Actor	Customer
Precondition	The ATM is 'In Service' AND Customer Proximity Detected.
Invariants	None
Minimal guarantee	Customer access is logged
Requirements	Banking Systems plc: ATM2000 SRS V1.0 : 2.3.1-2.3.2
Scenario	Primary
Name	Validated Session

(*cont.*)

Table 4.2 (*Cont.*)

Description	1 the system raises the 'shield'
	2 the customer to insert their bank card
	3 the card details are verified and the customer prompted to enter a PIN
	4 on entry of correct PIN details the system displays a list of transaction types and an exit option
	5 log the fact that the customer has been authenticated to the system
	Extension point 1
	6 on selection of the exit option the system ejects the customers card
	7 once customer proximity is absent for 30 s the ATM lowers the shield
Post condition	None.
Additional Notes	
Scenarios TBD	Invalid PIN
	Stolen card
	PIN not entered in time
	Service Exception

Note that the initiating actor is still the customer. With the specification of the extending use cases we omit the initiating actor, since this is derived from the extended use case. Also, we specify the extension points in the extended use case, but we do not specify which extension occurs (the extended use case is unaware of which, if any, use cases will extend it).

Table 4.3 An extending use case

Use Case ID	ATM10
Name	Make Withdrawal
Purpose	To allow a bank customer to withdraw a specified amount of cash
Precondition	The ATM0 Use Case is in progress AND the customer has selected the 'Cash Withdrawal Option' AND the Cash Supply Balance > Min Withdrawal Amount.
Invariants	
Minimal guarantee	
Requirements	Banking Systems plc.: ATM2000 SRS V1.0 : 2.3.3-2.3.4; 3.1; 5
Scenario	Primary
Name	Successful withdrawal
Description	1 the system prompts the user to select an amount of cash
	2 the machine delivers the requested cash
	3 returns to the ATM0 use case – menu selection
Post condition	The customer's account balance, daily withdrawal balance and the ATM Cash Supply Balance are reduced by the withdrawal amount.
Additional Notes	
Scenarios TBD	Insufficient funds in account
	Insufficient funds in machine
	Daily Withdrawal limit exceeded

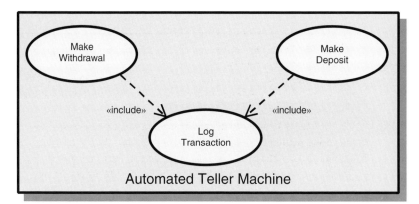

Fig. 4.3 ATM with «include» relationship

4.6.3 Include relationship

A complex use case model can be further simplified by the factoring out of common behaviour patterns into use cases that behave like shared subroutines. For example, in the ATM system, it may be a requirement that all customer transactions which affect the balance of a customer account should be logged to a separate non-volatile storage medium. This behaviour may be encapsulated in a common **abstract** use case called 'log transaction' and then invoked as needed by the existing 'Make Withdrawal' and 'Make Deposit' transactions in the manner of a computer program calling a common subroutine. We say 'Make Withdrawal' **includes** 'Log Transaction'. This is shown diagrammatically in Figure 4.3.

As with an extend, the behaviour of the special use case is inserted into the body of another use case. However with an include relationship it is the included use case that is ignorant of which use cases it is used by, whereas the users must be aware of their invocation of the common use case. This property is reflected in the diagrammatic notation in both «extend» and «include». The dashed arrow points to the use case that is ignorant of the use case that extends or includes it. Unlike an extend the including use case does not stand alone but requires the included behaviour in order to be considered complete.

4.6.4 Multiple use case diagrams

For systems with a large number of base use cases or where significant use has been made of extend and/or include relationships it may become impractical to show all the use cases on a single use case diagram. In these situations it is perfectly acceptable to produce multiple diagrams each showing a subset of the use cases.

Partitioning into multiple diagrams is typically done on the basis of similarity of purpose between use cases or use cases initiated by a common actor.

If necessary to aid clarity the same actors and use cases can be shown on multiple diagrams, although this should only be done if thought absolutely necessary, since it can cause confusion to the readers of the diagrams.

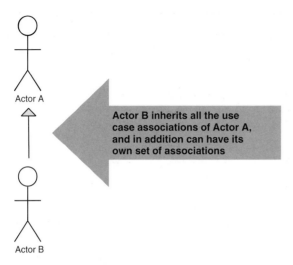

Fig. 4.4 Actor generalization/specialization

4.6.5 Actor generalization

Consider a software configuration management (CM) system. Users of a CM system can play a number of roles: developer, reviewer, team leader and administrator. If building a use case model of a CM system we would expect each of these roles to be represented by actors on the use case diagram. When considering the use cases associated with the actors we will undoubtedly find that each actor has certain use cases they can perform but we may also note that the actors are organized in a kind of hierarchy. Developers can perform basic create, extract, return operations, team leaders can do the same but can also action items through certain states and administrators can do all these normal operations plus a number of specialized administrative tasks.

As things stand we would expect to show each actor having associations with the relevant use cases. With the situation described, this would result in a messy use case diagram with a large number of association lines.

An alternative is to make one actor a specialization of the other. The use of generalization/specialization between actors allows an actor to inherit the associations of another. We can show this on the use case diagram as in Figure 4.4.

It is possible to consider instances of the supertype actor with no corresponding subtype instances being present (looking ahead, it should be noted that this interpretation of generalization/specialization between actors does not have the same meaning as this relationship type between classes in xUML, where, as we shall see in Chapter 6, the specialization/generalization relationship is treated as {`disjoint,complete`} and the supertype is {`abstract`}).

4.6.6 Configuration management use case example

Figure 4.5 illustrates the ideas of actor inheritance, «include», «extend» and multiple diagram views to attempt to aid clarity.

Use case modelling

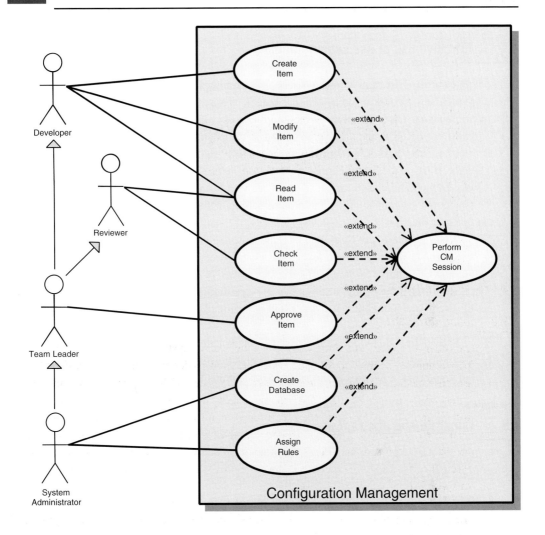

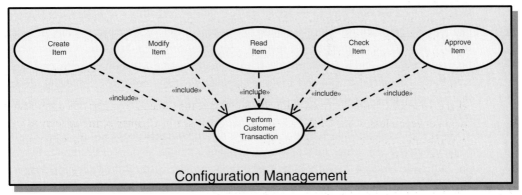

Fig. 4.5 Configuration management use case example

4.7 Effective use of use case modelling

In the previous section we met a number of techniques for managing complexity in use cases («include», «extend» and actor inheritance). Whilst these are helpful if applied with care, to make use cases comprehensible and reasonably concise, the process of decomposition must not be taken too far. As we stated at the beginning of this chapter, use cases are a start to the process of building models, not a fully-fledged modelling technique in their own right. If we go too far in refining our use cases it will, at best, prejudice and possibly distort our domain and class models; at worst we will fall into the trap of performing top-down functional decomposition, with domains acting as 'functions' that we have identified in the use case diagrams.

If you are new to use cases then it is advisable to concentrate on identifying and describing use cases at the appropriate level. The best source of guidance that we are aware of is Alistair Cockburn's 'Writing Effective Use Cases' (Cockburn, 2001) (which we meet again in Section 4.9). He makes the point that use case diagrams should only be used to summarize the use cases identified, that is don't start by trying to draw the definitive use case diagram – it won't be right!

The features that we have discussed, such as extend, include and actor generalization, should be ignored in early use case development. It is far more important and valuable to get a good set of use cases than using these features to organize a poor set.

4.8 Concrete and abstract use cases

When starting the process of identifying use cases, it is helpful to take each actor and ask what does this actor require from the system. It is often tempting to jump straight into a general form of the use case; for instance, in a telecommunications system, imagine we were considering a basic use case for a 'subscriber' actor as 'Make Telephone Call'. This is too generic to start with. We should start with a number of specific use cases, such as 'place national call to a second party', 'place a local call', 'make a conference call', etc. It may even be helpful to name the specific locations we have in mind, in order to fully explore the complexities of that case, for instance we might consider a phone call between Guildford and Keswick in order to explore the routing possibilities for this national call. Early generalization can be a route to skipping what may look like 'irrelevant' detail that actually turns out to be some of the interesting subtleties that we must eventually model.

Once we have modelled enough of these specific, or **concrete** use cases, we can perform some generalization, which may or may not lead to a limited amount of decomposition of our initial use case set. Usually a number of concrete use cases will be subsumed into a single abstract use case. By considering the specific use cases first we make sure that we limit the scope for ignoring detail that we must address and then make a conscious decision to generalize or elide it in the refined use cases.

4.9 Use case levels

Alistair Cockburn in his book (Cockburn, 2001) concentrates on the notion of the use case as being goal-driven. What is more, he gives a taxonomy for the levels of goals, reproduced in Figure 4.6. The idea is that we generally start the use case process with aspirational 'summary goals' (white), that then are refined into a number of user goals (blue) – these are the most important level of use case to document thoroughly. Next come the underwater subfunctions (indigo) that are derived from the user level goals. Some of these may be worth documenting particularly when we wish to «include» the same use case a number of times (logging is an example of a subfunction). Care must be taken not to try to document everything in the use cases (we shall fully model the system using xUML constructs such as classes, operations and statecharts), so we generally avoid the lowest level subfunctions (indigo moving to black).

Another way of contemplating the goal levels is to consider the amount of time it will take a user to achieve the goal. If the time is considerable (greater than, say, half an hour)

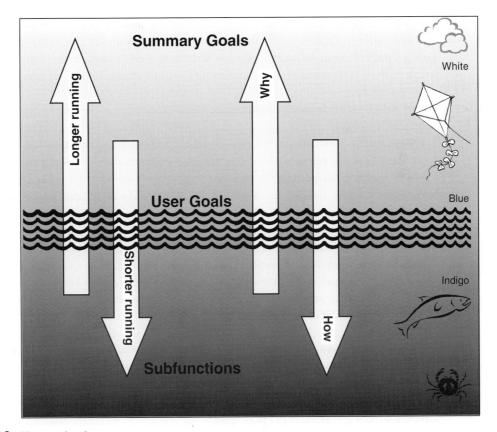

Fig. 4.6 Use case levels

then this is probably a white level goal. If the goal can be achieved in a shorter time, in what could be considered a single user session, then this is a good candidate for a user level goal. The white level goals act as a context for the sequencing of user level goals that they summarize.

Cockburn also points out that there is a continuum from the white to the black, through shades of grey (or blue rather!) that goes from concentrating on the why of the system to the how.

4.10 Specifying performance

Failure to satisfy performance requirements is a common risk for many software development projects. A first step to mitigating the risk is to raise the profile of performance issues on a project at an early stage and avoid the temptation to brush them under the carpet.

When developers are asked the question 'does it (the system being developed) meet performance requirements?' they often respond by pointing out the absence of any performance requirements to begin with and cite the consequent impossibility of verifying performance.

Use case modelling presents an opportunity to capture these elusive creatures and nail them down in a form amenable to analysis and verification by the development team.

Performance requirements are all about specifying how long the system should take to do the 'things' it does. As we have already discovered we can capture the 'things it does' as use cases so performance modelling becomes a matter of specifying how long each use case should take.

Performance is modelled by extending the specification of each use case with the following fields:

- **Trigger event** – this identifies the external unsolicited event which initiates the execution of the use case and whose arrival serves to bound the behaviour for which a performance requirement is to be specified;
- **Periodicity** – this specifies the nature of the trigger event (is it periodic, aperiodic or sporadic?);
- **Rate** – if the trigger event is periodic then this specifies the frequency.

The specification of each scenario is then given a:

- **Response event** – this specifies the event that marks the end of the critical performance region of the scenario. The performance requirement therefore relates to the period between the trigger event and the response event. This is not necessary if the response time is associated with all of the scenario behaviour;
- **Response time** – this specifies the required response time, measured as the time interval between the arrival of the trigger event and the occurrence of the response event or completion of the scenario as appropriate;
- **Response type** – this can be **hard** (each and every instance of the use case scenario must complete within the specified response time) or **soft**. The average of the actual response times for instances of the use case scenario must fall within the specified response time.

- **Source** – this specifies the source of the performance requirements. Examples include:
  - **Estimate** – the customer has not explicitly specified a requirement but one is suggested by common sense and experience;
  - **Exists** – an early version or competitor's product exhibits a specific performance, this system should be at least as good;
  - **Specification** – the customer has explicitly specified a requirement, refer to the Requirements field for traceability information.

4.11 Capturing other requirements types

Not all requirement types are easily or efficiently captured using use cases. Section 4.10 showed how a non-functional requirement, namely response time, can be handled in the use case. Other requirements are not so amenable to specification in the use case. These typically include:
- safety;
- robustness;
- throughput;
- redundancy policy;
- required uptime;
- required software defect density;
- response to uneven data arrival rates;
- security;
- *r.f.* emissions policy;
- distribution issues;
- persistence issues;
- development paradigm;
- adherence to standards (e.g. use of language standards such as MISRA);
- conformance with a quality standards (e.g. ISO 9000);
- application of formal methods;
- update or modification of the system whilst in operation;
- human factors (operational, maintenance, usability, etc.);
- compliance with the law(s);
- etc!

Such requirements must be documented and indexed using natural language so forming another source of requirement.

4.12 Conclusion

We have seen the use cases are a useful tool throughout the MDA process. They provide a way to start the formalization of requirements of the system we are to build, prior to modelling those requirements formally in xUML. Use cases are often a natural way to communicate requirements to stakeholders and to form the basis for requirements' refinement as the MDA process proceeds. They continue to be useful as the xUML models are developed providing a testing agenda for the executable models and finally, when we come to deploy and commission the system, they can be useful for documentation.

5 Platform-independent modelling with domains

5.1 Introduction

So far we have seen that our approach of using xUML in an MDA process is capable of bringing many benefits to software development. Use cases have been explained as one of the techniques suited to capturing stakeholder requirements. In this chapter we shall proceed with the explanation of the MDA process and show the strategy we take to partition the system into a number of separate Platform-independent Models (PIMs) that reflect such requirements.

5.2 System partitioning options

Few software development processes have much to say about large scale partitioning of systems. A large system may contain over a thousand classes and so some partitioning strategy for managing this is definitely required. Without a well-founded strategy the default approach is often functional decomposition, which is flawed in its own right, and definitely not a sound basis for an object-oriented development (see Section 5.2.1). Partitioning by use case allows behavioural partitioning but this needs to be harmonized with structural partitioning (we shall examine this approach in Section 5.6.4).

First, we examine the most common partitioning strategy and then go on to explore a strategy based upon the subject matters present in the system.

5.2.1 Partitioning based upon function

A common approach, particularly with code-centric software development processes, is to base the initial partition of the system upon the idea of **function**. Typically such high level functions comprise differing areas of expertise. For example, in an air traffic control system, a team working on collision alerts would have to understand the subject matter of air traffic control (e.g. the policy that states that if any two aeroplanes are at the same height (within a tolerance) and on constant relative bearings, then there is the potential for a collision), the subject matter of radar data processing (e.g. the way that chirp pulses emitted from the radar head are turned into plots), the way that these plots are correlated to form tracks and then, given these, the detection of the emergency situation, and finally what policy is used to display this to the controller (the user interface). The team will also have to understand how to use and best exploit the platform technology upon which they are to build the system (e.g. CORBA and C++ or Java and J2EE). Such a functional split intertwines expertise from air traffic control to computer science (going via radar data processing and correlation theory). If any of these aspects change (including the platform technology) then the likelihood is that the entire slice has to be re-engineered since there is no separation of the differing concerns. Such a 'vertical' slice of functionality in a system is often attempted and this leads to a number of unfortunate consequences:

- misuse of skilled personnel (there is a tendency for all developers to become 'jack-of-all-trades');
- sensitivity to changing technology (e.g. changing implementation language, user interface technology, database, etc.);
- instability of the system to changing requirements over time;
- high risk when integrating the functional areas to form the final system.

Such functionally oriented partitioning is frequently seen as an 'obvious' route because source requirements are often presented in a functionally oriented way or a naive use of requirements has been presented in a use case (which leads to a vertical slice view of the system).

However, to realize the full advantages of an MDA approach, we choose to use a partitioning strategy based upon subject matter.

5.2.2 Partitioning based upon subject matter

Subject matter partitioning is one of the best-accepted and mature partitioning strategies (Shlaer and Mellor, 1992).

The basis for subject matter partitioning is to recognize that any system comprises a set of subject matters or **domains**. These units will be used as a basis for partitioning the analysis effort for the system. A domain will comprise a number of classes and can be represented in UML using a **package**.

Figure 5.1 shows a **domain chart** for a simplified air traffic control system. Each domain (shown as a UML package) represents a distinct subject matter within the whole system, whilst the dotted arrows represent **dependencies** between the domains. A dependency indicates that the client requires some services that are fulfilled by the server domain. Each domain will be specified in the form of a PIM.

A domain is more formally defined as:

> **Method Definition**
>
> A **domain** is a separate, real, hypothetical or abstract world inhabited by a distinct set of classes that behave according to the rules and policies characteristic of that domain.

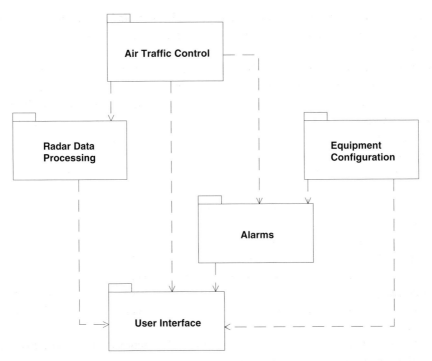

Fig. 5.1 A simplified domain chart for an air traffic control system

Sounds intimidating? Let's take the definition a step at a time:

A separate real, hypothetical or abstract world means that the domain might represent a **real world**, such as air traffic control, military command and control or patient administration, for instance. Such a domain usually reflects the end-user requirements directly. We typically have little or no discretion over these domains, we just formalize the requirements that impinge upon them, aided by any use cases or other requirements that have been documented. A **hypothetical world** might be a domain that performs mathematical transformations, such as 3D geometry or statistical analysis. Such domains serve to formalize the rules of mathematics – again, not much scope for imagination here. An **abstract world** is a world that we have invented for our own convenience, such as a user interface domain. In these domains, the requirements are invented by us to meet the overall needs of the system. For example, we need to establish a policy regarding unavailable menu items – are they 'greyed out' or are they not shown?

A distinct set of classes means that a class should appear in only one domain. Note, however, that the same real world thing can appear at different levels of abstraction on different domains. For example, a real aircraft might appear as an AIRCRAFT, FREIGHT CARRYING VEHICLE, SERVICEABLE ITEM, RADAR TRACK and ICON in various domains (classes are discussed in detail in Chapter 6). These would be correlated by means of counterpart associations (as discussed in Chapter 12).

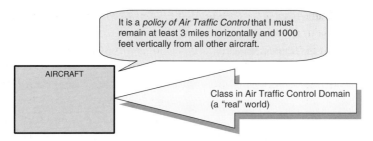

Fig. 5.2 A class from the air traffic control domain

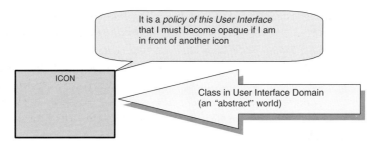

Fig. 5.3 A class from the user interface domain

Behave according to the rules and policies characteristic of that domain means that each class understands the context within which it exists. An AIRCRAFT class in the air traffic control domain embodies air traffic control rules about separation and so forth (Figure 5.2). It knows nothing about how it is displayed and its behaviour is only modelled from the viewpoint of its containing domain.

An ICON in a user interface domain knows only about policies for displaying icons. It knows nothing about the rules and policies governing the things it represents. It is this clean separation of concerns that is the hallmark of domain partitioning (Figure 5.3).

This approach exhibits a number of distinct advantages:

- **Reuse** – each domain is largely self–contained, forming a potentially massive reusable component;
- **Well-defined interfaces** – a domain presents a well-defined interface to other domains which may want to use its services;
- **Effective utilization of subject matter knowledge** – each domain is a subject matter in its own right, therefore staff holding suitable competencies may analyse the appropriate domain (or set of domains), unhindered by consideration of other knowledge areas unfamiliar to them;
- **Stability to changing requirements** – the domain that captures the purpose of the system, with regard to the end-users' point of view, is the application domain; it will typically be augmented as further requirements are modelled, whilst domains further down the domain chart are isolated from such changes;
- **Stability to changing technology** – as technology advances inevitably parts of a system will become obsolete. To avoid this we must ensure that new technologies are readily

accepted into the framework of the system. Domain partitioning recognizes this issue and allows service domains (those that represent highly reusable, technology-oriented subject matters) to be replaced in a highly modular fashion, without affecting other domains. Such domains are referred to in MDA as **pervasive services** (www.omg.org/mda);

- **Incorporation of third-party software** – many systems will incorporate legacy code, third-party libraries or services. Domain partitioning recognizes that this is a risk area for any project and defines such units as implementation domains. The domain approach therefore does not insist upon a homogeneous approach but rather manages the interfaces between differing components, promoting the use of COTS (Commercial Off-The-Shelf) software where appropriate;
- **Incorporation into a use case driven approach** – interaction diagrams readily document use cases on an inter-domain basis. A high-level view of system interactions is extremely valuable and can be used to help scope a domain and define its interfaces.

These are the principles that underpin the MDA process. We separate our system into subject matters, some application-oriented, some representing pervasive services such as communications and persistence, and some technology-oriented. A PIM is built with well-defined interfaces for each domain. It is the fact that each PIM is uncontaminated by other subject matters that makes it both simple and reusable.

A number of architectural frameworks exist, such as the three-tier and N-tier architecture (Sun Microsystem Inc., 2000), the pipe and filter architecture (Allen and Garlan, 1994) and others. Domain partitioning is a complementary rather than an alternative approach to using such frameworks. For instance, with the three-tier architecture (User Interface, Business Rules and Database), the Air Traffic Control example would map the user interface to the User Interface layer (obviously). The air traffic control, radar data processing, equipment configuration and alarms domains would all form sub-partitions of the Business Rules layer and the data access tier would be present in the derived PSMs.

5.3 The domain chart

Typically the first modelling activity on an MDA project is to build a domain chart that shows the domains (the distinct subject matters) that are required for that system. For each domain, we shall build one PIM.

The fundamental unit of system partitioning is the **domain**. A domain represents a large reusable component and is depicted using a domain chart using UML packages and dependencies. The dependency from a **client** to a **server** domain shows that the client requires some services, which are provided by the server. The dependency does not denote flow of control or data. From the client's perspective it simply requires some services and the server provides services to anonymous clients (Figure 5.4). We always maintain the view that domains are loosely coupled via their dependencies.

For each domain we write a **mission statement**, which provides a short summary of the purpose of the domain and its main responsibilities. Each dependency has a description that explains its role in this particular system (Figure 5.5).

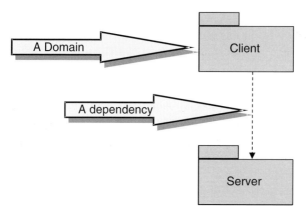

Fig. 5.4 Elements of the domain chart

Each *dependency* between a client/server domain pair has a *dependency description* that explains the purpose of the dependency in this particular system...

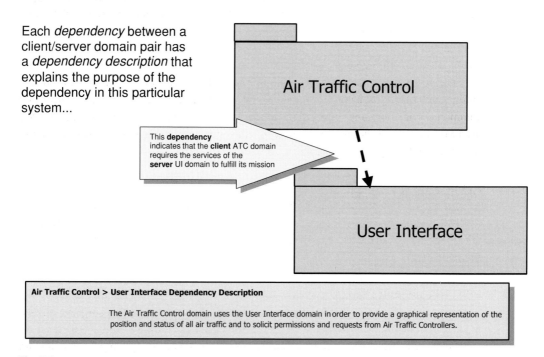

Fig. 5.5 The role of the dependency on the domain chart

5.4 Domains types

The process of identifying and documenting domains is based upon finding domains that fall into one of four categories.
- Application domain;
- Service domain;

- Architecture domain;
- Implementation domain.

5.4.1 Application domain

Modelling a real world – this represents the purpose of the system from the users' point of view. Typically this will form the closest and most direct match to any initial end-user requirements. This domain is normally the (or one of the) **real worlds** from the definition of a domain (Section 5.2.2).

5.4.2 Service domain

Creating a world – all systems require certain basic services, independent of the application. For example, the ability for classes to communicate with each other and to store persistent data. Such pervasive services are provided by what we have termed the software architecture domain. Other pervasive services, typically associated with a type of application, are represented by **service** domains.

Some of the more obvious service domains are discussed blow. For a few we present outline class diagrams, a concept we shall explore fully in Chapter 6. There are many other common service domains. If an MDA project is taking place in an establishment that has used MDA or xUML before, then a domain catalogue may exist – in which case browsing this may yield ready-to-go service domains for your project.

5.4.2.1 Process input/output
This deals with direct physical interfaces so isolating the domains that make use of such devices as sensors and actuators from the actual hardware types present in a particular release. The class diagram in the PIM for this domain looks something like that depicted in Figure 5.6.

5.4.2.2 Resource allocation
In many types of system there is a need to allocate resources of a particular type against a pool of competing resources, usually given some set of constraints and priorities. It also usual that the resources have to be connected in some predefined topology. Such a domain does not know or care whether it is allocating weapons against threats or patients to beds in a hospital; they are simply modelled as requests for resources against a particular resource type.

5.4.2.3 Logging
This domain creates time-stamped snapshots of significant system incidents. The logged data are typically read-only and have no associated application-dependent processing other than querying and garbage collection. The class diagram in the PIM for this domain may look like Figure 5.7.

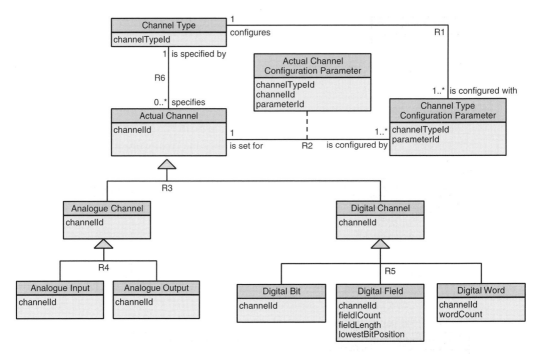

Fig. 5.6 A process input/output domain

5.4.2.4 Alarms

This domain manages alarm conditions as they arise, soliciting acknowledgements from the operators where required.

5.4.3 Architecture domain

Not much imagination required here – you need one! This is the domain that provides the execution environment for all the application and service domains that have been formalized using xUML. It also embodies the strategies for integration of the xUML implementation domains including legacy code and bought-in components. The architecture domain, and the domains that support it, provide the xUML virtual machine. It is often a bought-in domain, although it may be built or produced by modifying an existing one. Many of the requirements for this domain derive from the xUML formalism itself, while others come from the type of run-time environment that must be supported. The architecture domain is an important part of a project and usually has its own distinct development plan. It is closely related to the translation of xUML models into target code. As a result these two important topics are discussed thoroughly in Chapter 13.

Because all xUML-analysed domains depend upon the architecture domain, it is often omitted from the domain chart.

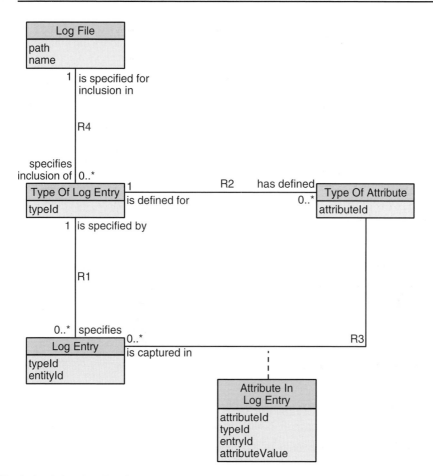

Fig. 5.7 A simple logging domain

5.4.4 Implementation domain

Implementation domains represent pre-existing components. They show:
- Pre-existing software components that are to be reused (often disparagingly referred to as legacy code);
- Software components to be built as part of the development process but outside the scope of the xUML work;
- Bought-in components such as compilers and databases.

A key system design decision is the choice of the software implementation technologies on which the system will be built. In general these will include:
- Programming language;
- Operating system;
- Data storage technology (e.g. database).

Where the system is to be implemented on a network of processors, we also need to choose technologies to support the invocation of operations, and the addressing of messages, across the network.

Implementation domains are not typically subjected to xUML, unless they are so poorly documented that some level of 'reverse engineering' is required to understand how they work and how they are to be integrated.

5.5 Organizing the domain chart

By convention we lay out the domain chart with the application domain at the top, followed by the raft of supporting service domains. Then comes the architecture domain (although we possibly omit that) and, below it, the implementation domain Figure 5.8 illustrates a domain chart for a Naval Command and Control System (dependencies are omitted for clarity).

The service domains are also layered, with the most generic and hence reusable, at the bottom of the service domain layer and the more specialized ones at the top of the layer. For instance it would be improbable to contemplate the reuse of the threat evaluation and weapons allocation domain in a non-military system, whilst a recording domain is highly reusable in a wide variety of differing applications.

5.6 Techniques for identifying domains

In this section we give a range of techniques (which are not mutually exclusive) that may be used to establish the domain chart for a system.

5.6.1 The system level object blitz

The idea of the **object blitz** was briefly introduced in Chapter 3. The **system level object blitz** involves identifying all classes in the system, without regard for domain boundaries, and then identifying clusters of classes that are coupled because they belong to the same subject matter. For example, in an air traffic control system, the object blitz might turn up these classes as in Figure 5.9.

Which would lead to the domain partitioning in Figure 5.10.

We shall see later (Chapter 12) that the relationship between Icon and Aircraft will be modelled as a counterpart association.

5.6.2 Domains by analogy

The system level object blitz technique is often used in the situation where we are modelling a type of system that is completely unfamiliar. If this is not the case, and we have encountered

Platform-independent modelling with domains

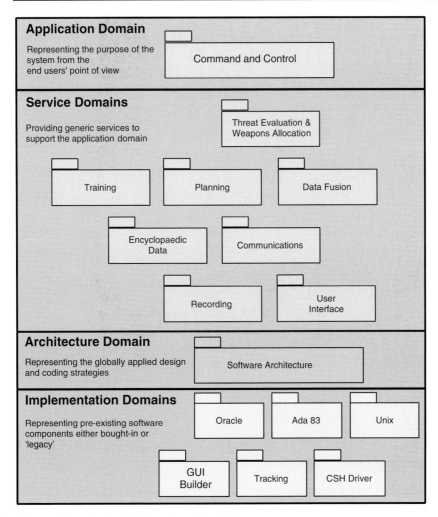

Fig. 5.8 A layered domain chart (without dependencies)

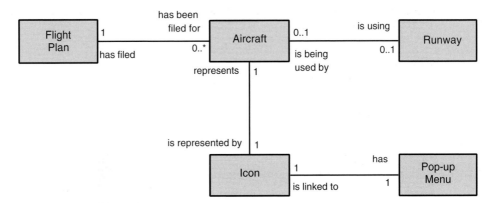

Fig. 5.9 A class model after a system level object blitz and before domain partitioning

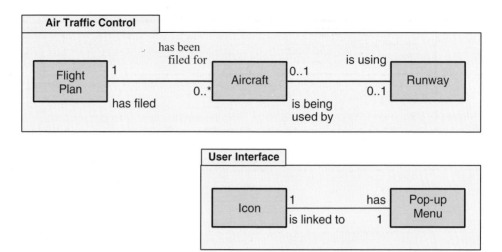

Fig. 5.10 A class diagram partitioned into domains

the type of system before, we frequently have a good inkling as to the domains to expect and use this along with domain interaction analysis (see Section 5.6.5) to derive the initial domain chart. The experienced analyst will ask whether he has encountered a system *like* this one before; the application domain may be different but the class of system may be generalizable. For instance, many real-time systems have common patterns of domain that support sensors, actuators, signal conditioning, failover, etc. (we give an example of a domain level pattern in Chapter 11).

5.6.3 Exploiting generic characteristics

The system level object blitz provides a way of bypassing any preconceptions that a development team might have about the way to partition a system. It also improves your chances of identifying common patterns that can be factored into in a service domain.

Consider an aircraft system (Figure 5.11), consisting of subsystems, such as Electrical Equipment, Hydraulic Equipment and so on. While building the electrical network domain, it will be necessary to get to grips with the distribution system on the aircraft. This will result in classes like POWER SUPPLY, CABLE, ISOLATOR and EQUIPMENT UNIT. At the same time, the hydraulic system domain analysis will reveal classes like COMPRESSOR, PIPE, VALVE and PISTON.

It is obvious that at the network topology level of abstraction these three sets of classes are the same so we can abstract a distribution network domain that can distribute electricity or compressed fluid without distinguishing between the two. The classes in this new domain would be as shown in Figure 5.12.

This highlights the importance of maintaining a system-wide view, and the danger of premature or functionally oriented domain partitioning.

Platform-independent modelling with domains

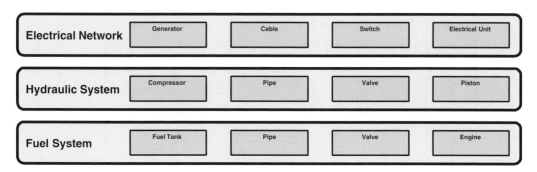

Fig. 5.11 Specific network classes

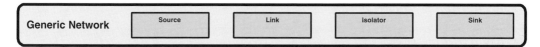

Fig. 5.12 Generalization of the network model

A word of caution, however; it is possible to over-generalize models, which may maximize the potential for reuse but make the models themselves so difficult to understand that analysts feel intimidated by them and so avoid their use – thus defeating the object of maximal reuse!

5.6.4 Common use cases

Each use case represents an operation from the users' perspective, therefore use cases, for the most part, correspond to application domain functionality. However use case diagrams may reveal more about the domains present in the system; **included** use cases are candidates for generic service domains. In Figure 5.13 the requirement to log an incident in the patient administration use cases leads to the realization that we require a logging domain (if we hadn't already identified this obvious domain!). Actors represent entities outside the system with which we must communicate. Therefore these imply the need for 'interface service' domains, either hardware or liveware oriented, so the fact that there is a human actor in the hospital system, the admissions clerk, leads directly to the user interface domain.

At this stage we should sound a note of warning. It is tempting to 'over factor' use cases, leading to top-down functional decomposition and hence yielding a functional rather than subject-matter oriented domain chart.

5.6.5 Domain interaction analysis

When we are considering domains and use cases, it is very common for a use case scenario to imply behaviour in several domains. This means that the domains must interact to

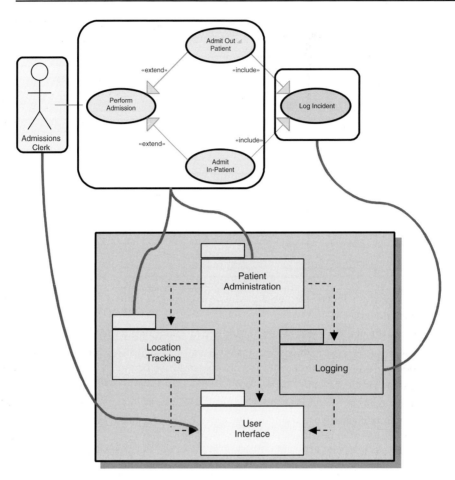

Fig. 5.13 Use cases and domains

realize the use case scenario. A valuable tool in bringing together the uses cases and the domains is a domain level sequence diagram. This is a UML sequence diagram in which the lifelines represent the domains rather than individual objects. We will return to the more conventional object level sequence diagrams later when we consider behaviour within a domain. The domain level sequence diagram for the 'Admit In Patient' scenario is shown in Figure 5.14.

The interactions between the domains are represented as arrows between the lifelines on the sequence diagram. These should be labelled to reflect the meaning of the message in the context of the use case. For each individual use case we identify the set of domain interactions that are required to support it. This will give us a set of domain level sequence diagrams. If we look at that set of sequence diagrams from the viewpoint of a single domain, that is one of the lifelines, the set of messages coming into the domain will represent the set of services that domain needs to provide. The set of messages going out of the domain

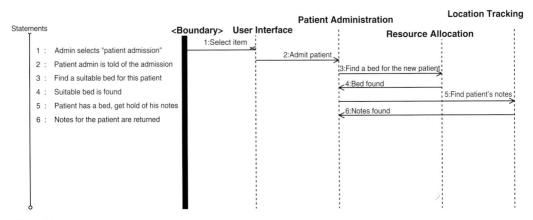

Fig. 5.14 Domain level sequence diagram

will represent the set of services that the domain requires other domains to provide on its behalf. Hence, we can build up a picture of the domain's interfaces. This can be repeated for each domain.

If we perform an object blitz on a domain and we use the set of domain level sequence diagrams to identify the domain's interfaces then we start to build up a good picture of what the domain's responsibilities are. We can then proceed to model the internals of the domain with some confidence that it fits coherently into the system as a whole.

Some styles of sequence diagram allow a commentary to be placed down the left-hand side of the diagram with the statements in the commentary linked to the interactions on the sequence diagram. If the commentary contains the use case scenario description then the correlation between use case and domain interactions is extremely clear. An example of this style of domain level sequence diagram can be found in the case study in Chapter 14.

One frequently asked question about domain level sequence diagrams relates to the labelling of the messages. The domains at either end of the message arrow are different subject matters so the meaning of the message out of one domain will not be the same as the meaning of the message into another – after all if both domains knew the same subject matter we would have a poor domain split. So the question is 'How do you label a message going between two domains?' Let us consider an example taken from Figure 5.14, the patient administration domain sends a message to the logging domain to indicate that a patient has been admitted to the hospital. From the patient administration domain's viewpoint this message is logAdmission, since the domain knows about patient admissions and discharges and knows that they have to be logged. From the logging domain's viewpoint this message means logItem. The logging domain is general purpose and knows nothing of admissions: it just logs things, that is it serves an anonymous client. Hopefully, you will agree that logAdmission is a better message name to put on the sequence diagram than logItem since it conveys more meaning. The general answer to the question therefore is 'Label the message with the name that is most closely related to the business of the system being analysed.' This means that the names are usually taken from the viewpoint of the domain furthest up the domain chart in any interaction. Note that this is not the same as saying label the message from the viewpoint of the sender (or receiver). A further justification for this approach is that the message names will more closely correlate with the use case descriptions.

Some tools allow us to label the interactions in three ways; from the viewpoint of the use case (as discussed above) and also to give both the required and provided message names. This trio of names allows us to fully document the use case and also to start he process of marrying required to provided services (more on this in Chapter 12, where we consider domain integration).

5.6.6   Common behaviour in a domain

When studying a domain, you may get that déjà vu feeling. If you think you have modelled this behaviour already in this domain then you have probably found a common service. For example, in the patient administration domain for a hospital system there will be the need to allocate patients to beds, consultants to operating rooms and so on. This points towards the existence of the **resource allocation** service, which resolves generic requests for different types of resource from different types of requester.

5.6.7   Embedded subject matters

Occasionally the process of asking detailed questions about an application domain leads to the 'discovery' of new domains. For example, when studying the problem of finding a path through a telecommunications network to set up a two party call, you will come across the need to understand the **network topology** and how to find the shortest/fastest route between two nodes.

This represents the discovery of a new subject matter – network topology. This can be made the subject of a separate analysis activity and the application domain will publish service requirements on this domain, such as 'establish route between two nodes'. Incidentally, the resource allocation domain can be reused in this system to resolve requests for timeslots in a multiplexed link.

5.6.8   Determining if a service domain is needed

The questions to ask when considering whether to separate part of the behaviour of a domain into a separate service domain are:
- *Can I envisage this domain in a different type of system?* – for example, network topology would be a domain required in a network management domain and train management domain;

- *Can I see more than one strategy for this part of the system?* – for example, the billing policies for a telecommunications system might vary widely by territory and customer. Although it would be desirable to capture the potential differences in a generic way, this will sometimes be impossible. In this case, a number of variants of the billing domain can be constructed, and the appropriate one included in each system build. Another way of thinking about this property is to consider so-called 'plug-and-play' with a service domain – is this required or likely to happen in the future?
- *Can I imagine a system of this type that includes this class but not this other class?* – for example, can I imagine an air traffic control system with RUNWAY class but no AIRCRAFT class? No – they are part of the same subject matter. Can I imagine an air traffic control system with a RUNWAY class and no ICON class? Yes – they are completely independent ideas and should be kept separate in separate domains.

5.6.9 Benefits of finding service domains

Identification of service domains yields a number of benefits, both immediate and longer term:
- The **application PIM becomes simpler** because it no longer has to deal with the problems of network traversal and resource contention;
- The **scope for concurrent analysis is increased** because we now have three domains (and therefore three PIMs) rather than one. Of course, this is only beneficial if there are sufficient analysts to exploit this concurrency;
- The newly identified service domains will become **generic, reusable components**.

5.6.9.1 Analysis reuse

Note that this approach represents a rather more strategic attitude to reuse than the classic 'reuse individual classes' approach taken elsewhere. Such a naive classed-based approach to reuse has two obvious drawbacks:
- Individual classes are rather small (perhaps containing only tens or hundreds of lines of code). Developers believe that the cost of finding them and assessing their fitness for purpose will probably take longer than writing them again from scratch. When it comes to reuse, size **is** important!
- The classes are expressed using an implementation-specific language. History shows that software engineers have a great track record in using a technology once and then moving onto another one. This is a proven technique for ensuring that the components developed during the previous project achieve obsolescence in the shortest possible time.

xUML captures expertise in an entire subject matter in an implementation-independent way so we have large components of reuse that can be mapped onto whatever technology is on the cover of the latest issue of *Class* magazine.

Of course, it is feasible to use the code generated by one project in the context of an implementation domain on a subsequent project.

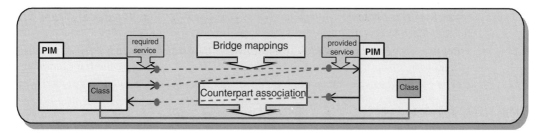

Fig. 5.15 Define mappings (correspondence) between PIMs

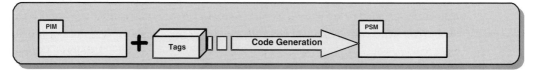

Fig. 5.16 Define mappings from the PIM to the PSM

5.7 The MDA process: a summary

The process of acquiring expertise, capturing it in a form that is uncontaminated by other subject matters and making it accessible to others is the essence of the MDA process. It is fundamental that if a component is to be reusable it does not intertwine separate subject matters, such as network management, user interface and C++ structures, since it has proven historically impossible to untangle these to reuse any one of them. This is why domain-based system development is primarily based upon **reuse of expertise** expressed in domain models rather than **reuse of implemented code**. The use of an implementation-independent action language is a critical component in this strategy otherwise we have tied a critical part of our PIMs to a specific platform, eliminating (or making extremely difficult) the transition to new technologies in the future. For a discussion of action languages see Chapter 10.

5.7.1 Mappings in MDA

The process of building a system now becomes one of:
1. assembling a set of domain PIMs;
2. specifying mappings between them;
3. selecting the most appropriate mappings onto the PSM as the basis of the code generator.
   We can summarize this by Figures 5.15 and 5.16. Of course we shall be examining these concepts closely throughout the book.

5.7.2 Summary of domain partitioning

In summary, the process of partitioning the system proceeds as follows:
1. identify the separate subject matters within the system (domains);

2. write a mission statement for each of the domains that captures its purpose and forms a basis for further modelling;
3. produce a domain chart that depicts the domains and identifies the dependencies between them.

5.7.3   The process of constructing a domain chart

Most systems are developments of existing systems and/or there are strong pressures from customers to use particular third-party software in the system. As a result, there is usually a set of components that must be used in the system. System designers should capture those components as the first domains on the domain chart as these components are givens for the system. Dependencies are used to represent the existing interfaces between the identified legacy domains and the existing interface specifications are used as the definition of those dependencies.

Secondly, the domain library should be searched for pre-existing domains which can be reused in this project. Selected, reusable domain versions should be the second set of domains added to the application domain chart and dependencies added to represent interactions and dependencies between reused domains and between reused xUML domains and legacy domains.

Finally, new domains and their associated dependencies should be modelled and included on the domain chart.

5.7.4   Documenting the domains

Each domain identified on each domain chart will be allocated a unique key letter and will be documented in a domain description as follows:

In the case of pre-existing domains (either xUML or code domains), the domain description will identify all relevant documentation defining that domain. The definition should also summarize any refinements that are required in order to make this domain suitable for reuse on this project;

In the case of domains to be developed as part of this project, the domain description should comprise:

- **Mission statement** – this section is for use by system designers who are identifying potential domains for inclusion into a system. It describes the purpose of this domain in the general case, that is, the role it can play in any system, not just the system under development. This requires that reference to specific client subject matters should be avoided. If it appears to be necessary to produce more than one mission statement for a given domain, perhaps to reflect a change of mission during the evolution process, then this may be indicative of the need for a different domain, rather than a different version of the same domain;
- **Domain version role** – this section is primarily for use by the analysts of other domains in this system who need to understand what the domain does in the context of this system. It describes the role that this domain plays in the system under development. In this case

it is legitimate to refer to ideas in other domains in order to make clear to the reader how this domain contributes to the overall system purpose.

5.8 How to do bad domain partitioning

First, please note the symbol in Figure 5.17. This section is **HOW NOT TO DO DOMAIN PARTITIONING!** – So that said there are lots of ways to develop perverse domain charts and improve your chances of failure. Here are a few of the more popular ones we have encountered:

1. **Build an upside-down domain chart** which shows a hierarchy of control rather than a hierarchy of abstraction. It is pretty easy to do – just make the **user interface** the application domain. If challenged, you can explain that, in a user interface centric system, this domain is the source of most stimuli and therefore must be at the top. You can now focus all your energy upon getting to grips with that lovely GUI builder technology and, with careful planning, you will hardly have any time available for understanding the application. Just move directly into the class-oriented hacking phase, writing lots of Visual BASIC behind each control. If anybody asks what happened to the analysis phase, just tell them you are prototyping – that will keep them away until you have got everything more or less working and there is no time left for analysis. Of course, you know that the **user** interface domain does not represent the purpose of the system. The purpose of the system is to **manage air traffic**, not to manage icons. So, if you want to build the domain chart properly, the **air traffic control domain** should be at the top;

2. **Identify functions, not domains**. Make sure your partitioning is based upon what the system does, rather than what it does it to. One straightforward approach is to take a really cohesive subject matter, such as **Navigation**, and split it into 'phases'. For example, **navigation planning**, which builds and verifies navigation plans made up of segments, legs and waypoints, and **navigation execution**, which guides a vehicle along an established navigation plan. Clearly, these two domains will embody the same set of classes. By taking two copies of each, and analysing them separately each class will have a state machine that specifies a subset of its complete life cycle. The domains are therefore inextricably coupled together, as there is no risk of either being of any value without the other. You can now soak up all that spare time and resource by specifying the same set of classes twice and then putting in place the bureaucracy necessary to ensure they are kept in step. The way others might spot this is by looking for the same noun in different domain names. For example, **navigation** planning **navigation** execution. Try to avoid this, otherwise you will be asked to merge them into one domain. The most

Fig. 5.17 Bad practice symbol

common justification for partitioning the system badly is that it is 'use case driven'. If you come up with a domain for each use case, or even a domain for a set of use cases, you will have a badly partitioned system which you can get away with for some time. The behaviour in the models associated with a single use case will typically, in a well partitioned system, run through many domains;

3. **Ensure domains correspond to software units**. The domain should represent a subject matter. In the general case, there does not need to be any correlation between a domain and a unit of software. It is the job of the software architecture to map the domains onto the target code structure according to specific system architecture requirements;

4. **Work breakdown-based domain partitioning**. It is convenient to have one analyst for each domain (especially if this is the only way you can get multi-users to actually work in your chosen UML tool!). Therefore, we split or combine our initial domains, that were based upon subject-matter, so that their number corresponds to the number of available analysts. Of course, this strategy does have the slight drawback that we have to refactor the domain chart every time an analyst resigns (as surely they will if we go down this route).

5.9 Conclusion

We have seen how domain partitioning fits into the MDA strategy of building separate PIMs linked by dependencies. The domain, capturing a distinct subject matter of the system, promotes reuse on a massive scale, provides 'islands' of stability in a sea of varying requirements and changing technology and allows specialists to work in areas of their own expertise unencumbered by baggage from other subject matters.

6 Class modelling in a domain

6.1 Introduction

So far, we have considered the way we use xUML in an MDA process and the concept of the domain (a separate subject matter within the system under consideration). For each domain that we analyse we build a Platform-independent model (PIM). The PIM is a crucial part of MDA, allowing us to reason about the problem space in a model uncluttered by irrelevant realization technology. Such separation of concerns allows us to change the realization platform independently from the business functionality we wish to support.

The key component of the PIM is the class diagram, one of which will be produced for each domain. A class diagram is a static view of the system. It describes classes and associations. It describes the abstractions of the system being analysed and how those abstractions are related. It does not describe when instances of the classes (**objects**) and association instances (**links**) are created and destroyed. It does not describe how objects are used or how associations are queried. In other words it is a declarative specification. In this section we shall look at some of the major facets of the class diagrams before moving on to describe the notation and form of xUML class diagrams.

6.2 An overview of the class diagram

The class diagram is a **structured declaration**. Everything that happens, happens to something in the class diagram. Behaviour in an object-oriented model is expressed as the:
- creation and deletion of objects;
- reading and writing of attributes;
- linking, unlinking and navigation of relationships;
- invocation of operations.

It defines the **key abstractions** in the form of classes. Each class has properties described that are its attributes and operations. Classes participate in relationships that are captured as associations and generalizations. A class diagram can be read as a set of **verifiable statements**: if the class diagram is correct all of the statements will be true. For example:
- A flight carries many passengers;
- A passenger can be on only one flight at a time.

Fig. 6.1 A simple class diagram

Fig. 6.2 A class with no attributes

```
         Match
         attributes
isSafety
burnTime
manufacturerName
length
typeOfWood
         operations
```

Fig. 6.3 A MATCH class

An example of a class diagram depicting this simple scenario is given in Figure 6.1. We shall introduce the notation used during the course of this chapter.

It is a **stable definition**. A good class diagram will remain stable throughout the life of the system under development. It is expressed in the terminology of the subject matter under analysis and anyone familiar with that subject matter should be able to understand the class diagram.

It is an **organized glossary**. Any informal glossary produced in the requirements gathering phase is likely to be superseded by the class diagram. The class names are not sufficient to define an abstraction. The attributes and relationships help qualify the precise meaning of the class. For example consider the MATCH class in Figure 6.2.

What is intended as the basis of the MATCH class? The abstraction becomes clearer when the attributes are added, as in Figure 6.3.

In addition, the naming of associations is critically important to convey the correct meaning in the model. In Figure 6.4 we see that there can be numerous ways that a person can be associated with a car!

Now that we have seen the central role that the class diagram has to play in the MDA process and some of its notation, we shall start by considering how we build xUML class diagrams.

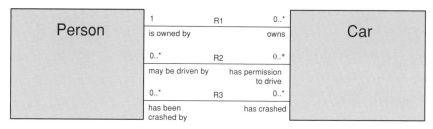

Fig. 6.4 Associations convey meaning

6.3 The Life cycle of the class diagram

The first cut class diagram is produced early in the analysis activity, typically in the object blitz, as discussed in Chapter 3. An object blitz is a short, intensive, group activity aimed at getting an agreed starting point for a domain class diagram. The object blitz will be performed early in the project, say during the inception phase, and is used to identify the scope, complexity and risk associated with the domains to be developed. An object blitz usually identifies 70–80 per cent of the final classes in the domain being analysed. The classes have some representative attributes and the key associations are captured. Subsequent work on the class diagram will add further attributes and refines the form, multiplicity and naming of relationships. Operations are added to the class diagram usually in the later stages of analysis. Detailed descriptions of the classes, attributes, associations and operations are added as the analysis progresses.

The class diagram need not be complete before it is used to develop early iterations of the system. Provided the class diagram has sufficient classes, attributes, associations and operations to support the use cases to be realized, development of the executable PIMs can proceed.

There is no single right answer. For any domain, there will be a number of possible class diagrams that will deliver compliance. They each will have benefits and drawbacks. There are standard patterns that can be adopted. Some of these we shall touch on in this chapter and they are considered in greater depth in Chapter 11. Ultimately, however, as in most things, the best solution is usually the simplest. Avoid the temptation to over-engineer a domain. If you create a monster of a domain, you will probably find yourself in the same position as Mary Shelley's Frankenstein. It seems like a good idea when you are creating it but it soon becomes very hard to manage.

The primary purpose of the class diagram is to depict the classes in the domain. Classes represent the conceptual entities that make up the subject matter under consideration. They each possess pertinent characteristics, which are depicted as attributes, and they participate in relevant associations, which are depicted as links between the classes.

The class diagram represents the **data required to support the computations of the domain**, nothing more and nothing less.

6.4 Classes

6.4.1 Definition

We start with a definition of a class:

> **Method definition**
>
> A **class** is an abstraction of a set of things in the domain under study such that:
> - all of the things in the set – the objects – have the same characteristics, and
> - all objects are subject to and conform to the same rules and policies.

The fact that a class is an **abstraction**, tells us that it is a model, a representation, of the entity under consideration, from the viewpoint of its containing domain. The class acts as a template for all of its instances, the **objects**, which must all have the same characteristics (their **attributes**) and behave in the same way (offer the same operations and, if applicable, have their state behaviour defined by a common state model).

The power of this definition, and its true significance, will become apparent as we build up xUML models throughout this book. The reader should always bear this central definition in mind when considering candidate classes in their own models.

6.4.2 Class descriptions

Like any graphical model, the class diagram provides a visual agenda of items about which we need to say more. Firstly, the classes are formally defined in terms of up to one state machine, zero or more operations and zero or more attributes. These are the ingredients that deliver the capability of executing models expressed in xUML and the potential to translate such models to the target system (see Chapter 13). Secondly, to make the models more approachable for people, such as customers and users, we also provide class descriptions. This covers that embarrassing situation where, while your model is being reviewed, somebody asks 'By the way, what is the TRACK class?' When you put it on the model, you knew exactly what it was, but now, with all these reviewers waiting expectantly, you can think of a dozen things it might be. So what do you do? You can take the humiliation now or you can bluff. If you go for the bluff, make sure you are ready to think on your feet, for you are about to fall victim to one of the great qualities of a precise formalism. It makes it easy to refute assertions. So the likelihood is that the conversation will take on the feel of the Spanish Inquisition, in which you will slowly but surely be backed into a corner. Humiliation is inevitable. The cure for this is to document the classes as you put them on the model. It makes life easier for you, and for those privileged to maintain your model in the future.

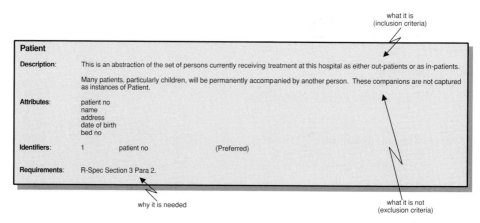

Fig. 6.5 A class description

A good class description is short, states the inclusion criteria (what objects can be considered as members of this class) and any exclusion criteria (what objects are not valid instances of the class). We may also choose to link the class description back to any formal statement(s) of requirement that we have available. Figure 6.5 gives an example of a class description of a PATIENT class, within a hospital administration domain.

6.5 Attributes

Finding attributes is considerably easier than finding classes. Unlike classes, they can be added and removed with minimal impact to the class diagram.

6.5.1 Definition

> **Method definition**
>
> An **attribute** is an abstraction of a single characteristic possessed by all entities that were themselves abstracted as a class.

From the definition we can see that all objects of a class must be capable of being described by an attribute. It is not acceptable to have only some of the objects of a single class that can take on a valid value for an attribute. If you find that you require a 'not applicable' value for some objects of a particular class, then this is a sure sign that another class (abstraction) is required in the domain model.

An attribute represents a value that has been assigned to an object. This value may be arbitrary, such as **patientNo**, or descriptive, such as **patientDateOfBirth**. It is common practice for analysts to allocate an arbitrary integer **id** attribute to classes. As will be seen

Table 6.1 xUML base types

Type name	Example values
Integer	1234 −1234
Real	12.34 1E27
Boolean	TRUE FALSE
Text	"Hello"
Date	1996.04.21
Time_of_Day	21:30:00

Table 6.2 User defined types in xUML

Type name	Example values
Traffic_Light_Colour	'RED', 'AMBER', 'GREEN'
CoolantTemperature	0.1 to 117.5

later, this simplifies the state machine views and can make debugging easier. However, this should not be used as an easy escape from the real job of specifying all the inherent candidate identifiers (see Section 6.10).

6.5.2 Data types

All attributes (and, as we shall see later, local variables and parameters) have a data type. The formalism incorporates a number of base types, which are as shown in Table 6.1.

The user may construct new data types based upon these base types. For example (Table 6.2).

6.5.3 Attribute descriptions

We refer the reader to the earlier section on class descriptions. An attribute description should also be produced contemporaneously with the modelling activity. It too provides an insight into the basis of the abstraction that is the class and its attributes.

6.6 Graphical representation of classes

The graphical notation used to depict classes on a class diagram is shown in Figure 6.6.
- The attribute and/or operation compartments can be suppressed.
- The types of attributes can be shown. (Figure 6.7)

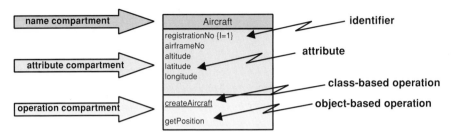

Fig. 6.6 Graphical representation of a class

Fig. 6.7 Graphical representation of a class with typed attributes

6.7 Tabular representation of classes

As you will see during our progression through the xUML formalism, tables are used in a variety of ways. They have the advantage that, unlike class diagrams and statecharts, they are in common everyday use and so provide an easy way to present your ideas to subject matter experts and solicit their feedback without subjecting them to a five-day technical training programme.

Use of tables can be particularly useful when discussing a system with non-technical experts, who will recoil at the sight of any form of class diagram, but will be very comfortable with tables.

It is good discipline to produce a few example objects in tabular form for each class. This often brings out problem-oriented detail that does not emerge when thinking about classes rather than objects. In a tabular representation of a class, each column represents an attribute and each row corresponds to an object. An example object table for a PATIENT class might look like that shown in Table 6.3.

We shall see later how populating object tables helps to tease out hidden classes and highlight problems that may be lurking in our classes. The use of tables to capture example

Table 6.3 An example object table for a PATIENT class

patientNo	name	dateOfBirth
1	Catherine	16:12:1485
2	Henry	28:06:1491

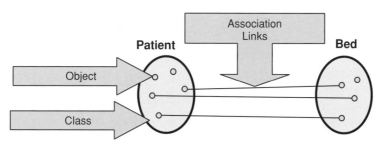

Fig. 6.8 A potato diagrams

objects of each class will also make the model easier to understand and check because it becomes much clearer what data are held and where.

6.8 Potato diagrams

Potato diagrams provide an informal way of graphically depicting classes, their objects and instances of associations (the links). They are helpful when trying to establish the number of objects that participate in each association. Each 'potato' represents a class, and each 'hole' (or 'eye' on the potato) represents an object of that class. Links can be depicted as lines from one object to another. In fact, potato diagrams are a very useful way of understanding objects and their association links (Figure 6.8).

Note that this type of information can also be represented (in a less compact form) using UML object diagrams.

The potato diagram is not part of the UML and is not presented here as a formal notation. Don't rush off and add 'Development of Potato Diagrams' to your project plans or add 'Potato Diagrams' to your list of project deliverables. This type of diagram is an unceremonious tool that helps you get to grips with the form and cardinality of associations. Potato diagrams can be useful when you are developing a class diagram to work out what type of association is required. They can also be useful when reviewing a model to understand what the model is stating about the related objects.

6.9 Associations

Classes do not exist in isolation, they need associations too. There is a prevailing view that classes are individually reusable. This view is not held in xUML. That is because the classes in xUML have intelligence. They are not just data structures with `get` and `set` operations for each attribute. A BED class in a hospital system is aware that a PATIENT can occupy it. This couples the classes together and means that individually they are useless. Hmmm..! Sounds like bad news. But think what we are saying if we make them independent of one another. The knowledge that beds are occupied by patients, and the rules about how they are allocated, and how many patients can be assigned to a bed, must be held somewhere. Of course, we could put it into an association class that formalized the association, and in this case that would not be unreasonable. Or we could put this knowledge in manager or controller classes. This is a feasible strategy but it yields a system in which the majority of the classes are stupid, passive data structures, with a few indecently complex manager classes that are coupled to dozens of stupid classes. These manager classes typically exhibit poor forms of cohesion since they contain knowledge of totally unrelated parts of the system and they do not correspond to anything in the problem. They typically have exactly one object and correspond to a unit of software that we envisage writing one day. This is not the intention of finding classes in xUML.

So in xUML, we identify the tightly binding associations between classes. This is easier than it sounds. Ask yourself 'Does a hospital bed make sense without patients?' Of course not, so there is little incentive to build a BED class that does not know about patients.

This approach to classes and associations yields a view of the real world, rather than an artificial view, tainted by our preconceptions about how to organize our software, and littered with meaningless associations like 'Hospital Controller manages Beds' and 'Hospital Controller manages Patients'. This reveals nothing about the problem.

Each association is specified in terms of:
1. Role phrases;
2. Multiplicity;
3. Conditionality;
4. Description.

6.9.1 Definition

> **Method definition**
>
> An **association** is the abstraction of a relationship that holds systematically between objects.

6.9.2 The role phrases

Each association can be navigated in either direction. In this case, we can start at a Bed object and navigate the association to find the object of Patient that is using it. Or we can go the other way. Each participating class has a perspective on the association, and we therefore assign two **role phrases** to each association.

Note that each role phrase should be based upon the same verb (e.g. 'is **using**' and 'is being **used** by') (Figure 6.9). Remember, there is only one association.

There is much debate about whether to use verb role phrases or noun role phrases (known as a **rolename** in UML). The benefit of the verb role phrase is that it allows a sentence to be constructed which is meaningful, and most importantly, refutable by anyone with knowledge of the subject matter of the domain. In short, verb role phrases make class diagrams easier to read and validate. With verb role phrases, read the phrase and cardinality at the far end

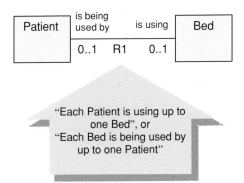

Fig. 6.9 Example role phrases

of the association; so we should interpret the association in Figure 6.9 as 'Patient is using up to one Bed' and is the opposite direction 'Bed is being used by up to one Patient'.

UML permits associations to be annotated to indicate that they are navigated in only one direction and therefore only one role phrase is required (this is done by placing an arrow on the association end). At this stage in the modelling it is too early to know how you will need to navigate an association. The PIM will define whether you need to navigate the association one way or both ways when we specify the actions. Associations are meaningful in both directions and adding role phrases to both ends makes the model easier to read. And, if you find later on in the modelling that you only navigate an association in one direction, DON'T remove the role phrase from the model – you never know what the next requirements change will mean. It is therefore our advice to never use navigability adornments on associations when building PIMs.

6.9.3 Multiplicity

The multiplicity indicates how many objects participate in each association. There are three choices:
- One to one;
- One to many;
- Many to many.

Note that 'many' means 'one or more'. Multiplicity is depicted by the cardinality string. Figure 6.10 shows how association multiplicity is depicted graphically on a class diagram and how it is manifested on potato diagrams.

6.9.4 Conditional associations

Figure 6.10 shows associations where all objects participate. It is possible to depict the fact that some objects do not participate by specifying '0' as part of the multiplicity (Figure 6.11).

Note that this conditionality cannot be established by taking a 'snapshot' of the situation under analysis. In the hospital system (Figure 6.12), it may be the case that no wards

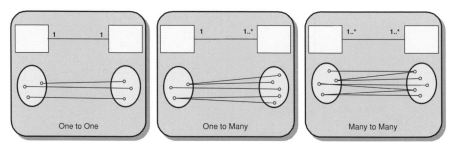

Fig. 6.10 Association multiplicity

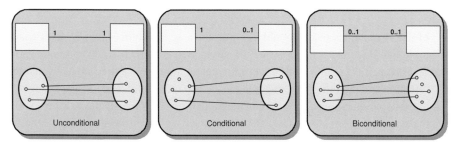

Fig. 6.11 Association conditionality

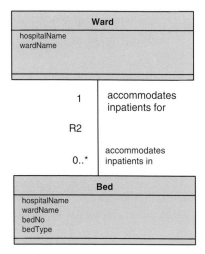

Fig. 6.12 A conditional association

currently have zero beds in them. This does not make the association unconditional. The question is 'Could a ward ever have zero beds?' If the answer is 'Yes, because a ward can be temporarily closed', then the association is conditional.

Figure 6.13 is a reminder of the notation that we have met so far when modelling classes.

Class modelling in a domain

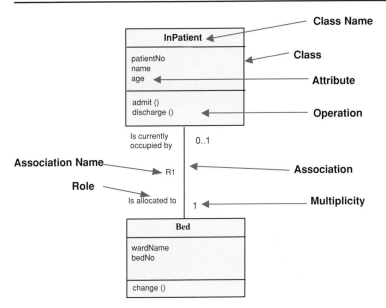

Fig. 6.13 Summary of association terminology

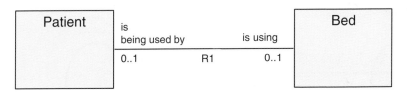

Fig. 6.14 An example association

6.9.5   Association descriptions

You should have already got the idea that we need to document our model. We have written descriptions for classes and attributes, so it should come as no surprise to discover that we also document associations. The description should give an indication of who maintains the links (if this is done dynamically) and the basis of the choice of objects to participate in the link (Figure 6.15).

Still don't like the idea of documenting stuff? Have you ever looked at somebody else's code and thought, 'Why was the code written like this? It is perverse, and this person is clearly in need of psychological help.' You have? Well, it is probable that somebody will one day look at your model and, finding no descriptions to explain your rationale, will reach similar conclusions regarding your level of competence and mental health. Remember, there is much more to a model than the graphics. They merely provide a visual index to the real expertise, which is described formally in state machines and operations, and informally in descriptions, to make the expertise accessible to normal people. An example of an association and its description is given in Figures 6.14 and 6.15, respectively.

> **R1**
> **Patient is using up to one Bed**
> **Bed is being used by up to one Patient**
>
> Description: This association records which Patients are in which Beds. Instances of this association are created and maintained by the staff on each Ward.

Fig. 6.15 An example association description

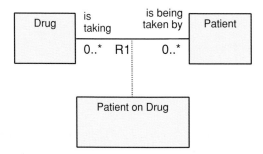

Fig. 6.16 An association class

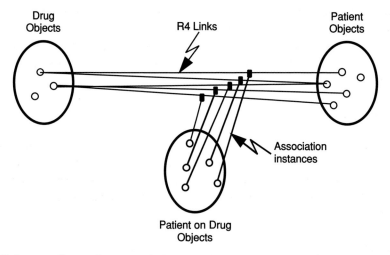

Fig. 6.17 A potato diagram for an association class

6.9.6 Association classes

Any association may be formalized using an **association class**. This is a class that associates the two related classes. It is depicted graphically as in Figure 6.16, to emphasize that the class is really an abstraction of the association itself.

In potato diagram language, an association class is depicted as in Figure 6.17.

Note that now there are two types of linkage: the normal links corresponding to the association instances and the lines connecting each association instance to an association

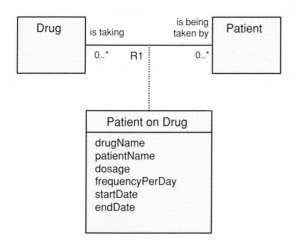

Fig. 6.18 Capturing characteristics of an association

class object. Note also that the notion of conditionality does not apply to the associations – if an association instance exists, then a corresponding association class object must also exist.

It is common that the analyst needs to describe characteristics about an association. In this example, we need to capture, for each PATIENT_ON_DRUG object, the dosage, frequency, start date and end date. This is formalized by attributing these facts to the PATIENT ON DRUG class, as in Figure 6.18.

Reading the association between PATIENT and DRUG works in the usual way – 'Drug is being taken by zero or more Patients', and the other way – 'Patient is taking zero or more Drugs'.

6.9.7 Multiple associations between classes

A pair of classes can participate in any number of associations. Each association typically involves a different set of objects. In Figure 6.19, we see that the current doctor (GP) for a PATIENT (if any) is given by R1, whilst R2 informs us of the historical registrations for that patient. Note that such multiple associations only make sense when well-thought-out role phrases are used.

6.9.8 Generalization/specialization hierarchies

The essence of abstraction is finding **the difference that makes the difference.** When studying the subject matter of air traffic control, we can observe that all aircraft are different. Some have two engines, some have four. Some have vertical take-off capability, some do not. Some are big, some are small.

The analyst's task is to decide whether these differences matter in the context of the domain under study. Specifically, do these differences manifest themselves as:
- Different attributes;
- Different associations;
- Different states?

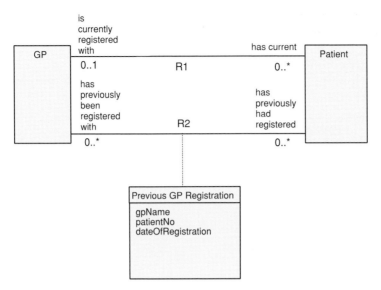

Fig. 6.19 Multiple associations between classes

The difference in the number of engines can be captured as a **number of engines** attribute. This would best be attributed to a specification class, such as TYPE OF AIRCRAFT, since the number of engines depends upon the aircraft type, not the actual aircraft. This is a difference, but it does not make a difference to the model. It does not justify abstraction of extra classes named TWO ENGINED AIRCRAFT and FOUR ENGINED AIRCRAFT.

However, the fact that some aircraft have vertical take-off capability would imply that they behave differently to horizontal take-off aircraft. This could manifest itself in a number of ways. For example, only the class HORIZONTAL TAKE-OFF AIRCRAFT would have:

- An attribute named **minimumRequiredTakeOffSpeed**;
- An association to an allocated runway;
- A state named `TaxiingToRunway`.

In other words, the fact that some aircraft take off vertically and some take off horizontally is a difference that makes a difference. We must abstract separate classes to reflect the differences. Importantly, whether the difference makes a difference depends on why you are producing the model. There isn't a single correct model for every application. If you are modelling Pilot Licensing then aircraft in different weight bands are an issue as is whether it is a single engine or multi-engined aircraft. These differences are not significant in the air traffic control case.

Of course, in such circumstances, there will typically be a number of attributes, associations and states that are common to all members of the set.

Consider this model fragment for a patient administration domain (Figure 6.20):

Clearly, all patients have a patient number, name and date of birth. But:

- **creditLimit** and **currentBalance** apply only to patients paying for their own treatment;

Class modelling in a domain

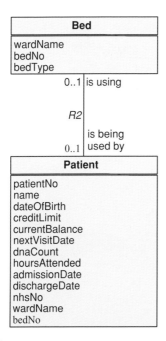

Fig. 6.20 A schizophrenic PATIENT class

- **nextVisitDate** applies to outpatients;
- **dnaCount** (nothing to do with genetics – this stands for 'Did Not Attend'!) applies to those outpatients attending (in fact, not attending) consultant clinics;
- **hoursAttended** applies to out patients attending day care centres;
- Only in-patients are allocated to beds, and have admission and discharge dates.

So we could refine the model to reflect this, as in Figure 6.21.

6.9.8.1 Subclass hierarchies

We still need to deal with the fact that the original **hoursAttended** attribute applies only to day care outpatients, while **dnaCount** applies to consultant-clinic outpatients. We can address this need by further specializing the OUT PATIENT subclass, as in Figure 6.22.

6.9.8.2 Multiple classification

Finally, we can address the fact that each patient, regardless of whether as in-patient or outpatient, is either being funded by the National Health Service (NHS) or by private funding. This can be achieved by introducing another subclass family for PATIENT. This ability of a class to have multiple subclass families is known as **multiple classification** (Figure 6.23).

What have we achieved here? We have **formalized expertise** about the subject matter of hospitals. The original PATIENT class did not conform to the definition of a class because it does not represent a set of things that have the same characteristics. Not all patients have a

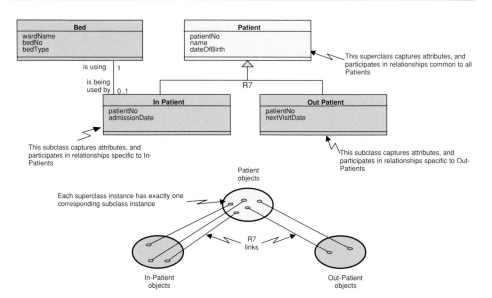

Fig. 6.21 Improvement 1: a specialization–specialization hierarchy

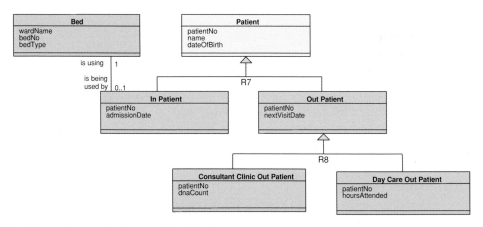

Fig. 6.22 Multilayered generalization–specialization hierarchy

'credit limit' or a 'dna count'. The reader is therefore left to make assumptions about which attributes pertain to which patients. The second model makes it clear which attributes pertain to the different types of patient. It also makes clear the policy that only in-patients are allocated to beds.

But what about size and speed? It is easy to jump to the conclusion that more classes mean more code and slower execution, but a closer examination shows the opposite may be true.

Let us suppose that we wish to find all the in-patients due to be discharged today. The process specification for the smaller model would look something like this (this is given in the Action Specification Language (ASL) which we shall consider in detail in Chapter 10):

Class modelling in a domain

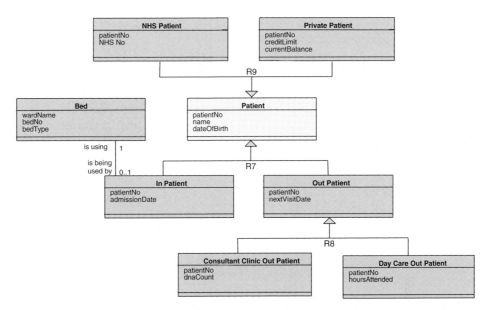

Fig. 6.23 Multiple classification

```
for each Patient
  # check to see if this is an in patient
  thebed = this -> R1
  if thebed != UNDEFINED then
    if dischargeDate = current-date then
      [daysStayed] = \
        noOfDaysBetween[admissionDate, dischargeDate]
      totalDaysStayed = totalDaysStayed + daysStayed
      totalInPatients = totalInPatients - 1
    endif
  endif
endfor
averageDaysStayed = totalDaysStayed / totalInPatients
```

Most of the code for this IN-PATIENT class will be infested with these 'if it is this type of patient, then this makes sense, otherwise it does not' tests. The predominance of if...then...else or switch statements in the code for a class is usually indicative of an effort to economise on classes. The price is more code.

With the larger class diagram, the code would look like this:

```
for each InPatient
  if dischargeDate = current-date then
    [daysStayed] = \
      noOfDaysBetween[admissionDate, dischargeDate]
    totalDaysStayed = totalDaysStayed + daysStayed
    totalInPatients = totalInPatients - 1
  endif
endfor
```

There is less code, occupying less space, and there are fewer objects to access, consuming fewer CPU cycles. The larger class diagram thus yields a system that is **faster** and **smaller**.

> **Modelling technique: subclasses to capture physical differences**
>
> One of the reasons that we abstract subclasses is to capture physical differences. Because these differences will typically manifest themselves as attributes, associations or states, the definition of 'class' requires that we abstract different classes.
>
> For example, an aircraft is either a 'Powered Aircraft', or an 'Unpowered Aircraft'. The powered aircraft is obviously physically different in that it has engines. This might manifest itself as a **fuel remaining** attribute, an association to the **Engine Type** class, and a state named All Engines Stopped.

6.9.8.3 Dynamic classification

We have seen generalization–specialization hierarchies used to capture differences between similar classes. There are some occasions when the differences do not pertain to what the abstracted entity **is**, but to **what it is doing**. For example, consider this AIRCRAFT class in Figure 6.24.

This AIRCRAFT class has all prime characteristics of lazy abstraction. It embodies attributes that only have valid values for part of each object's life cycle. For example, **plannedTakeOffTime** is not relevant once the aircraft is flying. It also participates in two associations, both of which are conditional (and, in this case, mutually exclusive). This means that an object will only participate in each association for some of its life cycle. But which

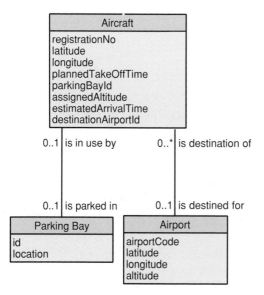

Fig. 6.24 A class model requiring dynamic classification

Class modelling in a domain

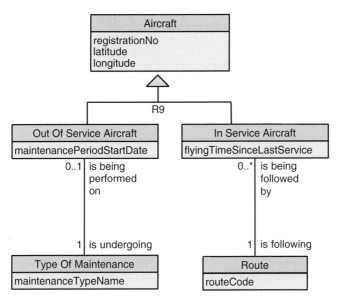

Fig. 6.25 Role subclasses

part? The reader is left to speculate about when each of these attributes and associations is relevant. This is not in keeping with the notion that a domain model should formalize expertise about a subject matter. One effective way to eliminate conditional associations like these is to abstract new classes that represent those instances that are currently participating. This will, by definition, make the associations unconditional so we could refine this class diagram as in Figure 6.25.

We have abstracted classes that represent **roles** played by the aircraft. These are referred to as 'Role Subclasses'. Each embodies the attributes that characterize objects performing that role and participate in associations that are specific to that role. Note that these typically correspond to one or more states of the superclass, but we are concerned at this stage about the different attributes and associations; that is why we need these extra subclasses.

At the point at which the aircraft is taken out of service for maintenance, the **flyingTimeSinceLastService** attribute and 'Route' association become irrelevant, and therefore the subclass IN SERVICE AIRCRAFT is deleted, and another subclass OUT OF SERVICE AIRCRAFT is created. This process is referred to as **role migration**.

It can be seen that, once the superclass AIRCRAFT object has been created, the creation and deletion of the role subclasses occurs independently (although the rules of one subclass object per superclass object must always be obeyed). This is known as **dynamic classification**. Conversely, subclass objects that represent physical differences are typically created and deleted at the same time as the corresponding superclass object. For example, a powered aircraft is always a powered aircraft (unless the analyst decides that an aircraft with all engines failed is an unpowered aircraft, in which case this begins to look more like role migration).

> **Modelling technique: subclasses to capture role migration**
>
> Another motivation for using subclasses is to reflect the fact that the class performs a number of different roles. This will be reflected by the presence of different attributes and associations. Note that a role subclass often corresponds to one or more states of the class. For example, an aircraft performing the role 'Flying Aircraft' has an attribute **time of take-off**, and an unconditional association with a destination 'Airport'. The state machine for this role subclass might include states like `Climbing Out`, `Cruising` and `On Final Approach`. In each of these states, the given set of attributes and associations exist.

6.9.9 Combining multiple and dynamic classification

The model in Figure 6.26 shows how aircraft can be partitioned into two ways. Each aircraft is either powered or unpowered, and, totally independently, each aircraft is either military or civil. This situation can be depicted using multiple subclass families.

Each family is identified by its relationship number. In this example, the two families are 'R8' and 'R9'.

6.9.10 Multiple superclasses

The rules regarding superclasses and subclasses are simple: a class can have zero or more subclasses, and zero or more superclasses. So the following model, Figure 6.27, can be built to depict the fact that a FLYING AIRCRAFT is a specialization of both AIRBORNE VEHICLE and AIRCRAFT.

This kind of structure is entirely legal but should be used with care. Experience has shown that it is not always the best modelling solution. A modelling problem which, at first

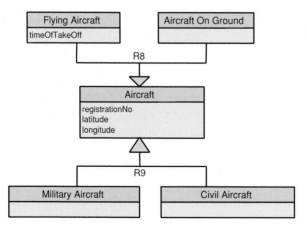

Fig. 6.26 Multiple and dynamic classification

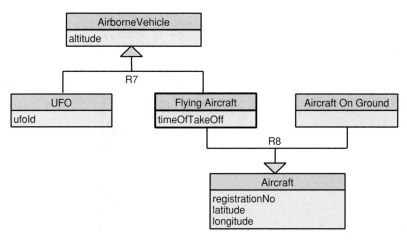

Fig. 6.27 Multiple superclasses

sight, may be best represented using multiple superclasses might be better expressed using multiple subclass trees.

(The authors appreciate that the attribute **ufoId** of the UFO class seems contradictory!)

6.9.11 The distinction between generalization and specification classes

Sometimes there is confusion over whether to use a generalizing class (superclass) or a specification class. The following draws the distinction:

Superclass versus specification classes

The basic distinction between specification classes and superclasses is:

A specification class captures *attribute values* common to a set of objects. For example, the *maximum altitude* for each type of aircraft.

A superclass captures *attributes* common to a set of objects. For example, the *current altitude* for each aircraft. The values of the attribute vary with each object.

Figure 6.28 depicts an example of both specification and generalization classes in use, showing how TYPEOFTORPEDO captures the attribute values common to all torpedoes of a given type, whilst the WEAPON superclass captures the attributes common to all weapon type subclasses.

6.9.12 Generalization–specialization in xUML

Now that we have discussed this type of relationship, we can express the way we use generalization–specialization in xUML by applying the UML property *{abstract}* to the supertype class (depicted by italicizing the class name) and the constraint *{disjoint, complete}*

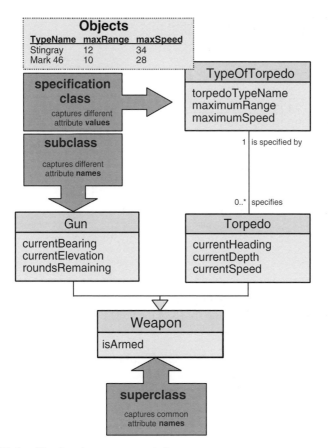

Fig. 6.28 Specification classes versus subclasses

to the discriminator (the separate named child hierarchies, R8 and R9 in this case, see Figure 6.29). The property abstract, for the supertype, means that we may not instantiate it by itself, so there must be at least one corresponding child in each of the hierarchies. Now to consider the constraint *{disjoint, complete}*; first *disjoint*, this stipulates that only one child may exist for any given parent in a single hierarchy; secondly, *complete* states that there are no missing, or elided children. The property of *{abstract}* and the constraint *{disjoint, complete}* give us the view of generalization–specialization presented in the preceding sections.

Since this is what we always mean my generalization–specialization in xUML models, it is usual to leave off the *{abstract}* and *{disjoint, complete}* adornments from the model, so Figure 6.29 is more usually presented in xUML as in Figure 6.26).

Another convention in xUML is to place additional semantic 'weight' to the use of the 'shared target style' for generalization discriminators. When we produce a model as in Figure 6.26, we could formally choose to write it as in Figure 6.30. This style soon becomes unwieldy and confusing, so we shall always use the 'shared target style' with a single discriminator name for each hierarchy.

143 Class modelling in a domain

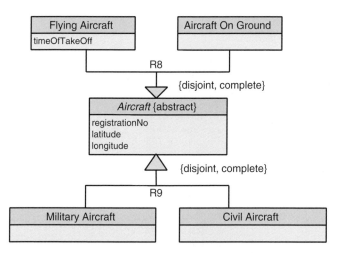

Fig. 6.29 Generalization–specialization with abstract keyword and discriminator constraints

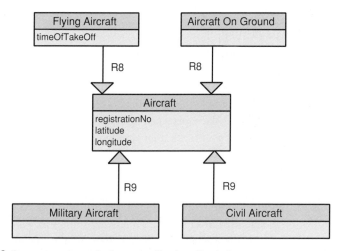

Fig. 6.30 Separate target style for generalization discriminators

6.9.13 Generalization–specialization versus inheritance

You may have noticed that we have not yet used the word **inheritance**. There is a popular misconception, sadly propagated in many textbooks, that specialization and inheritance are one and the same thing; they are not. There are two main differences between the notion of generalization–specialization in xUML and the much narrower view of inheritance, as taken in many OO programming languages. We have met all of these ideas in the preceding sections. Here we shall explain why these cause difficulties if one were to take a naive inheritance view of the world. The main point to stress here is that in xUML we consider the classes to be instantiated separately and linked via relationships (the generalization–specialization relationship) which have the specific semantics outlined.

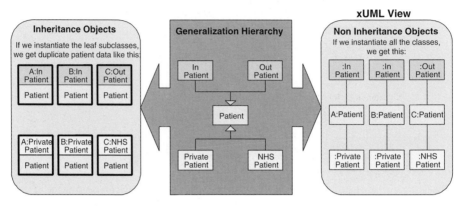

Fig. 6.31 Multiple classification and inheritance

First, consider **multiple classification**; the inheritance view of the problem leads to a dilemma – for each of the hierarchies we must somehow have a duplicate of the supertype (Figure 6.31).

Next we consider **dynamic classification**, where, for a given supertype, the corresponding child may change at run time. In the example in Figure 6.32 we consider an NHS patient deciding to go private. In the xUML view the supertype PATIENT object always exists and we link this to a newly created PRIVATE PATIENT after simply unlinking and deleting the old NHS PATIENT. In the inheritance view we have to delete the PATIENT and NHS PATIENT and recreate a new patient (with the same identifying attributes as the old one) that is now a PRIVATE PATIENT.

The reader may think that it must be impossible to implement the full richness of xUML generalization–specialization. It actually turns out to be a relatively straightforward task for most code generators, they simply do not use inheritance to implement this type of relationship; instead, constrained one to one binary associations are commonly used. Inheritance can be employed when we can determine that neither multiple or dynamic classification is present in all hierarchies from a given supertype, although it is often not worth the extra complexity in the code generation rules to exploit this situation.

As a final thought on the distinction been generalization–specialization and inheritance: the modelling techniques of multiple and dynamic classification are far too useful to abandon in order to allow a simple-minded implementation strategy of inheritance to be used.

6.9.14 Rules for objects

As we have seen, objects can be depicted using tables. It is useful to produce example object tables and then check them against the rules we introduce below:

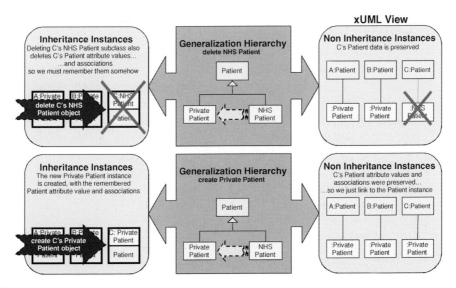

Fig. 6.32 Dynamic classification and inheritance

Patient Name	Number of Previous Ops	Date of Last Op
J Bloggs	3	29 FEB 1980
B Jones	0	N/A

↑ last patient in list

Fig. 6.33 'Patient on List' objects

> **Method rules**
>
> Each cell in an object table must contain exactly one fact.
> Row and column order are not significant.
> Each row must be uniquely identified by one or more identifying attributes.

For example, consider the object table in Figure 6.33, which captures objects of PATIENT ON LIST, an abstraction of a patient waiting for a particular type of treatment.

In terms of the rules, this set of objects is a disaster! It breaks all three rules:

1. The cell B Jones — > Date of Last Op contains a 'Not Applicable' value because he has had no previous operations. Note that the value '0' in **Number of Previous Ops** is fine, because this attribute is of type **Integer**, of which 0 is a legal value. But the **Date of Last Op** attribute is of type **Date**, and no meaningful value of that data type can be included in this cell. The solution is to recognize that an **Operation** is a class in its own right

and add it to the class diagram with a **Date** attribute and an association indicating which patients have undergone which operations.
2. The fact that 'B Jones' object follows 'J Bloggs' in the table cannot be construed to mean ordering. This must be formalized. The most straightforward way to do this is to include a reflexive association (an association between objects of the same class) that captures the fact that each PATIENT precedes another PATIENT. This will result in an additional referential attribute named **Next Patient**. Use of sequence numbers is undesirable because this makes it hard to insert new objects into the middle of the list.
3. Clearly, the patient name alone cannot be used to identify a patient – there is more than one B Jones in the world. The most common approach to this problem is for the analyst to establish an arbitrary, integer **id** attribute, over which the system has complete control.

6.10 Identifiers

It is the responsibility of the analyst to specify the way in which objects can be uniquely distinguished. This requires specification of identifiers. Each class has at least one identifier and each identifier comprises at least one attribute.

> **Definition**
>
> An **identifier** is a set of one or more attributes whose values uniquely distinguish each object of a class.

For example, the identifiers of the CONSULTANT class in a hospital system might be defined as in Figure 6.34.

There are three things to note about these identifiers:
1. Identifier 1 comprises a single, integer attribute whose values will be arbitrary. This is very common with classes that do not come with a 'natural' identifier. The values are under the control of the system and can therefore be guaranteed to be unique within the system;

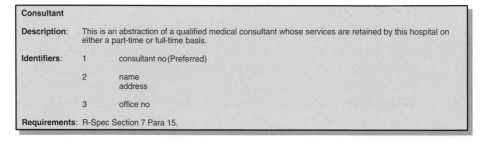

Fig. 6.34 Identifiers for a CONSULTANT class

Class modelling in a domain

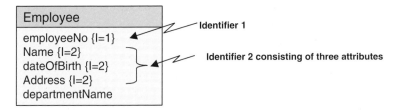

Fig. 6.35 Representation of identifiers in a class

2. Identifier 2 comprises two test attributes that are outside the control of the system. This identifier implies that there cannot be two consultants with the same name at the same address. This seems to tempt fate! Such identifiers are dangerous and should be avoided;
3. Identifier 3 is similar to identifier 2, except that the value it contains can be constrained by the hospital. This use of an identifier can be used to formalize a policy or constraint – in this case, that no two consultants can share the same office! Clearly, the hospital can enforce this policy, whereas forbidding two consultants with the same name from living at the same address may be perceived as rather draconian.

Specifying identifiers for each class is useful because it helps us think about how an object is recognized. Identifiers are frequently natural properties of the class being abstracted. Where there is no natural property an arbitrary identifier can be used, in which case only one identifier is justified.

Identifiers help formalize expertise about the domain, as in this example, where identifier 3 formalizes the fact that consultants don't mix well.

Identifiers are analysis artefacts – they do not represent a design decision. However, they do often map directly to the implementation. For example, **consultantNo** maps to the primary key of the Employee database table.

Identifiers are useful in formalizing associations and they help to identify missing associations. For example, identifier 3 (**officeNo**) of the CONSULTANT class implies that each CONSULTANT object participates in an association with exactly one OFFICE object. This should be formalized on the class diagram.

6.10.1 Representing identifiers

Identifiers are not supported explicitly by UML class diagrams so the convention is to use a UML tag to represent them (Figure 6.35).

Models can be built without identifiers in the classes but this makes them less formal and typically less useful. In any event, identifiers should be used for all classes or none.

6.11 Referential attributes

Referential attributes capture facts about an association in which the object participates. They correspond to identifying attributes of other classes.

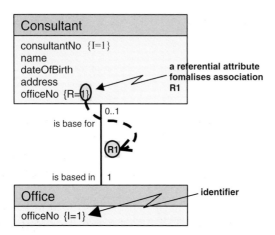

Fig. 6.36 Referential attributes

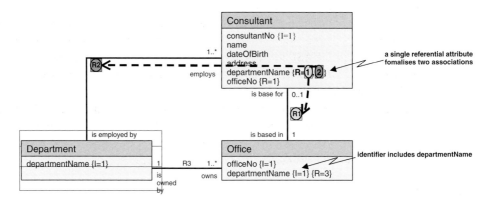

Fig. 6.37 Using referential attributes to constrain associations

Referential attributes, like identifiers, are depicted graphically using tags as shown in Figure 6.36.

Referential attributes are particularly useful where we need to formalize constraints about the associations in our class diagram. For example, suppose we wished to express the rule that consultants must be based in offices that are owned by the department that employs them. We could just write informal annotations on our model but this does not guarantee that they will be applied either by the xUML simulator or the executing system (since they cannot easily be picked up by a code generator).

We could formalize the constraint as shown in Figure 6.37. Here, we have included **departmentName** in the identifier of OFFICE. Naturally, this is also the identifier of DEPARTMENT. We can now use the referential attribute **departmentName** of CONSULTANT to refer to both the Office for that consultant and their Department. Clearly, this attribute can hold only one value for an instance of CONSULTANT and we therefore guarantee that the Office department name is the same as the employing department name.

Some CASE tools will automatically add these attributes as associations are added to the model.

6.12 Redundant attributes

There will be occasions where the analyst wishes to have an attribute that is redundant. From a purist viewpoint, it is never necessary to have an attribute whose value can be derived by accessing other objects. However, there are a few situations in which redundancy is permissible, mainly oriented towards improving performance. Examples of acceptable redundancy are:
- The value is also held in another domain but the analyst does not wish to incur the overhead of having to access it by invoking a bridge operation. In this case, the analyst must take steps to ensure that the redundant attribute is sufficiently up to date for the purposes of this domain. The attribute description should explain why the redundancy is necessary and how the redundant copies are kept up to date;
- The attribute is attached to a superclass and captures the class name of the corresponding subclass object. This is strictly speaking redundant but can be included in all superclasses in order to allow the ASL to know the destination class name when navigating from superclass object to the corresponding subclass object. No explanation is therefore required in the attribute description;
- The attribute represents the state of the active class to which it is attached. The values of this attribute may or may not correspond to the state names on the state machine. The attribute description for such attributes should explain why the state of this class is of interest to other classes and specify which classes they are.

6.13 Normalization

The primary goal of normalization is to eliminate redundancy in the data model. A side effect of this is that it typically brings to light new classes that were hiding in the original classes, resulting in a clearer model, as illustrated by the forthcoming example.

A normalized model allows the greatest discretion during the system generation phase. By establishing a non-redundant PIM, we are not committing to a non-redundant PSI. There are many good reasons to have redundant information in the PSI. It may be done for performance reasons or to improve reliability. However, every system has different non-functional requirements and it is highly undesirable to cast implementation decisions into your PIMs, as this compromises their clarity and reuse potential. It is far easier to take a non-redundant PIM and generate a redundant PSI than the converse.

Normalization is best explained using an example so consider this set of attributes (Figure 6.38) that have been postulated for an AIRCRAFT class.

Is there any redundancy in these attributes? It can be hard to tell without understanding what the instances will look like. So consider the object populations given in Figure 6.39.

Attribute
Manufacturer Name
Model Name
Serial No
Wingspan
No of Engines
Maximum Altitude
Current Altitude
Date of Last Engine Service
Time Since Take-Off
Current Parking Bay Id
Passenger Names

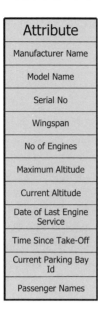

Fig. 6.38 Attributes of an AIRCRAFT class

Attribute	Instance 1	Instance 2	Instance 3
Manufacturer Name	Boeing	Boeing	Schempp-Hirth
Model Name	747-400	747-400	Nimbus 3
Serial No	1234	1235	6789
Wingspan	64	64	24
No of Engines	4	4	0
Maximum Altitude	13,000	13,000	4000
Current Altitude	10,000	100	2000
Date of Last Engine Service	25/10/1999	1/4/1999	N/A
Time Since Take-Off	85	N/A	20
Current Parking Bay Id	N/A	25	N/A
Passenger Names	A, B, C	D, E, F	G

Fig. 6.39 Unnormalized class with example objects

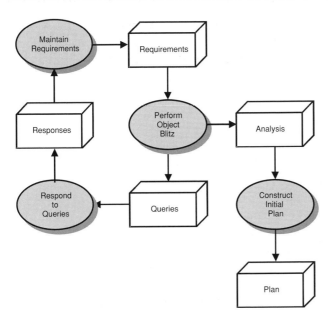

Fig. 6.42 The preliminary system design loop

6.17.2 Required inputs

The information required to perform the object blitz comprises any textual system requirements, the set of primary use cases and scenarios and a preliminary domain model. Also access to a subject matter expert, ideally as a participant, is good practice, but otherwise any relevant background documents should be at hand.

6.17.3 Who should attend

An object blitz should be attended by the analysts, one nominated as senior analyst, a product manager and (ideally) appropriate subject matter experts. The senior analyst shall be responsible for the conduct of the object blitz (facilitator). If the attendees are inexperienced then the use of an experienced mentor is recommended.

6.17.4 An example object blitz process

The following example illustrates the process of performing a blitz. Figure 6.43 shows the principle components of a managed telecommunications network. Each node comprises a cabinet containing a number of shelves, each having several slots that can accommodate up to one card. Only specific types of card can be inserted into each slot. Each connection and each node can generate alarms when faults occur.

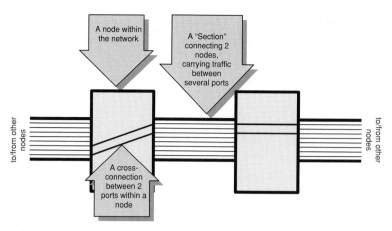

Fig. 6.43 A network management system

6.17.4.1 The system level class diagram

The first class model to be constructed provides a complete view of the system, in terms of the primary classes needed to satisfy the available requirements. No domain partitioning is performed at this stage, although the analysts will start to form ideas about candidate domains.

One of the biggest problems with this type of work is deciding where to start. This issue can be addressed by constructing the system level model in a series of stages:

1. Look for 'tangible' or 'real' things in the problem, such as **Actual Node** and **Actual Card.**
2. For each 'actual' class, identify any required specification classes, such as Type Of Node. This will capture common qualities of all nodes of a particular type, such as traffic carrying capacity.
3. For each 'actual' class, identify any required role classes. For example, Actual Card can be performing the role of Actual Out Of Service Card or Actual In Service Card. Such role classes will typically manifest themselves as subclasses, with subclass-specific attributes and associations.
4. For each class, identify any required incident classes. For example, Actual Connection will be the subject of a number of Actual Connection Requests, **which** must be queued and processed sequentially.
5. For each class, identify any required interactions with other classes. For example, each Type of Card can legally be inserted into one of many Type of Slot. Such interactions will typically manifest themselves as association classes (in this case Legal Type Of Card In Type Of Slot).

Note that this part of the process is not taking account of the use cases. By focusing on the problem space, the resulting class diagram should cover the scope of the problem and contain sound class-based abstractions. The use cases will be considered against the resulting class diagram(s) to explore the interactions between the domains and between the classes within the domain. In the authors' experience this produces far superior results to identifying

It can be seen that there are a number of problems with this class:
- The **Passenger Names** attribute contains a list;
- **Wingspan**, **Number of Engines**, and **Maximum Altitude** are stored several times for each type of aircraft. If there are 1000 Boeing 747–400s, we will store these three values 1000 times;
- Some attributes contain 'N/A' values.

This class is a disastrous abstraction. Here's why:
- It tries to capture information about lots of different things – Aircraft, Aircraft Models, Passengers;
- It fails to formalize the fact that an Aircraft can be performing a number of roles – Parked or Flying – each of which has its own attributes. For example, only Parked Aircraft have a **Parking Bay Id**;
- It fails to bring out the differences between Powered Aircraft, which have a **Date of Last Engine Service** attribute, and Unpowered Aircraft, for which this value is meaningless.

The world is littered with models that contain classes like this one. Their authors often defend them on the grounds of performance. To understand why this defence is invalid, see Section 6.9.8 about '*The difference that makes the difference*'.

The new set of classes, which brings out the full set of classes in this domain and enables a smaller, faster implementation, is depicted in Figure 6.40.

6.14 Static and dynamic classes

More exciting terminology – classes whose objects are not created or deleted at run time are known as **static classes**.

Classes whose objects are created and/or deleted at run time are known as **dynamic classes**.

Note that this is an orthogonal property to whether the class is **active** or **passive** (i.e. whether the class has stateful behaviour, see Chapter 9).

6.15 Improving model efficiency

With some effort, it is possible to build inefficient PIMs. In an xUML system, efficiency is achieved through two routes:
1. Efficient modelling of the required behaviour. This is the responsibility of the analyst building the PIM;
2. Efficient implementation of the xUML formalism on the target environment. This is the responsibility of the software architecture domain that embodies the mappings from the PIM to the PSM.

A good architecture can no more mitigate for incompetent modelling than can a good C++ compiler alleviate the problems of a poor C++ program. The analyst must take responsibility for efficiency. However, this does not mean building a model that exploits

Aircraft

Attribute	Instance 1	Instance 2	Instance 3
Manufacturer Name	Boeing	Boeing	Schempp-Hirth
Model Name	747-400	747-400	Nimbus 3
Serial No	1234	1235	6789
Current Altitude	10,000	100	2000

Aircraft Model

Attribute	Instance 1	Instance 2
Manufacturer Name	Boeing	Schempp-Hirth
Model Name	747-400	Nimbus 3
Wingspan	64	24
No of Engines	4	0
Maximum Altitude	13,000	4000

Powered Aircraft

Attribute	Instance 1	Instance 2	Instance 3
Date of Last Engine Service	25/10/1999	1/4/1999	

Flying Aircraft

Attribute	Instance 1	Instance 2	Instance 3
Time Since Take-Off	85		20

Parked Aircraft

Attribute	Instance 1	Instance 2	Instance 3
Current Parking Bay Id		25	

Fig. 6.40 Tabular representation of classes and their objects

some perversity of the implementation environment, which may change when the system is ported to another target. It means building models that deliver good performance at the xUML level of abstraction. Consider the following example. A car rental company routinely requests a list of available cars in a particular band. This could be modelled as shown in Scheme A (Figure 6.41):

153 Class modelling in a domain

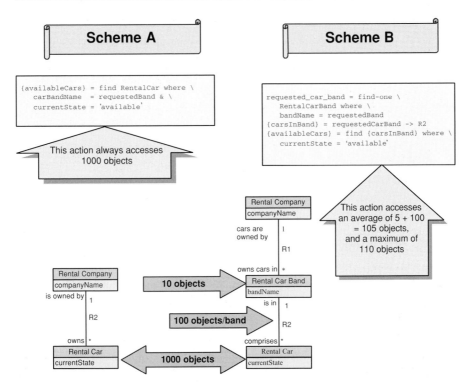

Fig. 6.41 Examples of 'inefficient' & 'efficient' modelling

However, additional abstractions, such as the Rental Car Band in Scheme B can make a model clearer and faster.

6.16 Attribute visibility

In OO in general there is much talk of encapsulation, a design concept intended to limit the unintentional coupling between classes in a system. In UML there are four possible visibilities that may be attached to an attribute, they are (OMG, 2001):

- **public** – any outside ModelElement can see the attribute, the attribute is prefixed with a '+';
- **protected** – any descendant of the ModelElement can see the attribute, the attribute is prefixed with a '#';
- **private** – only the ModelElement itself, or elements nested within it, can see the attribute. The attribute is prefixed with a '−';
- **package** – ModelElements declared in the same package (or a nested subpackage, to any level) as the given attribute can see the attribute. The attribute is prefixed with a '~'.

In building a PIM, expressed in xUML, our job is to expose requirements for one model element to read or write another's attributes, not prevent that by a design mechanism (encapsulation). Therefore, we do not use visibility adornments on attributes and we take all attributes to have public visibility. It should be noted that we may choose to use encapsulation in our platform-specific design, an issue we shall revisit in Chapter 13.

6.17 The object blitz

Now we have discussed the main features of class modelling, we turn our attention to one of the commonest techniques for starting the process of class modelling – the **object blitz**. We perform an object blitz to:
- obtain sizing information for estimation purposes;
- establish the level of abstraction for each domain;
- assist in identifying deficiencies in the requirements.

During the early stages of system development, a number of activities are performed on an iterative basis. The initial requirements, which will naturally be incomplete and ambiguous, form the starting point for a preliminary 'football field' class model. This model makes no assumptions about the system partitioning in terms of domains. It simply identifies the classes that the system must manipulate.

The concept of object blitzing is to provide a rough outline of the principal classes within each domain in the system, how those classes relate to each other and an outline of the primary class interactions. The object blitz is performed over a small number of days. This section is intended to convey a number of steps that could be followed when performing an object blitz. The process is informal, so it is not mandatory to follow all of the steps or guidelines.

During the construction of this model, queries will arise regarding the requirements, which will in turn elicit responses from the relevant subject matter experts. This will be used to extend and refine the requirements, which will then be captured in the model. The process is summarized in Figure 6.42.

6.17.1 The system level object blitz

The system level object blitz forms a framework within which to establish the preliminary 'system design' in terms of domains, key classes and counterpart associations. It is this type of work that can be performed by a 'system' group, prior to handing over to the 'software' groups, which will perform the detailed analysis of each domain.

When performing an object blitz it is important to remember that there are no rigid rules that must be applied, rather a set of guidelines that may aid the analysts in their approach to producing a preliminary class model. The process is akin to directed brainstorming. The senior analyst (see the roles defined in Section 6.17.5) has discretion on how long each step will take, but the object blitz usually lasts between one and three days.

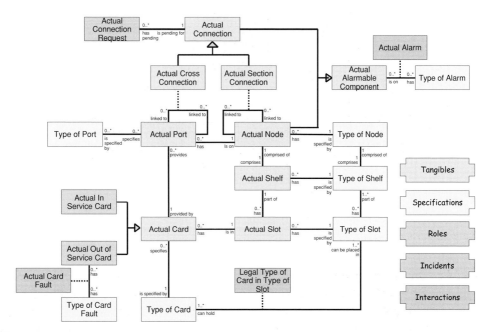

Fig. 6.44 The system level class diagram

the classes from the use cases. The latter approach produces functional abstractions with unnecessary controller classes.

The resulting class diagram will then look something like that presented in Figure 6.44.

Note that the associations have been labelled with multiplicity and role phrases at both ends of the association. At this stage we are capturing the real world associations that exist between the classes we have identified. We do not know (or care) how they will be navigated. It helps the reader of the diagram to understand it if the associations are all fully labelled.

Also the recording of attributes is encouraged. Whilst it is not necessary to have all of the attributes at this stage, it helps the confirmation of the class abstractions.

The next stage is to find the domain boundaries. This involves finding what the method refers to as **semantic shift**. Look for associations where the subject matter changes as they are traversed. For example ' ACTUAL CARD provides many ACTUAL PORT'. This is a shift from the subject matter of hardware to the subject matter of network management (Figure 6.45). This association therefore represents a **counterpart association**, linking two classes in different domains. This is discussed further in Chapter 12.

These counterpart associations can then be used to delimit the domains that they link and a set of candidate domain boundaries can be established (Figure 6.46).

The system will be clearer and easier to understand if counterpart associations have a multiplicity of one-to-one. That is, each object in one domain is represented by exactly one counterpart object in another domain. Therefore, the domain-crossing associations should be examined, and new classes introduced at one end to achieve a multiplicity of one-to-one where necessary. For example, the class ACTUAL PORT CONTAINER is an abstraction

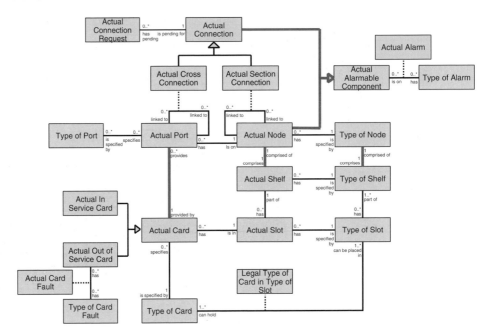

Fig. 6.45 Class diagram with ready for domain partitioning

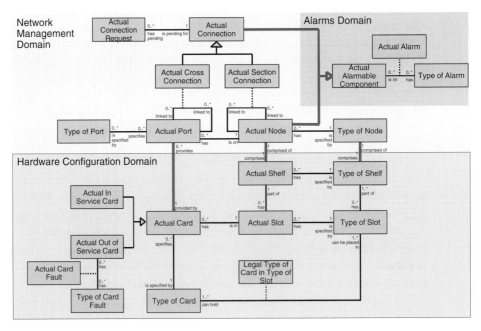

Fig. 6.46 Candidate domain boundaries identified

Class modelling in a domain

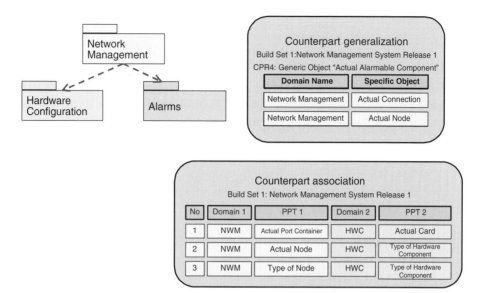

Fig. 6.47 The preliminary domains

of a set of one or more ports. This allows a one-to-one mapping with the ACTUAL CARD class.

The counterpart associations discussed are analogous to 'normal' binary associations in a class diagram. There will also be counterpart generalizations that are more like 'superclass–subclass' relationships. For example, each ACTUAL ALARMABLE COMPONENT 'is an' ACTUAL CONNECTION or ACTUAL NODE. See Chapter 12 for a fuller discussion of counterparting.

The system level PIM can now be partitioned into separate domains based upon the identified counterpart associations. These domains, with the preliminary class diagrams and counterpart associations, can form the basis of the initial system model (Figure 6.47).

6.17.5 Roles and responsibilities

The following list of roles identifies the responsibility of each attendee to the object blitz.

6.17.5.1 Analyst

This role can be fulfilled by an engineer on the project who has been tasked with producing an analysis of a domain.

The analyst is responsible for identifying the mission statement, threads of control, candidate classes, associations, attributes, asynchronous and synchronous behaviour between classes. The analyst is also responsible for building the class diagram and the sequence charts.

6.17.5.2 Blitz facilitator

This role can be fulfilled by an engineer on the project whose primary background is analysis and who has either extensive experience of the analysis on the project to date or has considerable experience of performing object blitzes on this or other projects.

The role involves moderating the object blitz, giving guidance to the analysts and acting as review authority on the final analysis. Particular attention should be paid to using the time effectively and avoiding side debates.

6.17.5.3 Verifier

This role can be fulfilled by a team leader responsible for the domain under work or by the development manager. It is the role that has responsibility for verification of a domain and giving approval for work to continue.

6.17.5.4 Subject matter expert

This role is fulfilled by the relevant subject matter expert or customer representative. The subject matter expert is not required to have experience of object blitzes but will be able to provide vital input to the blitz process through his/her knowledge of the problem space under analysis.

6.17.5.5 Product manager

This role is responsible for understanding and scheduling priorities of the customer requirements with respect to the resources available to the project. He or she should have a good understanding of the key features that are being requested by the customer and filter out less important features or features that do not add value to the overall product.

6.18 The dangers of premature partitioning

A system level object blitz provides a way of bypassing any preconceptions that a development team might have about the way to partition a system. It also improves your chances of identifying common patterns that can be captured once in a service domain.

Some years ago, we were invited to help in the analysis of an aircraft system. The organization building the system had a strong and successful background in structured methods. As is common in such organizations, the structure of the workforce matched the structure of the system. There were groups assigned to each of the 'subsystems', such as 'Electrical Equipment', 'Hydraulic Equipment' and so on. We were informed that this structure was tried and tested, and that it was to remain in place for the new, MDA-based project. We meekly complied and later regretted not putting up more of a fight. While building the electrical equipment domain, it was necessary to get to grips with the distribution system on the aircraft. This resulted in classes like POWER SUPPLY, CABLE, ISOLATOR and EQUIPMENT UNIT. At the same time, the 'Hydraulic Equipment' team were busy identifying classes like COMPRESSOR, PIPE, VALVE and PISTON. Only after significant effort had been replicated in understanding how to manage the distribution of power or

Table 6.4 Classes and their counterparts

Distribution network	Source	Link	Isolator	Sink
Electrical Equipment	Power Supply	Cable	Isolator	Equipment Unit
Hydraulic Equipment	Compressor	Pipe	Valve	Piston

compressed fluid did it become obvious that they were the same problem. At that point, we abstracted a distribution network domain that could distribute electricity or compressed fluid without distinguishing between the two. The classes in this new domain, and their counterparts in the other domains, were as shown in Table 6.4

This type of embarrassing incident highlights the importance of maintaining a system view and the danger of premature or functionally oriented domain partitioning.

6.19 Conclusion

We have seen in this chapter the central role that the class plays in the MDA process. Because the models that we build are executable and precise, the class diagrams that result from the process are expressive declarative statements of the requirements, rich with associations and meaningful role phrases; a reader of a good class diagram should be able to form refutable statements about the problem being modelled.

Having considered the domain and now the class, we have covered the static elements of the xUML. With this framework we can now proceed in the subsequent chapters to a discussion of how we express behavioural and processing requirements within our models.

7 Class behaviour and interactions

7.1 State-independent and state-dependent behaviour

We have so far built a class diagram for each domain. This is a static model of the classes that inhabit that particular subject matter, and their associations. It forms the cornerstone of the Platform-Independent Model (PIM) for the domain.

But these classes have behaviour. That behaviour falls into two basic categories: **state-dependent** and **state-independent**. State-dependent behaviour is expressed using the time-honoured formalism of **state models**, while state-independent behaviour is specified in **operations**.

For example, consider a diesel-powered train, which might have a state model like that shown in Figure 7.1.

Here we see a classic example of state-dependent behaviour. The signal named 'fireDetected' can occur in more than one state and causes different actions to be performed depending upon the current state – hence state-dependent behaviour. If a fire is detected while the train is stopped then the doors are immediately opened to allow a safe evacuation. However, if the current state of the train is 'moving', then opening the doors may be ill-advised. Therefore, we move into a state where the train is brought to a halt and then transition into the `Stopped For Evacuation` state.

Note also that the 'trainStopped' signal causes different actions to be performed depending upon the current state.

Let us now assume that the DIESEL TRAIN class has an attribute **fuelLevel** which is kept up to date by periodically updating its value at an appropriate rate. Let's examine the effect this has on our simple model if this is to be achieved by sending a signal 'fuelLevelUpdated' to the respective train object. Clearly, this signal could arrive in any of the above states. If we were to model this behaviour using a state model, it might look like Figure 7.2.

So we now have twice as many states and a whole stack of new transitions. Further, we can see that the action associated with the new states is always the same. In other words, the action taken in response to the 'fuelLevelUpdated' signal does not depend upon the state of the train. Modelling such behaviour on a state machine, as this example shows, results in a proliferation of artificial states. We should model such state-independent behaviour as operations provided by the class, which can be invoked regardless of the current state of the specified object.

Much of the behaviour within a domain is typically **state-independent.** That is, the processing to be performed does not depend upon the state of any particular object. As this

Class behaviour and interactions

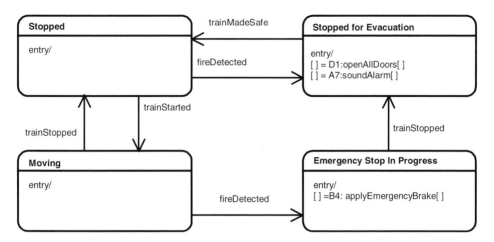

Fig. 7.1 State model for a DIESEL TRAIN class

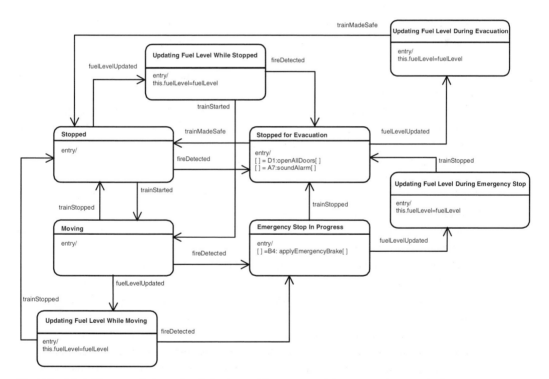

Fig. 7.2 Modelling state-independent behaviour with a state model causes undue complexity

example shows, such behaviour, if captured on state machines, results in abominations, in which every state has transitions for almost every signal.

Therefore, the method offers the ability to specify state-independent behaviour in terms of **operations**.

7.2 Operations versus states

In general, the circumstances under which operations should be used in preference to signals are easy to identify. For example, it is obvious that determining the number of **Bed** objects that are currently unoccupied in a hospital should be achieved using an operation – since the result is independent of the state of any particular bed.

The circumstances under which operations should be used are when:
- The required processing is **independent of the state of a particular object**;
- The required processing **does not affect the state of a particular object**;
- The required processing is **common to multiple state actions** and must complete before the completion of the action;
- It is necessary to invoke services provided by one or more other domains.

Do not use operations when:
- The processing is **dependent upon the state of a given object**;
- The operation body would need to read the value of the 'Current State';
- The operation body would need to write the value of the 'Current State'.

So classes can provide services in two distinct ways: as an **operation** or as a **signal**. When a class provides a service in the form of an operation it can be used in the following ways:
- A call to an operation is synchronous, that is the caller will wait for the operation to complete before it continues;
- Input parameters can be passed to an operation;
- Output parameters (more than one is permitted in xUML) can be returned by an operation;
- The behaviour that results is not dependent on the state of the object that performs an operation;
- Operations are typically defined to act on a specified object of the class – these are referred to as object-based (or sometimes object-scoped) operations;
- Operations can be defined to act over the objects of the class as a whole – these are referred to as class-based (or sometimes class-scoped) operations;
- Operations can be defined to act on the domain – these are referred to as domain-based (or sometimes domain-scoped) operations.

When a class provides a service in the form of a signal, which it is prepared to receive, it can be used as follows:
- The sending of a signal to an object is asynchronous, that is the sender carries on with its business immediately and is neither affected when the signal is acted upon nor by the (state-dependent) processing performed as a result;
- Input parameters can be sent with a signal;
- Output parameters are not allowed as they make no sense with an asynchronous interaction;
- The behaviour that results from sending a signal will depend upon the state of the object that receives the signal;
- Signals are always directed at a specified object.

Class behaviour and interactions

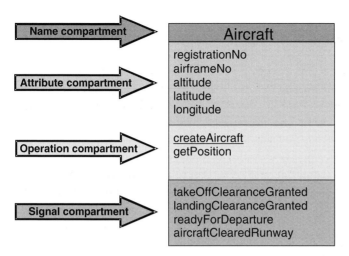

Fig. 7.3 Class with signal compartment

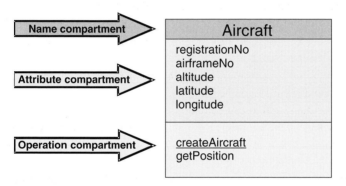

Fig. 7.4 Common class compartments

Note that the above is a simplified set for comparison since polymorphic behaviour associated with counterpart relationships is not considered until later chapters.

Operations and signals can be shown on the class diagram (Figure 7.3).

Unfortunately most UML tools do not support the signal compartment and so it is frequently omitted. It is most common to see the operations on the class diagram and the signals on the corresponding statechart for the class (Figure 7.4).

The state-independent behaviour for the class is thus defined in terms of the set of operations that it provides. The detailed processing associated with each such operation will be defined in its respective method using the Action Specification Language (ASL). Chapter 8 discusses the modelling of operations in more detail.

The state-dependent behaviour for the class is defined using a state model that, in essence, describes the states, the allowable transitions between those states and the signals that give rise to those transitions. Each state has an entry action that defines the processing that will be performed on transition into the state – again, such processing is defined using ASL.

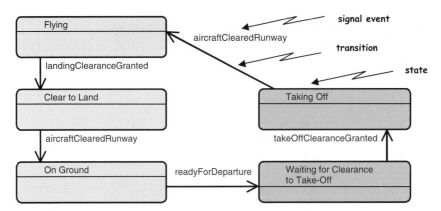

Fig. 7.5 Introduction to state model terminology

Figure 7.5 shows a simple example state model for the AIRCRAFT class expressed using a statechart and introduces the notation of statecharts. Chapter 9 discusses the modelling of state-dependent behaviour using state models in more detail.

7.3 Object and class interactions

When one object interacts with another either synchronously, by invoking an operation, or asynchronously, by sending a signal, this couples the two objects together. By implication the classes of these two objects also become coupled. The extent and form of these interactions therefore is of great importance to the quality of a model.

When we show object level interactions, the diagrams are showing particular objects of the classes. This is the way that use case scenarios are related to the class diagram. The use case scenario is an example sequence of events resulting from a given set of conditions. An object level interaction diagram illustrates the manner in which the objects in the model interact to achieve the scenario.

The UML provides us with two means of representing and viewing interactions; the **sequence diagram** and the **collaboration diagram**. Both forms show interactions but each emphasises different aspects. The sequence diagram emphasises the time ordering of the interactions. Figure 7.6 is an example of an object level sequence diagram for the primary scenario of the **Admit In Patient** use case for a patient administration system. It shows the interactions within a single domain, but where the interaction would provide services to or require the services of another domain then an interface class, stereotyped «terminator», is shown in the interaction (the use of the «terminator» stereotype is fully explained in Chapter 12). The interactions involving «terminator» classes must be consistent with the interactions appearing on the domain level sequence diagram described in Chapter 2.

The sequence diagram is time ordered from top to bottom. The interface corresponding to the Administrator initiates the sequence by sending the 'admitPatient' signal the

Class behaviour and interactions

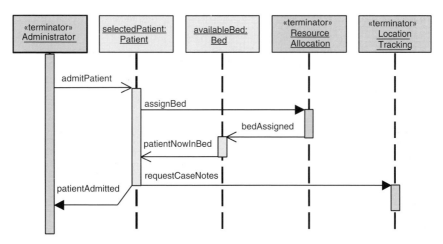

Fig. 7.6 Sequence diagram for **admit in-patient** use case

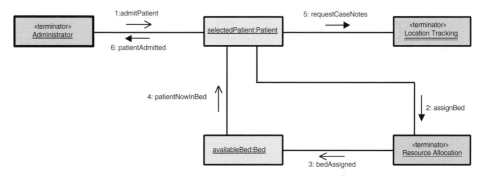

Fig. 7.7 Object collaboration diagram for **admit in-patient** use case

'selectedPatient' object. The sequence progresses with the 'selectedPatient' object requesting a bed to be assigned by the resource allocation domain, which in this scenario is successful. The response from the resource allocation domain identifies the 'availableBed' object of the BED class, which returns a message to the 'selectedPatient'. The respective patient's case notes are requested from the location tracking domain (this domain tracks case notes as well as items like mobile hospital equipment) and the Administrator is informed that the patient has been successfully admitted.

The collaboration diagram ignores time ordering in favour of emphasising patterns of interaction and closeness of collaboration. The same primary scenario for the **admit in-patient** use case again showing the interaction between the objects appears in Figure 7.7.

Both forms show synchronous and asynchronous interactions. The amount of detail represented is optional. Interactions can simply be labelled with a name. Alternatively, all of the parameters names and corresponding data types passed with the message can be included. Too much information can make the diagram cluttered and confusing; too little and the diagram does not convey enough useful information to the reader. For object level

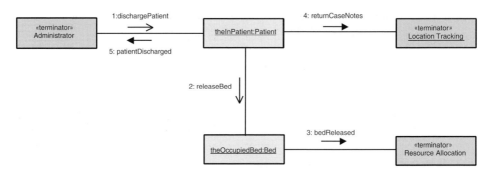

Fig. 7.8 Object collaboration diagram for **discharge in-patient** use case

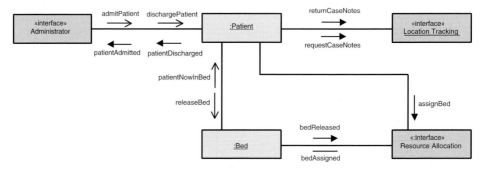

Fig. 7.9 Class collaboration diagram for multiple use cases

collaboration diagrams the message name with a sequence number is usually sufficient since details of the message's data can be found elsewhere in the model.

If we now consider a different use case scenario to 'Discharge A Patient' then, not surprisingly, the interactions are different but many of the same classes are involved (Figure 7.8).

The collaboration diagram can be used to show object interactions or, alternatively, it can be used to summarise sets of object interactions by representing class interactions. By combining these in a single diagram we are no longer showing the interactions involved in a single use case but the set of interactions involved in a number of or all use cases. The result reveals the patterns of communication between the classes rather than the interactions in a single use case scenario (Figure 7.9).

The class collaboration diagram (CCD) provides us with an important insight to the PIM that is not easily seen with object level interaction diagrams. The object level interaction corresponds to one thread of behaviour through the model. Typically, this corresponds to one scenario of one use case. A domain model in a typical system has behaviour corresponding to tens or hundreds of scenarios. The interactions viewed on a scenario-by-scenario basis may look completely reasonable but the overall effect of the coupling between the classes may be chaotic.

For example, consider an object of each of the classes PATIENT, CONSULTANT, X-RAYCLINIC, APPOINTMENT, X-RAYRESULT and CASENOTES. Let's imagine that

Class behaviour and interactions

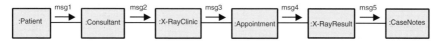

Fig. 7.10 Message sequence diagram for scenario 1

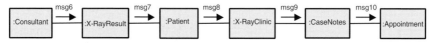

Fig. 7.11 Message sequence diagram for scenario 20

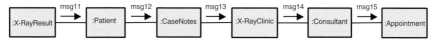

Fig. 7.12 Message sequence diagram for scenario 40

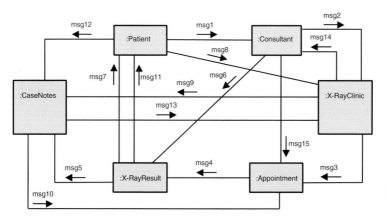

Fig. 7.13 Collaboration diagram for use case scenarios 1, 20 and 40

the message sequences defined in Figures 7.10–7.12 are constructed for their respective scenarios.

Whilst the interactions in the object level diagrams are simple, and maybe even justifiable, the overall pattern of communications between the classes is highly undesirable (Figure 7.13). It can be clearly seen that the coupling between the objects is unjustifiably high.

Considering use case scenarios and their sequence diagrams alone can mean we get too close and fail to see the big picture. The class collaboration model for a domain is the big picture for the dynamic behaviour within the domain and it provides a very valuable view.

Clearly, there is scope for redundant representation with these two similar notations so we will recommend their use within a domain in two distinct ways. We will use sequence diagrams to show object level interactions within a domain corresponding to a single scenario of a single use case. We will use the collaboration diagram to show patterns of interaction between classes. To reinforce this distinction we will refer to the collaboration diagram as the class collaboration model (CCM) throughout the book.

You should not infer from the above that we let the CCM slowly emerge as the analysis proceeds and then cry out in horror when it looks like a tangled web. Instead, we use this property of the CCM early in the analysis process to identify how to distribute responsibility between the classes.

We recall observing a PIM in which the CCM was produced rather late in the day. The model related to the control of a train. The class with overall responsibility for deciding on the motion of the train wasn't an abstraction of the driver, or any aspect of the signalling system; it was the DOOR class!

Whilst the doors represented an important interlock (the train shouldn't move with the doors open) it did not represent a very convincing abstraction of responsibility to have the doors in charge of the train! Had the CCM been used early in the analysis, the overall pattern of responsibility within the domain could have been developed on a sensible basis and the problem would not have arisen.

Recall that we have already used the sequence diagram notation to represent the interaction between domains. The lifelines on this style of sequence diagram represented domains and an interaction corresponds to the required service of one domain being mapped to the provided services of others. This is a more abstract interaction view for which the sequence diagram was not originally intended. However, it proves very effective in performing this additional duty.

7.4 Domain interfaces on the class collaboration model

Dependency can be shown on the class diagram. However, the primary purpose of a class diagram is a static declarative view in which associations and generalisations dominate. The way that an object uses an operation provided by another class or sends signals to another object is best described using the class collaboration diagram. We therefore adopt the style of not showing dependency on a class diagram. We use the CCM to show patterns of communication between classes. Both operation and signal messages appear on the CCM.

A domain represents only part of a system and a typical domain makes operations available for other domains to use and also depends upon operations provided by other domains. These are the domain's interfaces. One of the objectives of domain partitioning is to identify distinct subject matters and keep them separate. A result of this is that a particular domain must be built without any knowledge of other domains around it. A domain does not know which other domains will use the services it provides; this is quite normal since a server rarely needs to identify its client. A domain also does not know which other domains will provide the services it requires; this is less common since a client typically knows the identity of the server it depends on. This is, of course, the well-founded principle of encapsulation but at the domain level.

When we build a domain we adopt an abstract view of the 'external entities' with which the domain interacts. Take the lift example, as shown in Figure 7.14. The request to move

171 Class behaviour and interactions

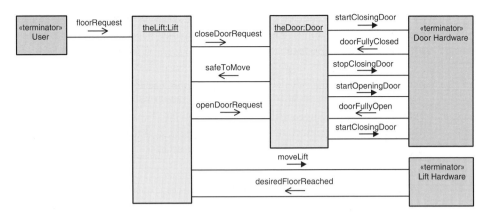

Fig. 7.14 Class collaboration model for the lift control domain

the lift comes from the User. In our final system there may be a user interface domain that knows all about buttons and lights, but to the lift control domain the request comes from the user. The lift control domain will issue commands to open and close the doors and to start and stop the lift. As far as this domain is concerned these commands are dealt with by the lift hardware and the door hardware. (The term hardware is included to distinguish from the class abstractions of LIFT and DOOR in the model.) In the full system there may be complex domains, which deal with the physical interfaces to the lift and door hardware. These 'external entities' are the sources and sinks of interactions with classes in the domain. Therefore they are usefully represented in the class collaboration diagram. Note they do not appear in the class diagram.

From the single domain viewpoint these external entities are the actors that interact with the domain. Caution is required using the term actor because its meaning is reserved for the external entities for the system as a whole and so using the term in this context could cause confusion. We have used the term **terminator**.

> A terminator is an abstraction of an entity external to a domain with which a domain interacts. A domain may have many terminators.

A domain can have any number of terminators. Each terminator can be the source of messages into the domain and sink of messages out of the domain. Messages into the domain from its terminators define the services that the domain provides to its clients. The messages from a single terminator define a **provided interface**.

> A provided interface is the set of services a domain offers on behalf of a single terminator. A domain may have many provided interfaces – zero or one per terminator.

Messages out of a domain to its terminators identify operations that this domain requires to be performed by some other (unknown) domains. The messages to a single terminator define a **required interface**.

> A required interface is the set of services associated with a single terminator that a domain will not realize itself, but the domain requires the services to be realized by some other domain. A domain may have many required interfaces – zero or one per terminator.

When we put a set of domains together to form a system, the job that has to be done is to 'bridge' the services in the required interfaces to the services in the provided interfaces. This will be discussed in detail in Chapter 12. In this section we will focus on the view within a single domain looking out.

Terminators are shown in CCDs and, because they represent the sources of messages entering and sinks for messages leaving the domain they traditionally appear at the edges of the diagram. As a helpful hint, terminators that represent external entities, which are clients of the domain, such as our USER in the lift model, will appear on the left-hand side of the diagram. Terminators that are servers to the domain appear on the right-hand side of the CCD. We also layer the classes in the CCD with the most responsible classes appearing to the left and the least responsible appearing to the right. We will explore this idea of layering the class collaboration diagram in more detail later.

7.5   The process of dynamic modelling

The process of dynamic modelling involves these steps:
1. Build a CCM to show the classes and how they interact to deliver the required behaviour of the domain. When building this model:
   - Identify stimuli from and responses to clients of the domain. The clients may correspond to human users interacting through some form of user interface or simply other parts of the system. For example, 'floorRequest' is a signal from the user interface (Figure 7.14);
   - Identify stimuli from and responses to service providers for the domain. Many domains require other domains to provide services for them. The domain being modelled views these as required services represented on «terminator» classes. Domains providing these services may also stimulate the domain being modelled. Both of these types of interactions need to be captured. For example, 'doorFullyClosed' is a signal from the door hardware (Figure 7.14).
2. Using the interfaces established on the CCM:
   - build statecharts to show the state-dependent behaviour of each class, and;
   - define operations on the classes to capture the state-independent behaviour.

So, to remind us, the CCD is just one way of representing interactions within a domain. Interactions can be represented at the object level or the class level using either a collaboration diagram or a sequence diagram. The decision about which to use will be driven by the subject matter being modelled. If time ordering is dominant, then sequence diagrams are most helpful. However, there is benefit to using a class level diagram, the CCD, because this allows establishment of a consistent set of communication patterns for the whole domain. We have seen projects run into difficulties when many sequence diagrams are built, with behaviour shared across many scenarios being realized using conflicting communication patterns. CCDs help to mitigate this risk.

7.5.1 The role of sequence diagrams for CCM construction

The construction of a CCM before describing the detailed behaviour of a domain (in terms of the operations and state models defined for its respective classes) will allow the analysis team to:
- Demonstrate the viability of the domain;
- Establish the interface for each class;
- Show how the domain delivers the interface shown in the system use cases.

In Chapter 5 we introduced the domain level sequence diagram to show the interaction between domains. We can use this diagram to develop the CCM for each domain. The incoming and outgoing messages for each selected domain define the interactions that the required and provided interfaces for that domain must support. As such, the domain level sequence diagrams provide a clear focus for establishing the preliminary CCM. Consider the domain level sequence diagram shown in Figure 7.15.

When building the preliminary CCM for the patient administration domain, clearly the CCM must support the interface specified for this system sequence diagram. This provides

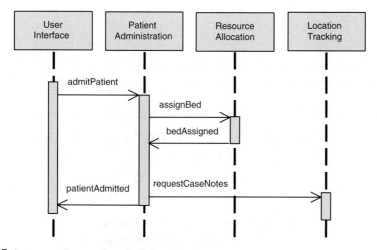

Fig. 7.15 Sequence diagram for **admit in-patient**

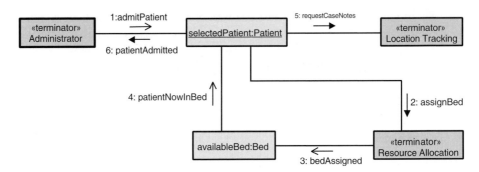

Fig. 7.16 Preliminary class collaboration model

the staring point for CCM construction. The basic process to be followed when building a CCM in this way is:
1. Use the sources of the incoming messages and the destinations for the outgoing messages to identify the «terminator» classes, reflecting either:
   - external entities, such as ADMINISTRATOR, or;
   - services provided by other domains, such as RESOURCE ALLOCATION and LOCATION TRACKING.

   Note the terminator name represents the domain's perspective of the source or sink, rather than simply using the domain name. For example the 'admitPatient' message comes from the user interface domain, but from the patient administration domain's perspective 'admitPatient' comes from the ADMINISTRATOR role that performs in-patient admissions.
2. Examine the 'unsolicited stimuli' (such as 'admitPatient') and direct it to the class that is best placed to deal with the stimulus (in this case, PATIENT);
3. Continue to invoke operations and transmit signals to other classes and «terminator» classes, in accordance with the interface established by the use case;
4. Repeat this process for each unsolicited stimulus identified by the system use cases.

Performing this process for the sequence diagram 'Admit In Patient' might yield a preliminary CCM for the patient administration domain like the one shown in Figure 7.16.

The annotations on this CCM show the sequence of interactions for the domain level sequence diagram 'Admit In Patient'. These will ultimately be used as the expected results for a single domain simulation run.

As each use case scenario is considered, the CCD builds up. We do not want the resulting diagram to look like a plate of spaghetti and the primary way of achieving this is to layer the CCM.

7.5.2 Layers on the CCM

Each class in a domain contributes to the overall mission of the domain by collaborating with a limited number of other classes. This is achieved by organising the classes into layers based upon how much intelligence they encapsulate about the domain in which they reside

Class behaviour and interactions

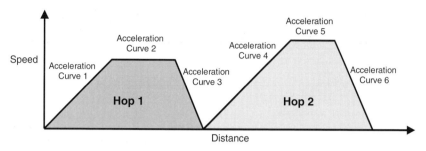

Fig. 7.17 Example profile for a journey through two stations

or how much responsibility they have for coordinating the behaviour of other classes. Using an example that we will revisit in Chapter 10, consider a train control domain, whose basic mission is to move a train that follows a route comprising a sequence of 'hops'. Each hop typically involves periods during which the train:

1. Accelerates at a constant rate to reach a predefined speed at a specified distance along the hop;
2. Maintains that speed until it has travelled a specified distance along the hop;
3. Decelerates at a constant rate to stop at a specific distance along the hop.

This is illustrated in Figure 7.17.

Part of the class diagram for this domain is illustrated in Figure 7.18.

The train's responsibility is to negotiate a number of hops, each of which consists of a number of acceleration curves. It seems reasonable to assert that the most 'responsible' and 'intelligent' class in this domain is the TRAIN, while the least intelligent is the ACCELERATIONCURVE. This might result in a preliminary CCM like that shown in Figure 7.19.

The layout convention is to place the more responsible classes to the left, and the least responsible classes to the right. A class to the left on the CCM will typically have broad responsibilities, be aware of some major capability provided by the domain, and achieve its task by coordinating the behaviour of other classes. A class on the right of the CCM will typically focus on a specific task and have no responsibility for how that task fits into the behaviour of the domain as a whole. An alternative view might be to think of this as a management hierarchy, with the senior managers on the left and the workers on the right. It will be seen that the 'manager' classes are characterised by the fact that they have simple, and principally coordinating, actions. These actions generally involve sending commands to their juniors to achieve some significant operation. For example, the TRAIN sends a signal 'negotiateHop' to the HOP class. Think of this as a command from a manager to a subordinate. Note that the subordinate in this case further delegates work to the 'AccelerationCurve' by sending the signal 'performCurve' to the ACCELERATIONCURVE class. Once a subordinate has completed a command, it will typically send an acknowledgement to its manager, such as the signal 'hopNegotiated' from Hop to Train. Note that although we have used the term 'manager' class this is still a class abstracted from the problem space (Train in this example) and not an arbitrary singleton 'manager' class invented for the task.

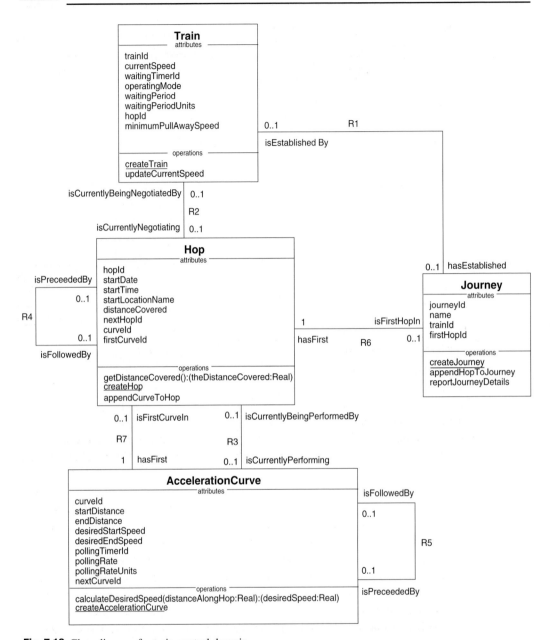

Fig. 7.18 Class diagram for train control domain

In our experience it is always possible to find classes abstracted from the problem domain that can take on the management responsibility. It is never necessary to invent a manager class.

It is sometimes helpful to annotate the classes to indicate their respective responsibilities, in particular:

Class behaviour and interactions

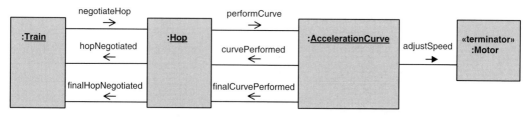

Fig. 7.19 Preliminary class collaboration model for **train control** domain

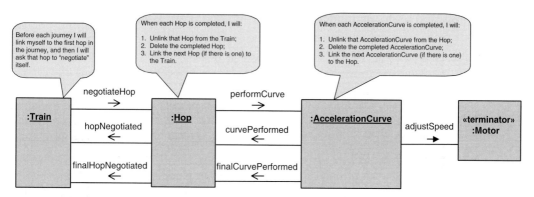

Fig. 7.20 Assigning class responsibilities

- Create and delete objects;
- Link and unlink objects of each association.

This will minimise the risk that some behaviour is specified twice, or not at all! It allows the safe concurrent development of different state machines. Figure 7.20 illustrates this.

This process of establishing a **consistent** set of policies for the entire domain, prior to building any state machines, will have devastating effects on the comprehensibility of the finished work. Once the reader understands the policies, questions like 'why did the analyst build the model this way?' will arise far less often. The fundamental policies that have been established for the part of the model shown in this preliminary CCD are:

- Dynamic classes are created 'top-down'. For example, the responsibility for creating objects of the ACCELERATIONCURVE class will be given to its immediate 'superior' – in this case the HOP class[1].
- Deletion is performed 'bottom-up'. For example, as each ACCELERATIONCURVE is completed, it deletes itself and links the next ACCELERATIONCURVE object to the current 'HOP' object.
- Commands are sent 'top-down'. Whenever a class needs to achieve a task, it delegates by transmitting signals to its junior classes. For example, the HOP class sends signal 'AC1:performCurve' to the next ACCELERATIONCURVE object.

---

[1] The 'creation' interactions have not been shown in our preliminary CCD.

- Responses are sent 'bottom-up'. Whenever a class has completed a task for its manager, it sends one of a number of 'completed' signals to it. For example, when the ACCELERATIONCURVE object has completed the task of moving the train up the acceleration curve, it informs the managing HOP object by sending either a 'H2:curvePerformed' or a 'H3: finalCurvePerformed'.

As will be seen later, these domain-wide policies lead to clearer state machines with common patterns throughout the domain.

7.6 Capturing state-dependent and state-independent behaviour

We capture state-independent behaviour using operations. In the next chapter we will explore how operations are described and how different types of operation can be used for different purposes.

State-dependent behaviour is captured using state models and these will be covered in detail in Chapter 9.

What makes xUML special is that we have the opportunity to state what actually happens in an operation and in state actions. We do this using the Action Specification Language that allows us to reference model elements, manipulate objects and links, send synchronous and asynchronous messages, and generally specify behaviour at the level of abstraction of the object model. The Action Specification Language will be described in Chapter 10.

8 Operation modelling

8.1 Operations

In this chapter we will look at the specification of state-independent behaviour using operations. Operations are most commonly associated with classes and visible in the operation compartment of the class but there are some other uses for operations in xUML. We shall look at the various types of operation in this chapter. The details of how the method for an operation is specified will be described in Chapter 10, which covers the Action Specification Language (ASL) (Wilkie *et al.*, 2002).

Within xUML the term 'operation' is used to indicate an action that can be invoked via a parameterised interface. The caller waits until the action is executed and resumes execution after the point at which the call is made. If desired, values may be 'returned' through the interface. The method for the operation can of course invoke further operations. It may also provoke asynchronous behaviour by generating signals, but in this case the caller will not wait for the results of those signals being processed.

In an xUML PIM for a single domain there will be one 'definition' of an operation and potentially many 'invocations' or 'calls' to that operation.

There are four types of operation:
1. Object-based;
2. Class-based;
3. Domain-based;
4. Bridge.

An operation provides a convenient way to encapsulate some desired behaviour, expressed as a set of ASL statements, which can then be invoked from within the processing of any xUML thread of control (Figure 8.1).

An operation is **similar** in many ways to the concept of a procedure, operation, or subroutine found in many programming languages.

Generic operation rules

1. At run-time the ASL operation is executed **synchronously**.
2. ASL operations may have zero or more **input** parameters.
3. ASL operations may have zero or more **output** parameters.
4. Input and output parameters may be of any valid data type.

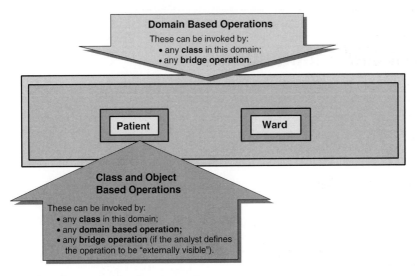

Fig. 8.1 Domain-, class- and-object based operations

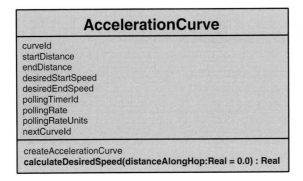

Fig. 8.2 Class with operation parameters

The standard UML syntax for declaring an operation is:

```
OperationName(ParameterName : ParameterType = DefaultValue,...) :
ReturnType
```

Figure 8.2 is an example.

However, since an operation can have more than one output parameter, we have extended this syntax to allow a list of output parameters. How these are actually represented will be tool dependent. Figures 8.3 and 8.4 show two examples of how this might be shown on a class diagram for the operation `reportPosition` of the AIRCRAFT class (note that the default value of each output parameter has been omitted for simplicity).

181 Operation modelling

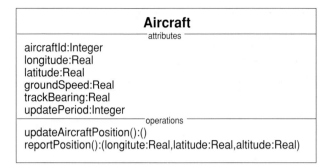

```
                    Aircraft
─────────────────── attributes ───────────────────
aircraftId:Integer
longitude:Real
latitude:Real
groundSpeed:Real
trackBearing:Real
updatePeriod:Integer
─────────────────── operations ───────────────────
updateAircraftPosition():()
reportPosition():(longitude:Real,latitude:Real,altitude:Real)
```

Fig. 8.3 Operations with multiple output parameters shown in 'UML' style

```
                    Aircraft
─────────────────── attributes ───────────────────
aircraftId:Integer
longitude:Real
latitude:Real
groundSpeed:Real
trackBearing:Real
updatePeriod:Integer
─────────────────── operations ───────────────────
[]=A1:updateAircraftPosition[]
[longitude:Real,latitude:Real,altitude:Real]=A2:reportPosition[]
```

Fig. 8.4 Operations with multiple output parameters shown in 'ASL' style

8.2 Class-based and object-based operations

8.2.1 Object-based operations

Object-based operations are the most common form of operation in xUML and are used to describe behaviour pertaining to a single object of a class.

Invocations of these operations are directed to a specific object of the class with which they are associated. The invoker of an object-based operation is required to supply the identification of the object to which the invocation is directed.

Object-based operations are distinguished from class-based operations in that they are not underlined on the graphic and have access to the variable `this`, which is a caller-specified instance handle of a specified class.

Figure 8.5 below shows an example of an object-based operation together with its method[1]:

---

[1] This example is put into the context of a more complete model in Chapter 10.

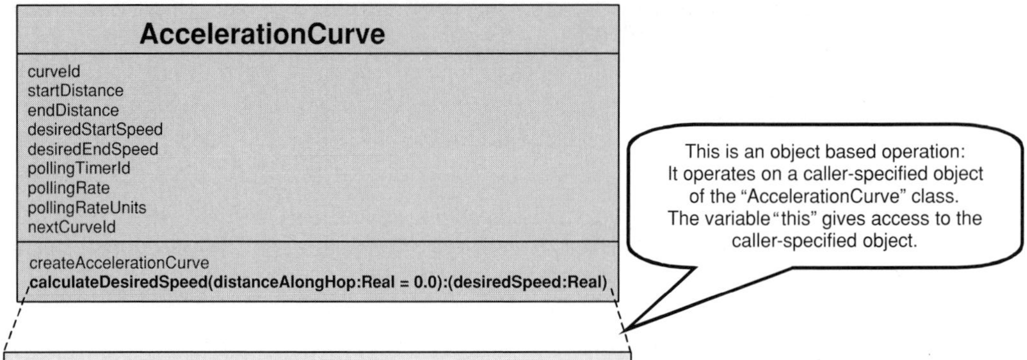

AccelerationCurve

curveId
startDistance
endDistance
desiredStartSpeed
desiredEndSpeed
pollingTimerId
pollingRate
pollingRateUnits
nextCurveId

createAccelerationCurve
calculateDesiredSpeed(distanceAlongHop:Real = 0.0):(desiredSpeed:Real)

> This is an object based operation: It operates on a caller-specified object of the "AccelerationCurve" class. The variable "this" gives access to the caller-specified object.

```
# The desired speed is determined by looking up the speed for
# the specified 'distanceAlongHop' (as defined by the linear
# acceleration profile defined by this Acceleration Curve).
# - except for the case when the train is at the start of a
# curve and is about to pull-away from rest. In such cases,
# the desired speed will be the respective train's
# 'minimumPullAwaySpeed'.

if (distanceAlongHop = this.startDistance) & \
   (this.startSpeed = 0.0) then

    # Must be about to pull-away from rest - so adjust the
    # speed to that of the train's minimum pull-away speed.
    theTrain = this -> R3 -> R2
    desiredSpeed = theTrain.minimumPullAwaySpeed

else

    # Not about to pull away, so control speed according to
    # this Acceleration Curve's profile.
    lengthOfCurve = this.endDistance - this.startDistance
    requiredSpeedDifference = this.endSpeed - this.startSpeed
    speedGradient = requiredSpeedDifference / lengthOfCurve

    distanceAlongCurve = distanceAlongHop - this.startDistance
    deltaSpeed = distanceAlongCurve * speedGradient

    desiredSpeed = deltaSpeed + this.startSpeed
endif
```

Fig. 8.5 An object-based operation

The invocation of an object-based operation in ASL takes the form:

> **Object-based operation rules**
>
> 1. Operation names are unique within the context of a class in a domain.
> 2. Operations act on an object that is referenced in the 'on' clause of the invocation.

```
[theDesiredSpeed] = AC2:calculateDesiredSpeed[hopDistance] on
theAccCurve
```

where:
- `theDesiredSpeed` is the returned output parameter mapped from desiredSpeed;
- `AC` is the key letter for the ACCELERATION CURVE class (an ASL short form);
- `2` is the operation number within the class (an ASL short form);
- `hopDistance` is the input parameter mapped onto distanceAlongHop;

- `theAccCurve` is an instance handle that refers to the object upon which the operation is invoked.

It is the `on` keyword in the operation invocation that indicates the operation is object-based.

Note: In ASL, key letters are assigned to domains, classes and terminators. Each signal and operation is assigned a number so that a short name for each can be derived. For example, the ACCELERATIONCURVE class can be given the key letter AC, and the operation `calculateDesiredSpeed` could be given the number 2, then the operation label is AC2. This label is pre-pended to the operation name as shown in the preceding example. The following rules are specific to ASL:

ASL object and class-based operation rules

1. Class key letters are unique within a domain.
2. Operation numbers are unique for a class.
3. Operation numbers can be the same as signal numbers for the same class.

8.2.2 Class-based operations

A class-based operation is an operation that is provided by a specific class but does not act upon a caller-specified object. For example, an operation that searches all objects of the class to find those meeting certain criteria or the classic example of an operation that creates a new object of the class.

Class-based operation rules

1. Operation names are unique for a class within a domain.

The invocation of a class-based operation in ASL takes the form:

`[newAccCurve]=AC1:createAccelCurve[startDistance,endDistance, ...]`

where:
- `newAccCurve` is an output parameter which, in this case, is an instance handle which refers to the newly created object;
- `AC` is the key letter for the ACCELERATIONCURVE class (an ASL short form);
- `1` is the operation number within the class (an ASL short form);
- `startDistance` and `endDistance` are (some of) the input parameters (the ... indicates that there are further input parameters that are not shown here – it is not valid ASL).

A class-based operation does not operate on a specified object. It is the absence of the `on` keyword in the operation invocation that indicates the operation is class based.

8.3 Domain operations

A domain-based operation is an operation that is provided by the domain as a whole and does not relate to any specific class in the domain. The most common use of domain-based operations is to define services the domain makes available to others, that is the realization of its provided interface. A good domain-based operation realises the service it provides by invoking operations on objects and classes within the domain or sending signals to objects in the domain. It will not have complex logic but rather has the knowledge of which object and class-based operations within the domain will achieve the operation it provides. One commonly adopted style of using domain-based operations is to provide one for every element of a domain's provided interfaces. This means that bridge functions 'using' the domain only call domain-based operations. As a result the bridge functions have no dependence on any of the classes within the domain. There are some specific ASL rules relating to domain-based operations:

> **ASL domain-based operation rules**
> 1. Domain key letters are unique within a system model.
> 2. Operation numbers are unique for a domain
> 3. Operation names are unique for a domain

The invocation of a domain-based operation in ASL takes the form:

```
[position] = TM1::reportTrainPosition[theTrainId]
```

where:
- `position` is the returned output parameter which in this case is a value of type Real;
- `TM` is the key letter for the train management domain (an ASL short form);
- `1` is the operation number within the domain (an ASL short form);
- `theTrainId` is an input parameter.

A domain-based operation does not operate on a specified object. It is the `::` that indicates the operation is domain based.

8.4 Bridge operations

Bridge operations are used to connect domains together. They are defined within a domain as operations provided by 'special' classes referred to as terminator classes – they can readily be identified since they are stereotyped «terminator». The methods for such operations do not really belong in a particular domain – rather they belong to a particular 'build' of the system. They are included here for completeness because they are a form of state-independent operation that can be specified using ASL. Essentially, bridge operations are used to map the service requirements of one domain to the services provisions of one or more other domains – they will be covered in detail in Chapters 10 and 12.

8.5 Where do operations live?

Every operation is associated with exactly one xUML element, which is:
1. A **domain**, for domain-based operations;
2. A **class**, for class- and object-based operations;
3. A «**terminator**» class, for bridge operations.

Associating an operation with a domain has no obvious graphical representation in UML. If it is not explicitly supported in the CASE tool you are using, then add a class to the class diagram with the same name as the domain. This will be a single object class (singleton) that holds the domain-based operations. It has no attributes and it is not associated with other classes. In fact, in the class diagram it just sits there on its own. The class implements the operations that are defined on it and generally will call operations on other classes and/or send signals to other classes. These interactions will be represented in the CCM but are not shown on the class diagram.

8.6 Polymorphic operations

As with signals directed at superclasses, it is useful for analysts to invoke operations on superclasses and have these implemented by the correct subclass.

Synchronous operations provide similar behaviour to polymorphic signals (Figure 8.6), with one important difference: **exactly one** implementation of the invoked service will execute at run time. Since the synchronous operation can return data in the form of output parameters, it makes no sense to say that the service will run multiple times; there could be no sensible rules for the return of the data.

Here are the detailed rules:
1. An object-based operation can be defined for a superclass. The implementation of this operation can be either in the superclass, or in any of its subclasses in **any** generalization hierarchy descending from it.
2. The analyst must ensure that there is exactly one implementation that can be executed at run time for any possible pattern of subclass objects. Thus:
   - Implementations must not exist in more than one generalization hierarchy;
   - Within one generalization hierarchy, implementations must not exist at more than one level in any descending branch.
3. At run time, when the service is invoked using the superclass handle, the implementation for the superclass or the corresponding subclass (or subclass of subclass) is invoked depending on how the model is specified.

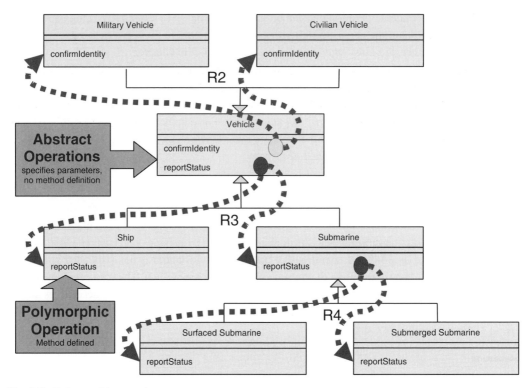

Fig. 8.6 Polymorphic operations

4. The handle `this` is accessible in the implementation of the operation and will be of the same type as the class with which the implementation is associated.

8.7 External visibility of operations

It is desirable to formally control the visibility of object-based and class-based operations from outside a domain. If an analyst intends that an operation be invoked from outside then the operation must be shown as being invoked by a «terminator» class on the CCM. Domain-based operations are always externally visible since their primary role is to define the operation provided by the domain.

The restricted visibility of object- and class-based operations allows the domain analyst to restrict access to the domain to known interfaces and thus reduce undesirable coupling. If this is not dealt with explicitly then a bridge operation could have an unseen dependency on a class. A developer working on the class may be unaware of its use by a bridge and if the class were modified or deleted then the domain would no longer work correctly. This really comes down to the general rule in software engineering of keeping coupling to a minimum but where coupling is necessary make sure it is explicit.

8.8 Summary

Each of the types of operation described in this chapter has its detailed behaviour specified using the ASL. This is covered in detail in Chapter 10. Operations allow us to specify the state-independent behaviour of a domain. In the next chapter we will look at the definition of state-dependent behaviour with state models. The state actions in state models are also specified using ASL.

9 Dynamic modelling

9.1 Introduction

In the previous chapters we have seen that the key concept of xUML in the MDA process is the **class** and in Chapter 7 we met the concepts of state-independent and state-dependent behaviour. In this chapter we turn our attention to state-dependent behaviour which can informally be thought of as that behaviour of a class which, given a signal, depends upon the state in which the instance receives the signal. In other words, to work out how an instance of a class will behave we need two facts, the signal that it receives and the state in which it receives the signal. If the class always behaves in the same way for a given signal then this is state-independent behaviour, as discussed in Chapter 8.

9.2 Definitions

We start with the definitions of the concepts of state modelling in xUML:

> **Method definition**
>
> A **state** represents a condition of the class subject to a defined set of rules, policies, regulations or physical laws.

> **Method definition**
>
> A **signal** represents an incident that causes a state transition.

> **Method definition**
>
> A **transition** specifies the state that an object will enter from a given state on receipt of a specific signal.

> **Method definition**
>
> An entry **action** is the sequence of processing which takes place on entry to a given state.

Dynamic modelling

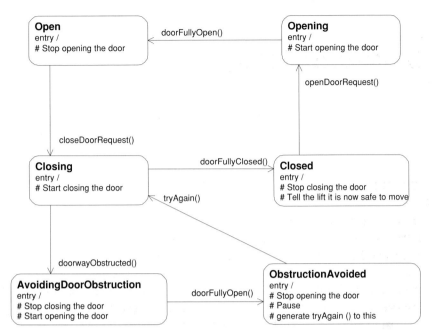

Fig. 9.1 Complete statechart for the DOOR class

The following sections explain how these definitions are put into practice. Diagrammatically we use a statechart to show states, transitions, signals and actions. As a first example of these definitions at work consider Figure 9.1, which gives an example of a simple statechart for a DOOR class in a lift control domain. The states (drawn as 'rounded' rectangles) represent a condition of the lift door in which certain rules apply. For instance, in the Open state, when the lift receives a 'closeDoorRequest' signal, the DOOR moves to the Closing state (defined by the transition between the two states, labelled with the 'closeDoorRequest' signal) and performs the entry action of that state (in this case it starts closing the door). We shall consider these constructs in greater depth in the coming sections.

It is important to realize that the statechart is a diagrammatic representation of a state machine. In xUML a state machine is always associated with one class (we do not attach state machines to any other type of model element). Those classes that exhibit state-dependent behaviour have one state machine and are termed as **active**. We can think of each object of an active class running concurrently, not only with objects of other classes but also with objects of its own class. The state machine acts as a 'blueprint' for the behaviour of all objects of the active class.

9.3 Statecharts

In Chapter 7 we saw that the class collaboration model (CCM) provides a graphical representation of the interactions between the classes in a domain, in terms of the operation calls

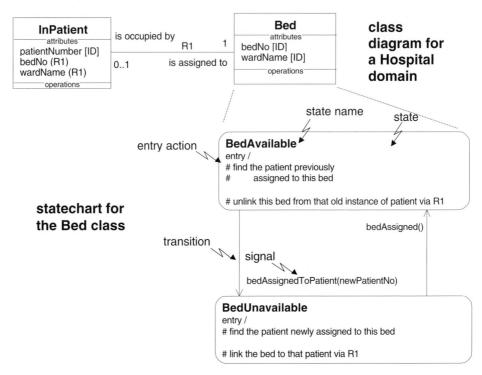

Fig. 9.2 Statechart diagram components

and signal transmissions. It is a **black box** view of the classes. The statechart provides a graphical representation of the way in which each class responds to the signals. It is a **white box** view of the class.

The statechart is the graphical representation of the state machine of a class. We shall see in Section 9.4 that we can also represent the state machine in a tabular form.

9.3.1 Basic structure

Not surprisingly, the basic structure of a statechart shows the states and transitions for its class. Each state has exactly one associated action, which is executed upon entry to that state. Each transition has one signal attached to it, which causes that transition to be made (we shall see an exception to this in Section 9.3.4).

Each class has up to one state machine, which describes the behaviour of every object of that class (Figure 9.2).

In the notation for a state, it is optional whether or not to show the state name in a compartment, that is there may be a line depicted below the state name as in some of the following examples. In the preceding examples we have shown the state action in the form of comments; to build a fully executable model we shall use the Action Specification Language (ASL) to specify the required processing in each state (see Chapter 10).

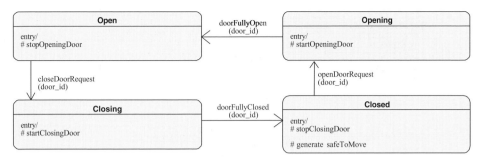

Fig. 9.3 Preliminary statechart for the DOOR class

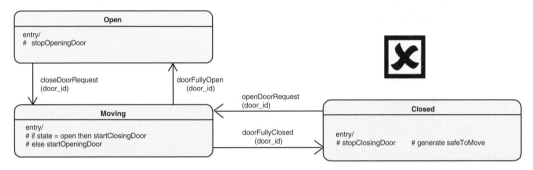

Fig. 9.4 A stupid state machine

To start the state modelling process, identify the states that represent normal behaviour (Figure 9.3). This will help ensure that the model reflects the expected system behaviour.

Once the analyst is satisfied that the normal, or fundamental behaviour has been captured, add states that represent abnormal behaviour. For example, if while the DOOR is in the Closing state it becomes obstructed by somebody entering the lift, we must transition to the AvoidingDoorObstruction state, and reopen the door. This leads to the more realistic model given in Figure 9.1.

9.3.2 Intelligent versus stupid state machines

The state machine captures expertise regarding the behaviour of a class in a domain. How much expertise should we include?

Consider the example of a lift door. It will respond to requests from a LIFT class to open and close the door at specific times. Only when the DOOR is closed should we indicate to the LIFT class that it is safe to move. We could build a state machine as in Figure 9.3, or we could take the view that the Opening and Closing states are basically the same, and economise on states by combining them into a Moving state, giving a model like that in Figure 9.4.

Which is the better model? Well, the developers of a system like this will probably be concerned with:

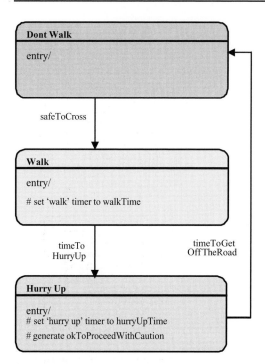

Fig. 9.5 A cyclic state machine

1. Safety, since this is clearly a system in which lives will be at risk if the lift ever moves with the door open. While the DOOR class is in the Moving state, it does not know whether it is opening or closing. So if the door is currently opening, what will happen if a spurious 'D2:doorFullyClosed' signal arrives? The DOOR will make a transition into the Closed state and inform the LIFT that it is safe to move. The lift may then start to move with the door still open, causing a safety hazard;
2. The degree to which the system can recognize and inform an operator of the occurrence of faults. The smaller state machine does not capture enough expertise about the subject of lift doors to recognize that a sequence of 'openDoorRequest' followed by 'doorFullyClosed' is impossible.

So the 'smaller' model results in a less safe and less 'intuitive' system. The moral of this is not to economise on states. There is little correlation between the number of states and the size and speed of the resulting system.

9.3.3 Basic forms of statecharts

There are two basics 'shapes' of statecharts that frequently occur. First, consider that some state machines just go round and round and round... (Figure 9.5).

Note that this does not necessarily mean that the objects are fixed; objects might be created and deleted using operations rather than signals. This pattern simply reflects the fact that while they exist, objects repeat the same set of actions cyclically.

Dynamic modelling

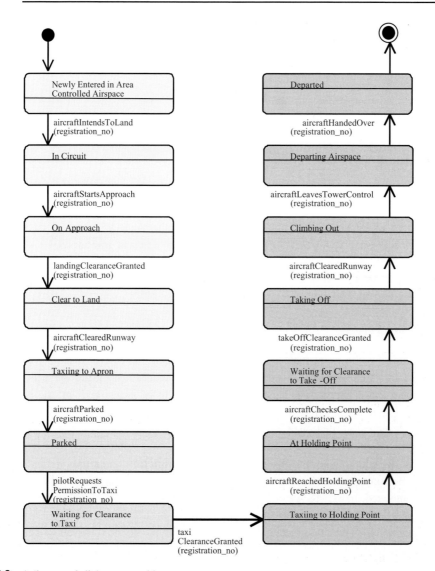

Fig. 9.6 A 'born-and-die' state machine

The other primary form of state machine shape is the 'born-and-die' life cycle. Classes that exhibit this form are definitely dynamic (they have instance populations that are created and deleted at run time). Such a 'born-and-die' state machine has one initial vertex and at least one final state vertex (we shall discuss the semantics of these in Sections 9.3.4 and 9.3.5). As an example, a state machine of an aeroplane approaching and leaving an aerodrome is given in Figure 9.6.

Here, the instance of the AEROPLANE class is created when it enters the controlled airspace (remember this in an **abstraction**, that is a model, of an aircraft under air traffic

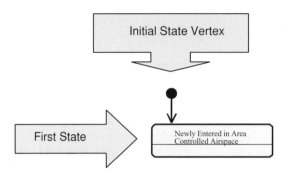

Fig. 9.7 The initial state vertex

control!). In this case the instance is created in the `NewlyEnteredInAreaControlledAirspace` state. It then goes through a series of intermediate states before it is finally deleted when it leaves the controlled airspace. Note that this model does not yet cater for failures, such as the aircraft failing an engine check at the take-off holding point and so going back to the hanger.

9.3.4 Initial state vertex

This indicates the point that an object of a class, that has a state machine, comes into being and starts operating. We create the object synchronously using the ASL `create` statement (see Chapter 10). This causes the object to be extant and 'in' the initial state vertex. The transition from the black 'blob' of the initial vertex, to the first state is unlabelled, indicating that once we have made the call to the `create` statement, the state machine automatically executes the transition to the first state and awaits signals as normal. In UML an unlabelled transition is called a **completion transition** and indicates that the transition will occur automatically the moment the state action for the source state has completed. In xUML we only use completion transitions from the initial state vertex (Figure 9.7) and transitions to the final state vertex.

Creation of a new object with a state machine will typically be achieved by a class-based operation (constructor) of that class, which contains the ASL `create` statement. This operation will:
1. Create the new object (which is mandatory);
2. Assign attribute values (every identifier must have values assigned to all their attributes, other attributes are optionally given values here);
3. Find unconditionally related objects;
4. Link new object to related objects.

9.3.5 Deletion states

A deletion state is a state in which the object is deleted. Obviously, there will never be a transition out of a deletion state to another state, since by the time the action has completed,

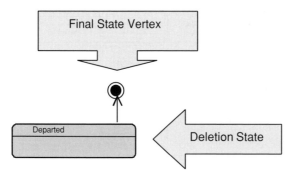

Fig. 9.8 The final state vertex

there is no object to receive further signals. Once the state action is complete there is a completion transition to the final state vertex where the instance is considered to be deleted. Note that at the end of the action in the deletion state the only thing that the state machine can do is to automatically transition to the final state vertex (Figure 9.8).

The action for a final state must always:
1. Find all currently related objects;
2. Unlink currently related objects;
3. Check for unconditional association violations;
4. Delete unconditionally related objects;
5. Delete this object.

9.3.6 Timers

It is often important to model time in state machines. This is addressed by the notion of a **delayed signal**. These may be used to model watchdog timers, periodic actions such as reading a sensor value or simulate continuous processes by small discrete time intervals. Once a delayed signal is generated it is queued and handled just like any other, as previously discussed. Three basic forms of ASL operation provide the required functionality and illustrate what may be achieved by the use of delayed signals:

1. Generate <signal> to <instance handle> after <delay> <delay units>. This will cause the specified signal to be queued to the specified instance after the specified amount of time. Note that this guarantees only that the signal will be received at or after the specified delay time. We may also want to set an 'absolute' timer, that is to generate a signal to an instance at a given time and date. In this case we use the form generate <signal> to <instance handle> at <date> & <time>. Time units can be 'millisecond', 'second', 'minute', 'hour' and 'day'. If we somehow specify the generation of a signal in the past, or with a negative delay, then the signal is generated straightaway;
2. Delayed signals can be cancelled using cancel-next <signal> to <instance handle>. If there are multiple instances of the specified delayed signal

Signal\State	closeDoor Request	doorFully Closed	doorFully Open	openDoor Request	doorway Obstructed	tryAgain
1. Open	Closing	Unknown	Unknown	Unknown	Unknown	Unknown
2. Closing	Unknown	Closed	Unknown	Unknown	AvoidingDoor Obstruction	Unknown
3. Closed	Unknown	Unknown	Unknown	Opening	Unknown	Unknown
4. Opening	Unknown	Unknown	Open	Unknown	Unknown	Unknown
5. Avoiding DoorObstruction	Unknown	Unknown	Obstruction Avoided	Unknown	Unknown	Unknown
6. Obstruction Avoided	Unknown	Unknown	Unknown	Unknown	Unknown	Closing

Fig. 9.9 Preliminary state transition table for the DOOR class

outstanding the signal which will be generated soonest will be cancelled. The form `cancel-all <signal> to <instance handle>` will cancel all delayed signals currently outstanding for an instance. If there are no delayed signals of the specified type outstanding, nothing happens. Note that, if the signal has already been placed on the queue for the target instance, then cancel does not remove the signal from the queue. This is typically addressed by making reception of the signal in other states 'ignored';

3. `<enquiryvariable> = time-until-next <signal> to <instance handle> in <units>`. This returns the amount of time remaining until the specified delayed signal is to be queued. If there are no such signals waiting, a time of zero is returned.

These operations may be invoked by any operation or state action of any domain.

9.4 The state transition table

The statechart provides a clear graphical representation of the state machine. It enables the reader to see the sequences and cycles of states entered by a class. However, the statechart does not require the analyst to address every possible combination of state and signal. So that this can be achieved, xUML provides an alternative view of the state machine, known as the state transition table (STT).

In this table:
- Each row corresponds to a state;
- Each column corresponds to a signal;
- Each cell represents a potential transition. These cells are referred to as effects.

The preliminary STT for the DOOR class statechart (Figure 9.1) would look like that shown in Figure 9.9.

197 Dynamic modelling

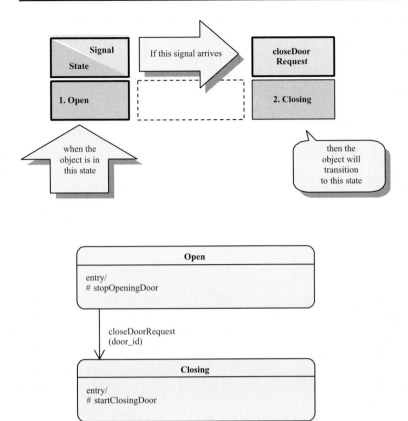

Fig. 9.10 Transition effect

Notice that most of the cells currently contain the value 'unknown'. This means that the state machine is incomplete. In order to complete the STT we must replace the 'unknown' effects with one of the those discussed below.

9.4.1 Transition effects

These correspond to the directed arc on the statechart. They show the state that is entered when the specified signal is received while in the specified state (Figure 9.10).

9.4.2 Ignore effects

These indicate that it is expected and acceptable to receive the specified signal in the given state and that no action is required (Figure 9.11).

Note that an 'ignore effect' is quite different to making a transition into the same state. In the case of such **reflexive transitions**, the action for the entered state will be executed. Remember; entry actions do not care which state you were in previously.

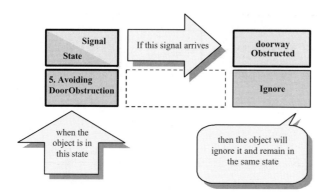

Fig. 9.11 Ignore effect

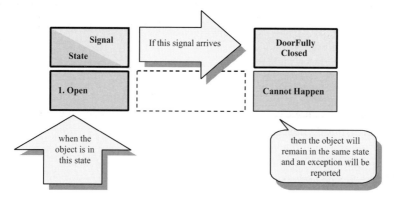

Fig. 9.12 Cannot happen effect

9.4.3 Cannot happen effects

These indicate that the given signal cannot occur in that state **if the system is behaving as intended**. It can happen – but if it does something has gone wrong and the analyst has explicitly decided not to attempt any recovery action (Figure 9.12). The behaviour is basically the same as for an 'ignore effect'; no transition occurs, so no action is executed, but the software architecture domain will typically add an entry to the system error log or take some corrective action. When the models are being tested in a simulator, these cannot happen occurrences will be flagged as serious errors and must be fully investigated by the analyst.

9.4.4 The complete state machine

Once all the cells have been given a value other than unknown, the STT is considered to be complete (Figure 9.13).

Signal / State	closeDoor Request	doorFully Closed	doorFully Open	openDoor Request	doorway Obstructed	tryAgain
1. Open	2. Closing	Cannot Happen	Cannot Happen	Cannot Happen	Ignore	Cannot Happen
2. Closing	Cannot Happen	3. Closed	Cannot Happen	Cannot Happen	5. Avoiding DoorObstruction	Cannot Happen
3. Closed	Cannot Happen	Cannot Happen	Cannot Happen	4. Opening	Ignore	Cannot Happen
4. Opening	Cannot Happen	Cannot Happen	1. Open	Cannot Happen	Ignore	Cannot Happen
5. Avoiding DoorObstruction	Cannot Happen	Cannot Happen	6. Obstruction Avoided	Cannot Happen	Ignore	Cannot Happen
6. Obstruction Avoided	Cannot Happen	Cannot Happen	Cannot Happen	Cannot Happen	Cannot Happen	2. Closing

Fig. 9.13 Complete state transition table for the DOOR class

Note that most cells are 'cannot happen'. This is often the sign of a good STT. If most cells are 'ignore', the implication is that the other classes are sending signals that cause no action. This is typically indicative of poor or lazy modelling.

9.4.5 The STT and the statechart

In UML an STT can be regarded as a stereotyped notation of a state machine, designed to address completeness of the analysis of a state-dependent behaviour of a class. It would be possible to capture all the information that is held in an STT on a statechart, using internal transitions with no action to indicate an 'ignore effect' or an internal transition that raises an exception for a 'cannot happen'. Whilst such statecharts are conceivable, modelling a problem of any complexity in this way would soon lead to a statechart that was incomprehensible.

The STT form also has the benefit that it is glaringly obvious when the analyst has omitted consideration of receiving a signal in a certain state, since that cell will have the effect 'unknown' indicating that the analysis of the state machine is not complete. Patterns are also evident, such as a signal from a timer being ignored in every state where it does not cause a transition.

9.4.6 Meaningless effects

Any cells appearing in a row corresponding to a **deletion state** (a state in which the object is to be deleted) are automatically populated with 'meaningless' (Figure 9.14). This is to reflect

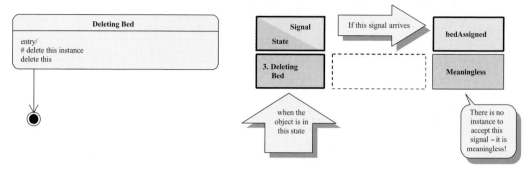

Fig. 9.14 Meaningless effect

that by the time the action for the deletion state has completed, the object will transition to the final state vertex and so no longer exist. It is therefore meaningless to specify the effect of the object receiving further signals.

9.5 The non-existent state

Despite its name, the non-existent state does exist! It represents the state of an object before it is created. When a signal is received in the non-existent state, it is handled by the class, since the instance to which it is bound is not extant. This effect is useful for specifying what should happen if a signal is generated to a 'stale' instance handle. Such occurrences may happen when timers are used, in which case we may want to set the effect to ignore. Mostly the effect is 'cannot happen' in the non-existent state. Note, we never allow a transition out of this state.

9.6 Execution semantics

The execution semantics of an xUML model are straightforward. They specify the basic rules that must be adhered to by all state machines in every xUML model. These rules allow analysts to model and interpret state machines in an unambiguous way. They are explained in the following sections.

9.6.1 State machines

The unit of concurrency in an xUML model is the **state machine**. Recall that a state machine belongs to exactly one active class (i.e. a class that exhibits state-dependent behaviour).

Each object of a class notionally runs it own 'copy' of the state machine. These concurrent state machines are able to communicate with any other state machine or themselves. Each state machine instance maintains a queue of signals that have been sent to it. Each state

Dynamic modelling

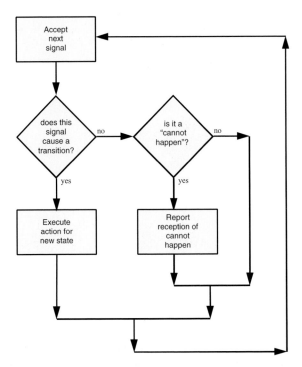

Fig. 9.15 The State machine loop

machine instance executes simple logic to process signals that have been placed on its queue (Figure 9.15).

9.6.2 Action execution rules

The execution rules for a state action may be summarized as:
1. Each action takes a finite time to execute (eventual termination of the state action must be ensured);
2. Once complete, the state machine instance will accept the next signal waiting on its queue. This signal is then deleted from the queue;
3. State machines are assumed to run concurrently.

Figure 9.15 reflects these rules and they are presented diagrammatically in Figure 9.16.

It is important that the model maintains the integrity of the relationships and objects as given in the class diagram. This means that actions must:
- Leave association instances consistent;
- Leave subclass and superclass objects consistent.

Consistency may be achieved either directly by the action, or by generating signals that will, in due course, ensure it is achieved.

In this model (Figure 9.17), for example, when the analyst specifies deletion of an object of 'IN PATIENT', he or she must perform the following as part of the same action:

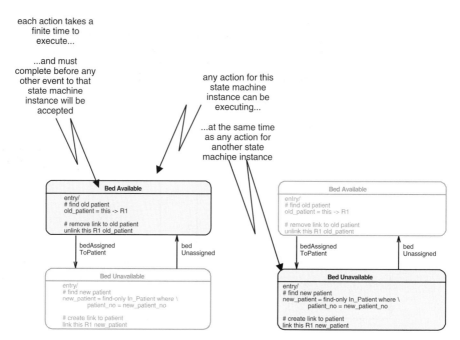

Fig. 9.16 Action execution rules

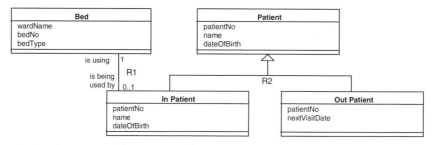

Fig. 9.17 Part of a patient administration domain

1. Unlink the 'IN PATIENT' object from its related 'BED' object across R1;
2. Unlink the 'IN PATIENT' object from its 'PATIENT' superclass object across R2;
3. Delete the 'PATIENT' superclass object.

9.6.3 Self-directed signals

Objects often talk to themselves. Such a signal, from an object to itself, is called a **self-directed** signal. Any signal generated by a state machine instance to the same state machine instance will be processed before signals from other state machines. This is why, on the DOOR state machine presented earlier (Figure 9.1), the effect of the 'doorObstructed' signal for the `ObstructionAvoided` state is 'cannot happen'. This means that the

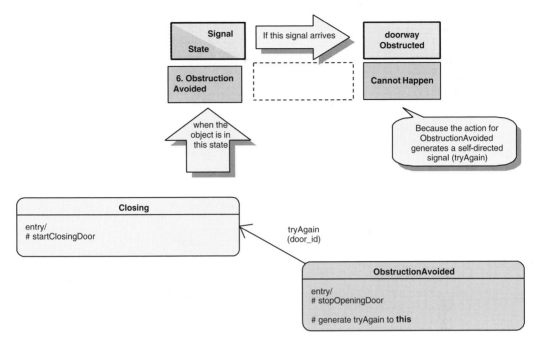

Fig. 9.18 State transition table entry for a self-directed signal

'doorObstructed' signal could have been queued in this state, but cannot possibly be at the front of the queue because the self-directed signal would have pushed past it (Figure 9.18).

9.6.4 Signal transmission rules

The rules that govern the way that signals are processed are:
- Signals received from the same state machine must be processed in the order in which they were generated;
- Signals received from different state machines may be processed in any order;
- Self-directed signals are processed before any others.

In the scenario depicted in Figure 9.19, the TheSwitch state machine can legally process the three signals shown in any of the following sequences:
- 1–2–3;
- 1–3–2;
- 3–1–2.

That is, the only thing that is guaranteed is that the two signals generated by the 'Alf' state machine to THESWITCH state machine will be processed in the order generated. This seems a reasonable rule, since in a distributed system it is highly likely that transport delays will result in signals being received at THESWITCH in a sequence that does not correspond to their generation time. A simple window-based protocol, in which signals from a given

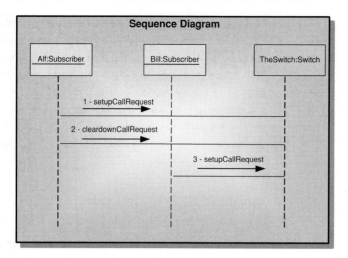

Fig. 9.19 Signal transmission rules

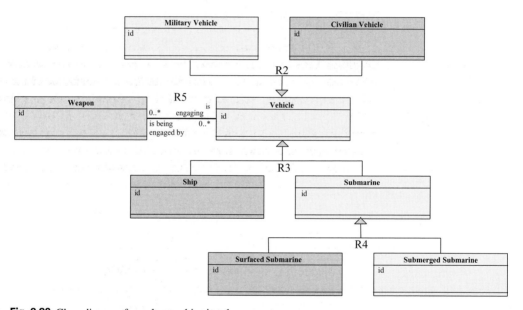

Fig. 9.20 Class diagram for polymorphic signals

state machine are tagged with a sequence number, will allow the software architecture to detect missing signals from each state machine.

9.6.5 Polymorphic signals

Consider a class diagram and corresponding object populations (Figures 9.20 and 9.21).

Now assume that a 'WEAPON' instance generates a signal to a 'VEHICLE' instance number 5. The signal is propagated, top-down, from the target (superclass) state machine (if it exists) to all corresponding subclass state machines. If the superclass is

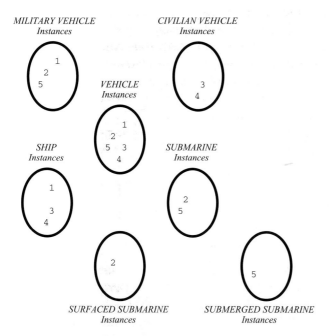

Fig. 9.21 Potato diagram for polymorphic signals

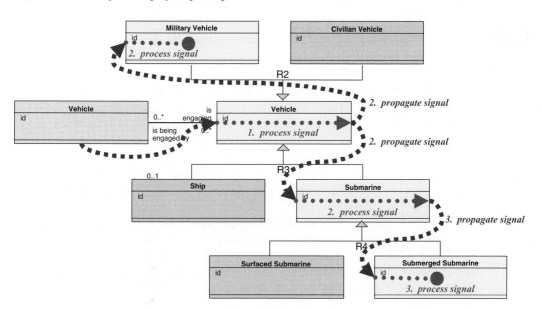

Fig. 9.22 Propagation of a polymorphic signal

passive (does not have a state machine) then the signal is still propagated down the specialization hierarchy. At each level, the associated action is completed before the signal is propagated to the corresponding subclass instance (or the signal is simply immediately passed on if the class at that point in the hierarchy is passive) (Figure 9.22).

Polymorphic signals behave in the following way:
- Any signal which is directed at a superclass is automatically available to all subclasses on all hierarchies descending from the superclass;
- A subclass may choose to ignore such available signals (by setting the whole response column in the STT to 'ignore') in which case the signal will not be processed by the subclass at run time;
- Any subclass which has not completely ignored the signal will receive and process the signal at run time if the subclass object is related to the superclass object to which the signal was directed when the signal was generated;
- A single signal generation (to the superclass) may result in multiple state machines responding at run time. This will happen, for example, if there are state machines at both the superclass and subclass level, neither of which ignore the signal.

As it turns out, in line with the 'keep it simple' philosophy of xUML, it is not really necessary to think of signals as being polymorphic or not. The models simply follow these rules:

> **Method rule**
>
> Each class has up to one state machine.

> **Method rule**
>
> Each signal is **directed to** exactly one class and is **available to** all subclasses of the class to which it is directed.

All signals can be 'ignored' on a class-by-class basis.

The benefits of polymorphic signals are:
- The sending state machine does not need to work out where to send the signal;
- The model is more stable in the face of changes in the subclass hierarchy.

9.7 Controlling state machine complexity

As discussed earlier, economizing on classes has a cost. The cost will become apparent at the state modelling phase when the 'big' classes begin to look decidedly schizophrenic. Is the AIRCRAFT in the model depicted in Figure 9.23 arriving or departing? The rules and policies for these two roles are quite different but now they have to be formalized on one state machine. It will obviously be complex.

To address this problem, we can abstract role subclasses to represent the sets of aircraft conforming to the different rules and policies for arriving and departing aircraft (Figure 9.24).

We now have the choice of building a single state machine for the superclass, as in Figure 9.25, or partitioning the state behaviour up, as in Figure 9.27. First, let us examine the single state machine approach.

Dynamic modelling

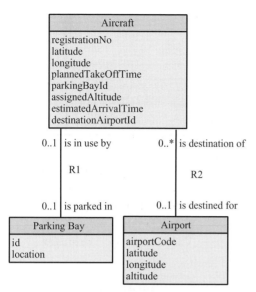

Fig. 9.23 A suspect class diagram

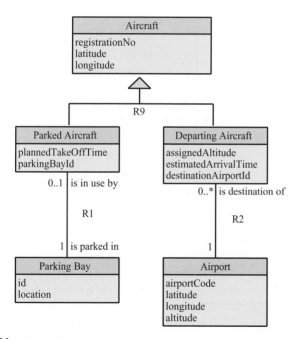

Fig. 9.24 A better class diagram

This is fine, but it is bordering on the level of complexity that will inhibit effective review and maintenance. An alternative is to partition these states across the role subclasses, and build a number of smaller, simpler state machines. If we extended the class diagram in Figure 9.25 to include the complete set of role subclasses, it would look like that shown in Figure 9.26.

Model Driven Architecture with Executable UML

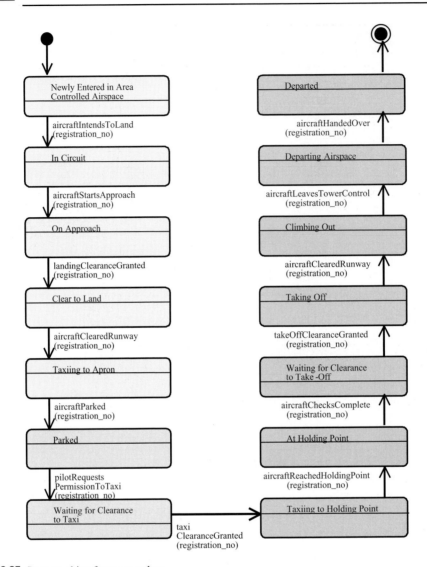

Fig. 9.25 State machine for a superclass

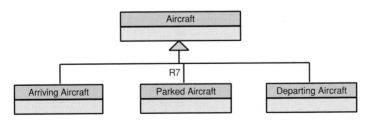

Fig. 9.26 Class diagram showing all role subclasses

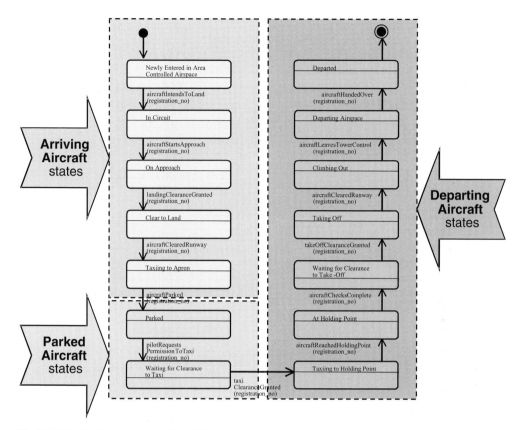

Fig. 9.27 Partitioning a state machine into roles

The states can be partitioned as in Figure 9.27.

Note that in this case, each role state machine needs to include a state that will migrate to the next role at the end of its life cycle. For example, the PARKED AIRCRAFT state machine would include the additional state Departing (Figure 9.28).

The class-based operation createDepartingAircraft would look like this...

```
createDepartingAircraft [theAircraft: Aircraft]

newDepartingAircraft = create DepartingAircraft with \
         aircraftId = theAircraft.aircraftId \
         & currentState = 'Taxiing_to_Holding_Point'
link newDepartingAircraft R7 theAircraft
```

9.8 How to build bad state machines

Here are some tips for building really bad state machines:
- Ensure that the model embeds no state-dependent behaviour at all. This can easily be achieved by designating signals to allow other classes to perform read accesses, or even

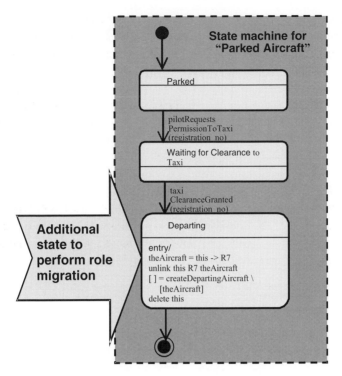

Fig. 9.28 Role migration action

write accesses, that do not depend upon or affect the state of the target object. Using this technique, you can quickly achieve an explosion in the number of states, since in every state, you need to respond to every signal. This will give the impression that you have analysed a really complex class. It will take orders of magnitude longer to develop in this way rather than with the rather humdrum technique of specifying one operation for each state-independent operation;

- Use polymorphic signals that are always queued to the same class. This is a useful technique for introducing unnecessary complexity to an CCM. If you are building a subclass state machine, and need to send a self-directed signal, avoid the obvious approach of sending a signal to this class; send it to the superclass instead. Lots of extra 'available signals' will appear in the state transition tables for the other subclasses in the family, which will help to compromise their readability. This approach will buy you lots of credibility in reviews, as your fellow analysts will murmur things like 'Ooh – he's used polymorphic signals – he must know what he's doing'.

9.9 Other forms of state modelling in UML

We have covered the basic form of state modelling in this chapter, the Moore machine, where actions are performed on entry to a state. This is sufficient to build executable models in xUML for all domain types. The UML provides other forms of state models, such as Mealy machines (where actions are associated with transitions) as well as decomposition of states using Harel statecharts. Hopcroft and Ullman (1979) provide a discussion of such **finite state automatons**. These forms may all be used in xUML (though non-executability may be a problem with Harel statecharts). If these additional form of state machine are employed then the analysts must be sure that the added complexity of these forms of statechart is justified in terms of ensuring that the models are still comprehensible and maintainable.

9.10 Conclusion

We have seen in this chapter how to build executable models that have classes whose behaviour is state-dependent. The use of the state machine is a powerful tool that allows us to capture in an elegant and readily comprehensible way the essential nature of class behaviour that is dependent upon the objects' prior history.

10 Action specification

10.1 Where are we?

So far, we have partitioned our system into:
- **Domains** – representing a subject matter to be studied;
- **Classes** – representing a set of like things;
- **States** – representing a condition of a class;
- **Operations** – representing state-independent behaviour.

These are depicted in Figure 10.1.

When an operation is invoked, or upon entry to each state, an action is performed. The language that is used to specify these actions is imaginatively named the **Action Specification Language** (ASL) (Wilkie *et al.*, 2002).

10.2 The Action Specification Language (ASL)

ASL is a platform-independent language for specifying processing within the context of an xUML model. The aim of the language is to provide an unambiguous, concise and readable definition of the processing to be carried out by a system. The UML does not define an action *language* but action *semantics* have been added to the UML standard.

> ASL complies with the UML action semantics.

ASL was used in the UML action semantics standard submission as an example of a compliant language.

Why do we need ASL? Why can't we use Java, C++ or another programming language? These are good questions. Programming languages are not designed to manipulate the elements of an object model. They do not provide the facilities that we need to be able to express the actions in a model in a clear and precise, yet abstract, manner. However, they do allow the developer to manipulate all sorts of implementation-specific things that are wholly inappropriate in a PIM. ASL is a small language compared to most programming languages, yet it has powerful abstractions that are easy to use and that do not bias the implementation.

Action specification

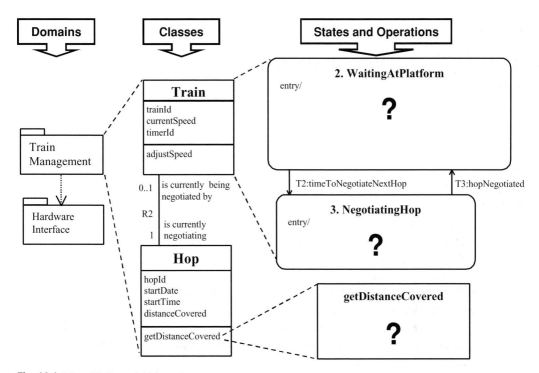

Fig. 10.1 The xUML model hierarchy

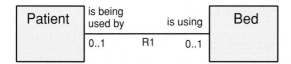

Fig. 10.2 Simple class diagram

For example, in Chapter 6 we introduced the class diagram with classes and associations as shown in Figure 10.2.

The association labelled R1 tells us that a particular patient may or may not be using a particular bed. It is commonplace in actions to want to find associated objects, such as which bed a patient is in, so we need to **navigate** the association. We use the verb 'navigate' to describe the action of finding the associated object (or objects) at the other end of an association. With a programming language we would need to know how the association is going to be implemented, say with pointers, and therefore navigate the association using pointers. This immediately makes the model implementation specific. In ASL we can simply navigate the association:

```
theBed = thePatient -> R1
```

This line of ASL can be read as 'theBed is found from thePatient by navigating through the association R1'

The ASL allows us to navigate the association simply and concisely, and it does not restrict the ways in which associations can be implemented.

A full definition of ASL can be found in the 'ASL Reference Manual' on the accompanying CD-ROM. This section contains the highlights.

ASL is used to specify all processing within the system, including the processing needed to test the system. Specifically, ASL is used to define:

- **Class and object-based operations** – for state-independent behaviour;
- **State actions** – for state-dependent behaviour;
- **Domain operations** – for the services made available by the domain;
- **Bridge operations** – to specify a mapping between a required service in one domain to one or more provided services in other domains;
- **Initialisation segments** – to define initial object populations;
- **Test methods** – to provide drivers to stimulate the model in a simulation environment.

Examples of several of these may be found in Chapters 7, 8 and 9.

10.2.1 The requirements for ASL

The purpose of building a platform-independent model is to understand a problem domain and to specify behaviour that will meet a set of requirements. To this end xUML allows the analyst to describe a system in terms of **classes** with **operations** and interacting **state machines** with well-defined processing actions. The domain models produced are sufficiently detailed and precise that they can be verified against external criteria (such as external reality or desired system behaviour) and, as such, these models are executable.

The xUML formalism achieves this detail and rigour by providing a set of notations with certain qualities:

- A class modelling notation that meets criteria of consistency, completeness and rigour;
- A dynamic specification formalism that provides clear statements on the nature of time in a concurrent environment, leading to rules about synchronization and interruptability of processing.

The requirements that drove the language definition for ASL were:

- The ASL must be **detailed and precise** enough that the resulting models can be executed without any ambiguity or use of assumptions. Of necessity, this will result in an ASL that has the appearance of a programming language, although not necessarily a procedural one;
- The ASL must be **rich enough** to specify all the processing that will be required. If the language is not sufficiently versatile, then analysts will be forced to resort to other, perhaps vague or platform-dependent forms of expression;
- The ASL must be **easily readable**. It is one thing to create a language that is precise. It is another to produce one that can be quickly and easily scanned by the human reader. There must always be a place for reviews and other such procedures and the ASL must not hinder that activity;
- The ASL must be **simple**, and **rapid to create**. Although a similar problem to the previous requirement, different issues come into play, such as the need to avoid long complex keywords or a wealth of different special characters;

- The ASL must be sufficiently rich that **automated execution** of architectural mappings becomes feasible. For example, to support 'instance handle' based architectures, it must be possible to recognise relationship manipulations clearly and unambiguously.

These requirements have implicit contradictions, for example:

- Short, easy-to-type keywords can produce cryptic names;
- Many analysts prefer to think algorithmically. However, while an algorithmic specification may be correct, there may be other, more appropriate algorithms that can be used in the implementation.

In defining this ASL, an attempt was made to form the best possible compromise between the competing requirements.

Many organisations have, for many years, been routinely specifying their entire system behaviour using this action language and then generating the target code from this specification.

10.2.2 Implications of using an action language

It is important to distinguish between being precise and being implementation-oriented. In the past, software engineers have been forced into the latter to achieve the former. The only formalisms available to them for expressing behaviour at the necessary level of precision have been traditional programming languages with their attendant obsessions with computer-oriented trivia.

So developers were forced to adopt low-level languages, such as C++, whilst knowing that if such languages ever fell from grace, their code would be consigned to oblivion. Of course, to many programmers, this is a rather attractive idea. It means they can continue to reimplement the same behaviour using the most marketable languages – a kind of 'CV++' mentality. After all, we have managed to pull this off for the past three decades and hardly anybody seems to have noticed. But the times they are a-changing.

The headlong rush to adopt the UML has brought a fundamental issue onto sharp focus. 'How do we specify the behaviour of all these classes?' Well, it seems we need a language to allow the specification of actions. This was the catalyst for the development of the ASL.

Once you have a platform-independent way of specifying your entire system, all sorts of opportunities surface. The most obvious is wholesale code generation. Once this is possible, it becomes immediately apparent that special code generators can be used to produce instrumented code to support testing and symbolic debugging of the domain models themselves, before any real target code is generated.

10.3 Key features of ASL

We are not going to attempt to cover ASL in detail in this chapter, but to give a flavour of its key features. These are:

- Instance handles;
- Object manipulation;
- Association manipulation;

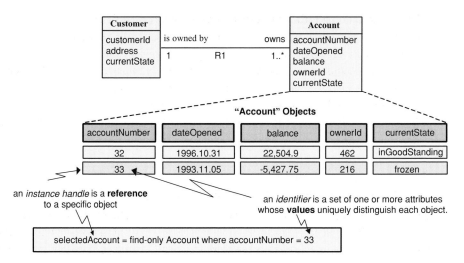

Fig. 10.3 Obtaining instance handles

- Invoking operations;
- Sending signals.

10.3.1 Instance handles

One of the most common things we need in ASL is the ability to refer to an object of a class. We do this using **instance handles**. An instance handle provides a reference to an object, and can be used to:
- generate a signal to that object;
- invoke an object-based operation on that object;
- link an association to that object;
- navigate a series of associations starting from that object;
- read and write attributes of that object;
- delete that object.

In Figure 10.3 the line of ASL states that an instance handle called 'selectedAccount' will refer to the object of the ACCOUNT class that has an **accountNumber** attribute with the value 33. What is more the statement asserts that there will be **one and only one** object that has an **accountNumber** with the value 33. This seems reasonable since **accountNumber** should be a unique identifying attribute for ACCOUNT. There is an important distinction here:
- **accountNumber** forms the identifier, namely an attribute that identifies an object because its value will be unique. It is not an instance handle;
- 'selectedAccount' is an instance handle, namely an object reference that allows us to refer to that object. It is not an attribute.

When executing a state action or an object-based operation, we are always executing in the context of a known object:

Action specification

- For a state action it is the object to which the signal was delivered;
- For an object-based operation it was the object upon which the operation was invoked.

To refer to such an object, ASL provides a special instance handle named 'this'.

An instance handle is a reference to an object of a particular class and therefore it gets its **type** from that class. 'selectedAccount' is of type 'instance handle of Account'.

Sets of instance handles are also useful. These allow us to refer to and to manipulate collections of objects. Using the example in Figure 10.3, consider an operation to return all overdrawn accounts. The ASL would take this form:

```
{overdrawnAccounts} = find Account where balance < 0.0
```

The `find` keyword is different from the `find-only` in the previous example in that it returns all of the objects that meet the condition. The condition being tested obviously may not resolve to a single object. There may be none, one or many objects of ACCOUNT where the condition is met. The returned instance handle set is denoted by {overdrawnAccounts}. The braces indicate that the variable is a set.

The instance handles in a set must be all of the same type. {overdrawnAccounts} is a set of instance handles of ACCOUNT.

10.3.2 Object and class manipulation

Objects can be created and deleted. Their attributes can be read and written. The creation of an object returns an instance handle that is the reference to the new object 'newAccount' in the example:

```
newAccount = create Account with accountNumber = 12345678 \
                          & balance = openingBalance \
                          & dateOpened = currentDate
```

`create` and `with` are keywords. The\denotes continuation of the statement on the following line and the & can be read as 'and' and denotes that the list continues. It is not essential that all attributes are specified, although it is necessary to define the values for all identifying attributes. You can assume that `openingBalance` and `currentDate` are local variables of type Real and Date, respectively.

Sometimes the identifier has an arbitrary value and in these circumstances a unique value is all that is required. ASL supports a variant of the `create` statement which will assign an arbitrary identifier.

```
newAccount = create unique Account with balance = openingBalance \
                                  & dateOpened = currentDate
```

In this example, the **accountNumber** attribute, being the identifier, will be set automatically.

The value of an attribute can be read. The result is returned as a variable.

```
theBalance = newAccount.balance
```

The value of an attribute can also be written. In this case, the value of the object's attribute is assigned from the variable `updatedBalance`.

```
newAccount.balance = updatedBalance
```

When an object is no longer required it can be deleted by reference to its instance handle.

```
delete selectedAccount
```

This also applies to sets of instance handles.

```
delete {overdrawnAccounts}
```

10.3.3 Association manipulation

One of the most powerful and elegant features of ASL is its capacity for manipulating associations. Associations link objects together. We need to be able to create and remove these links and navigate them so that we can find the related objects. We don't need to know how the association will be implemented to be able to do this. ASL provides a number of special primitives allowing the analyst to:
1. navigate an association (the -> construct);
2. create an association instance (the `link` keyword);
3. delete an association instance (the `unlink` keyword).

Referring to Figure 10.4, to navigate from an object 'selectedAccount' of class ACCOUNT to the instance of the class CUSTOMER that owns that account, you would simply write:

```
owningCustomer = selectedAccount -> R1
```

Note that because we are navigating along the association from ACCOUNT to CUSTOMER, we know that the result will be a single object. The multiplicity on the association says so: 'an account is owned by exactly one customer'.

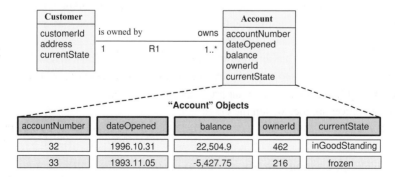

Fig. 10.4 Manipulating associations

To determine the set of accounts owned by a given customer, indicated by an instance handle 'theCustomer', we could simply navigate the association in the other direction.

```
{customersAccounts} = theCustomer -> R1
```

Note that now there is a set of objects returned by the navigation.

One of the most common errors committed by new users of ASL is to under-use the navigation features. Instead, new users often resort to unnecessary `find` operations. This results in clumsy ASL.

Before associations can be navigated they must be created. This is done using the `link` operation. To perform a link, instance handles for the objects to be linked together must have been obtained.

```
link theCustomer R1 theAccount
```

The `link` operation is often used just after a `create` operation, that is a new object is created and linked to its associated objects.

The `unlink` operation deletes the link between a pair of objects and takes the same form. The `unlink` operation must be performed before an object is deleted since the deletion of an object which is still linked to other objects is illegal.

```
unlink theCustomer R1 theAccount
```

There are refinements to these association manipulation primitives that allow association classes and reflexive associations to be covered. However, the principles remain the same.

10.3.4 Invoking operations

In Chapter 8 operations were introduced. The bodies of operations, that is their methods, are defined using ASL. Operations can also be invoked using ASL. The form of the invocation depends upon whether the operation is domain-based, class-based or object-based. Refer to Chapter 8 for the details of the different forms of invocation. Operations can have any number of input and output parameters. Consider an object-based operation, 'reportLocation', defined for the TARGET class that returns the location of the target in terms of range, bearing and elevation.

```
[range, bearing, elevation] = TGT1:reportLocation [] on theTarget
```

- `[range, bearing, elevation]` is the set of output parameters;
- `TGT` is the key letter for the class that provides the operation;
- `1` is the operation number within the class;
- `[]` defines the set of input parameters – in this case there are no input parameters;
- `on` is an ASL keyword indicating that the operation is object-based and will be performed on the following instance handle;
- `theTarget` is the instance handle referencing the object upon which the object-based operation will be performed.

ASL is unusual in allowing multiple explicitly declared output parameters in this way. ASL takes this form because it avoids the need to create unnecessary structured types just to return multiple values. Of course, in the mapping to implementation a structured type could be used.

10.3.5 Sending signals

Invoking operations provides a synchronous form of message sending. Sending signals provides an asynchronous form. The semantics of signals means that there is no return and so no return parameters are possible. Also signals must always be directed to a specific object.

```
generate D2:doorFullyClosed () to theDoor
```

- `generate` is a keyword used to indicate the sending of a signal;
- `D` is the key letter for the class in the same style as operation invocations;
- `2` is the signal number within the DOOR class;
- `doorFullyClosed` is the signal name;
- `()` contains the parameters to be sent with the signal, if any;
- `to` is a keyword denoting the object to which the signal is directed;
- `theDoor` is the instance handle referring to the object to which the signal is directed.

10.3.6 Other ASL features

There are numerous other features of ASL. It is not the purpose of this chapter to cover ASL in great detail but to provide a flavour of its key features and state its purpose. Refer to the ASL reference manual available on the CD-ROM or from the website (www.kc.com) and also take a look at the case study example in Chapter 14. The CD-ROM also contains further examples of ASL in use.

10.4 An example of ASL

In order to illustrate the use of ASL we take a very simple example class diagram from a banking domain, see Figure 10.5. In it we assume that an account is operated by one person only (so no joint accounts are allowed as yet) and that a bank customer must have at least one open account.

The following is the ASL for the `createAccount` operation of the ACCOUNT class.

```
# First create a new account with a unique id and set the opening
# balance to that passed in as a signal parameter
newAccount = create unique Account with balance = openingBalance

# Next we set the dateOpened attribute to today's date using
# the predefined ASL service current-date
today = current-date
newAccount.dateOpened = today
```

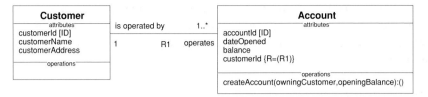

Fig. 10.5 A simple banking domain example

```
# Link the newly created account with the owning customer
# via association R1
link owningCustomer R1 newAccount
```

Note that comments are introduced by a hash (#) symbol. The first executable line of ASL creates a new object of the ACCOUNT class. `create unique` provides a value for the integer identifier of ACCOUNT (`accountId`) that is guaranteed to be unique. In the act of creating a new ACCOUNT object we also choose to set the `balance` to the value that was passed in as the `openingBalance` parameter. The next executable line uses `current-date`, which is a reserved word in ASL and returns the date in the local variable, which is typed, on first usage. Next comes an attribute assignment, which sets `dateOpened` for the new ACCOUNT to the value in the local variable `today`. It should be noted that ASL is strongly typed, so that a model compiler could check that the attribute and local variable types agree. We finally link the newly created ACCOUNT object to the object of CUSTOMER that was passed in to the operation as the parameter `owningCustomer`.

10.5 ASL and platform independence

This section offers the reasons why an implementation-independent action language is needed. The ASL is used, in this section, to provide concrete, real-world examples of how an action language is used to deliver a fully specified domain model.

The use of domain modelling with ASL has a dramatic effect on the way we think about software development. It gives a developer the ability to build and integrate genuinely reusable components that can be automatically mapped onto software units coded in any language, distributed across any target architecture.

Domain partitioning was covered in depth in Chapter 5, however, in order to explore how ASL supports domain modelling let us first revisit the concept of the domain.

10.5.1 Pollution control

Earlier, we observed that reuse is not about object-orientation; it is about avoiding pollution. Pollution control is the key to the success of the MDA process[1].

Pollution is a barrier to reuse.

---

[1] Reminder – by 'MDA process' here we mean the application of an MDA approach with the models expressed in xUML and translated into one or more target realizations.

If behaviour has been expressed in C, it is not readily accessible to somebody implementing a system using Java.

If an aspect of application behaviour also embeds knowledge of the technology, such as human or device interfaces, it will not be reusable in a system that deploys different technologies.

This is the price to pay for polluting one subject matter, like telecommunications, with a totally unrelated subject matter, like the C programming language.

Two strategies are needed to address the problem of pollution:
- **Domain partitioning** to keep separate subject matters apart;
- **Action language** to allow implementation-free (or platform-independent) specification of the behaviour within each domain.

10.5.1.1 Containing subject matter pollution

Talk to any member of the Green Party and they will tell you 'Pollution cannot be eliminated but it can be contained'.

Despite years of painful experience, developers continue to march down a road to sure and certain death. They take a perfectly unpolluted application class, which embodies only application level behaviour, and systematically pollute it with knowledge of:
- the user interface technology;
- the interface protocols;
- the target platform;
- the target language;
- the process partitioning and threading scheme;
- the database technology.

For example, take the simple AIRCRAFT class in our air traffic management system example with the operation `updateAircraftPosition`, as shown in Figure 10.6:

Let us assume that the air traffic control (ATC) domain in which this class resides provides two services that calculate a new latitude and a new longitude given current latitude and current longitude together with the direction and distance travelled. As such, the `updateAircraftPosition` operation might be defined as follows:

```
distanceTravelled = this.groundSpeed / this.updatePeriod

[newLongitude] = ATC1::calculateNewLongitude [ this.longitude, \
                                                this.trackBearing, \
                                                distanceTravelled ]

[newLatitude] = ATC2::calculateNewLatitude [ this.latitude, \
                                              this.trackBearing, \
                                              distanceTravelled ]
this.longitude = newLongitude
this.latitude = newLatitude
```

It is a PIM. It is delightfully implementation-free. It knows only about the subject matter of air traffic management. So it is reusable in any other air traffic management system regardless of the technology upon which it is based.

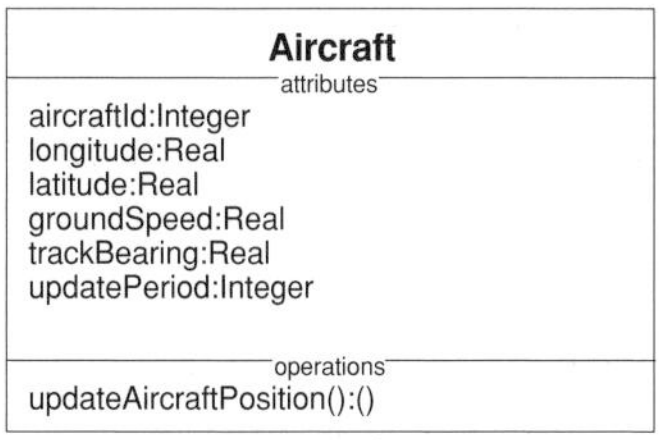

```
┌─────────────────────────────────────┐
│              Aircraft               │
│─────────────attributes──────────────│
│ aircraftId:Integer                  │
│ longitude:Real                      │
│ latitude:Real                       │
│ groundSpeed:Real                    │
│ trackBearing:Real                   │
│ updatePeriod:Integer                │
│─────────────operations──────────────│
│ updateAircraftPosition():()         │
└─────────────────────────────────────┘
```

Fig. 10.6 A simple AIRCRAFT class

But the road to code is long. We are about to enter the **design zone**. From now on, the class is doomed. It must endure a slow and painful process of contamination with subject matters that are totally unrelated to air traffic management, like user interface technology, messaging and database management – the result of which is illustrated in Figure 10.7:

By the time this methodical abuse is over with, we have a component that is completely schizophrenic. Is it a specification of the problem or a statement of the solution? Well, it is both, and they are now inseparable. The class is unfit for anything other than the system for which it has been specifically implemented. Reuse is not an option.

The superbly reusable analysis class will now fall inexorably into obsolescence, for it is likely that, from now on, any changes will be reflected in the code and nowhere else.

There is an evolutionary principle at work here:

> The more highly adapted an organism is to its current environment, the lower its chances of survival in a different one.

This principle applies to software too. So what is the message we want to emphasise? Do not adapt your PIMs to the target environment. Optimise your PIM to PSM mapping rules. Domain modelling with ASL makes this separation of concerns possible. The following sections show how.

10.5.1.2 Avoiding target language pollution

ASL assumes no particular target language and can therefore be mapped to any language. We have seen mappings from ASL to a diverse set of technologies, from low-level assembler, through C, Ada 83 and SQL to C++, Ada 95, Java and EJB's.

Use of an action language leaves a specification immune to changing fads and fancies of the software engineering world. Yesterday it was Structured Design and C or Ada 83; today it is OOD with C++ or EJBs. But what about tomorrow? What will be the 'next big thing'? If we continue to ignore historical precedent and specify our application level behaviour using the currently fashionable programming language, we must brace ourselves for the continued bleating of disillusioned managers. After all, they were told they would be able to reuse something one day but have yet to see it happen on any significant scale.

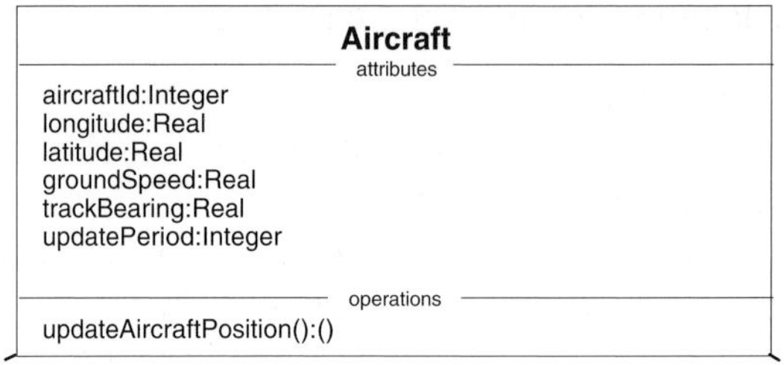

Aircraft

attributes
- aircraftId:Integer
- longitude:Real
- latitude:Real
- groundSpeed:Real
- trackBearing:Real
- updatePeriod:Integer

operations
- updateAircraftPosition():()

Aircraft:updateAircraftPosition()

```
distanceTravelled := this.groundspeed / this.updatePeriod;
this.longitude := ATC1::calculateNewLongitude(this.longitude, this.tackBearing, distanceTravelled);
this.latitude := ATC2::calculateNewLatitude(this.latitude, this.trackBearing, distanceTravelled);

screenXCoordinate := this.latitude * currentDisplayScale *...
screenYCoordinate : =  this.longitude * currentDisplayScale *...
GLOOM_updateIcon(iconId, screenXCoordinate, screenYCoordinate);

DOOM_saveState('COMPRESS', this.latitude, this.longitude);

if proximityViolation (...) then
   GLOOM_displayOKDialog("COLLISION WARNING", ..., Priorty > 3);
   GLOOM_makeIconFlash(...);
   Printf("COLLISION WARNING", ...);
 endif;

myStatusMsg[0] := 54;
myStatusMsg[2] := this.longitude;
myStatusMsg[4] := this.latitude;

POOP_sendMsg (myStatusMsg);
```

> I can be reused in any air traffic control system that:
> - Is based upon a user interface built using GLOOM;
> - Stores persistent data using the DOOM database;
> - Uses the POOP message protocol;
> - Produces a hardcopy incident log;
> - Is written in this language.

Terminology
- DOOM — Database for Object-Oriented Methods (Database)
- GLOOM — Graphical Language for Object-Oriented Methods (GUI Builder)
- POOP — Parallel Object-Oriented Protocol

Fig. 10.7 After the 'design phase'

10.5.2 How does an action language save development time?

Where do all those brain cycles go?

When a developer writes software in a traditional language such as C or C++, how does he or she expend intellectual energy? When specifying behaviour using traditional languages, such as C and C++, the developer must deal with these issues:

- How should I package my code headers and bodies into files?
- Should I use call by reference or call by value for this function?
- Should I make this function in-line?
- Is this method public, protected or private?
- Is this invocation crossing a process or processor boundary – do I need an RPC (remote procedure call)?
- Should I keep pointers for this relationship at one end or both ends?
- What data structure should I use to store objects of this class?

All of these concerns detract the developer from the fundamental developer issues of the application under development.

A key benefit of using an action language is that the poor developer is relieved of the burden of addressing these implementation-oriented concerns, leaving him or her to address only the problem in hand.

10.6 Use and benefits of an action language for the UML

This section looks at a whole, although simple, model and shows where ASL is used within the model. This provides a catalogue of the places in a UML model that can be more fully specified using ASL. At the same time that we are looking at these uses of ASL we can consider the benefits that ASL brings. This is done with reference to the document that specified the requirements for well-defined action semantics for the UML. The section also draws upon the Object Management Group's (OMG) *Request for Proposal on Action Semantics for the UML* document. It highlights the benefits cited in the OMG document and illustrates how they have been realised for the past eight years by users of the ASL. The text boxes in the following sections are passages from the OMG document (OMG, 2002) thus:

> **OMG problem statement**
>
> The UML is a rich and powerful language that can be used for problem conceptualization, software system specification, as well as implementation. The UML also covers a wide range of issues from use cases and scenarios, to state behaviour and operation declaration. However, the UML currently uses uninterpreted strings to capture much of the description of the behaviour of actions and operations. To provide for sharing of semantics of action and operation behaviour between UML modellers and UML tools, there needs to be a way to define this behaviour in a well-defined, interoperable form.

10.6.1 Scope of ASL in xUML models

The entire behaviour of a system can be captured using xUML with action language statements. The following example model illustrates all of the uses of ASL in an executable model. ASL is used to specify:
- Operations provided by classes and domains;
- Actions on state models;
- Interfaces between domains – bridges;
- Initial conditions for simulation and testing;
- External stimuli for simulation and testing.

The first three of these are specified in the executable model itself and will be covered in the following section. The last two of these form the model execution environment and are covered in Section 10.6.3.

The use of ASL in these various places within a model delivers the following benefit:

> **OMG stated benefit**
>
> Such action specifications can be used to build complete and precise models that specify problems at a higher level of abstraction than a programming language or a graphical programming system.

Model precision and level of implementation detail are two separate things. Platform-independent action specifications, in conjunction with the UML, can completely specify a computing problem without programming it.

10.6.2 The train management model

To illustrate how an action language works, an example system is presented in this section. The example is based upon the train management system (introduced in Section 7.5.2) in which we explore the train management application domain. To recap, the basic mission of this domain is to move a train that follows a route comprising a sequence of 'hops'. Each hop typically involves periods during which the train:

1. accelerates at a constant rate to reach a predefined speed at a specified distance along the hop;
2. maintains that speed until it has travelled a specified distance along the hop;
3. decelerates at a constant rate to stop at a specific distance along the hop.

This is illustrated in Figure 10.8:

The domain chart illustrated in Figure 10.9 separates the system into different **domains**, or subject matters.

A brief explanation of the responsibilities of each of the domains together with some example classes now follows.

Action specification

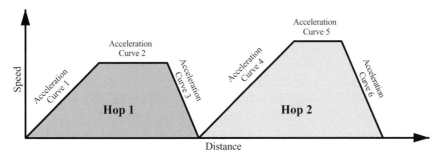

Fig. 10.8 Example profile for a journey through two stations

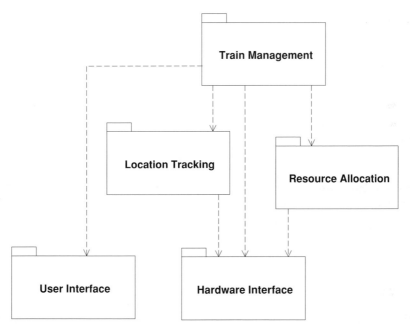

Fig. 10.9 Train management system domain chart

10.6.2.1 Train management

This domain is responsible for controlling the movement of trains along selected journeys in accordance with the defined speed profiles between each station on the journey.

Example classes are:
- TRAIN;
- JOURNEY;
- HOP;
- ACCELERATION CURVE.

The train management domain is now considered in more detail in order to illustrate how ASL is used to specify the behaviour within a domain – see Figure 10.10.

This domain is dependent upon a number of other domains in order to fulfil its mission:

- the location tracking domain to provide reliable positional information for each train that it is managing;
- the resource allocation domain for reserving sections of track for each train in order to ensure that no two trains occupy the same track section at the same time (this effectively provides a measure of collision avoidance);
- the user interface domain for providing a means of viewing the state of the rail network and for network administrators to interact with the system;
- the hardware interface domain for controlling the speed of the train by reading and writing to hardware devices that affect the speed of the train motor(s).

10.6.2.2 Location tracking

The location tracking domain is responsible for maintaining an up-to-date view of the location of a number of 'locatable items'. In the train management system this domain will be used to locate the position of each train in the network.

Example classes are:
- LOCATABLE ITEM;
- LOCATION.

This domain is dependent upon the services provided by the hardware interface domain in order to get raw position information from monitoring and reporting devices distributed around the rail network.

10.6.2.3 Resource allocation

The resource allocation domain is responsible for satisfying requests for resources by assessing their availability and then subsequently apportioning them in accordance with a number of allocation policies that might include request priority, maximum number of concurrent resources, requested duration, etc.

Example classes are:
- RESOURCE REQUEST;
- RESOURCE;
- RESOURCE SPECIFICATION;
- RESOURCE PRIORITY.

In our train management system this domain will be used to allocate the track section resources from the rail network such that no two trains are concurrently allocated the same track sections.

This domain is dependent upon the services provided by the hardware interface domain in order to check the availability of track sections using monitoring devices located around the rail network.

10.6.2.4 User interface

This domain provides a graphical user interface used to report system status and elicit commands and responses from users of the system. It employs colour, flashing and other techniques to draw attention to particular parts of the interface.

Example classes are:
- ICON;
- ICON TYPE;
- MENU;
- BUTTON.

In the train management system this domain provides a means for network administrators to view the state of the rail network and to interact with the system.

10.6.2.5 Hardware interface

This domain is responsible for providing services for interacting with a number of types of hardware device.

Example classes are:
- DEVICE;
- DIGITAL-TO-ANALOGUE CONVERTER (DAC);
- ANALOGUE-TO-DIGITAL CONVERTER (ADC);
- DIGITAL INPUT DEVICE;
- DIGITAL OUTPUT DEVICE.

In the train management system this domain is responsible for interfacing to hardware devices that control the speed of the train and report the status of monitoring devices distributed around the rail network such that the location of trains and the availability of track sections can be determined.

10.6.3 Specifying the actions in ASL

The class diagram in Figure 10.10 is for the train management domain. It specifies the classes with their attributes and operations and the associations between them. In the interest of brevity we have not shown the class diagrams for the other domains. Notice that some of the operations have their parameter signatures visible – these will be used later to help illustrate the specification of their respective methods using ASL.

Each journey is typically made up of a number of hops (the part of a journey between the stations at which the train will stop). The first hop in a journey is specified by the association R6. The sequence of hops on the journey is defined by the reflexive association R4.

Each hop is itself made up of a number of acceleration curves (typically three – a period of constant acceleration from rest to a cruising speed, followed by a period held at the cruising speed, followed by a period of constant deceleration which returns the train to rest at the next station). The first acceleration curve within a hop is specified by association R7 and the sequence of acceleration curves within a hop is specified by the reflexive association R5.

The ACCELERATIONCURVE class is an abstraction of a distance over which a continuous operation (accelerating, cruising, or decelerating) must be performed. Each object of this class defines:
- The start and end distance into the journey between which the operation must be performed;
- The speed at the start and the end;

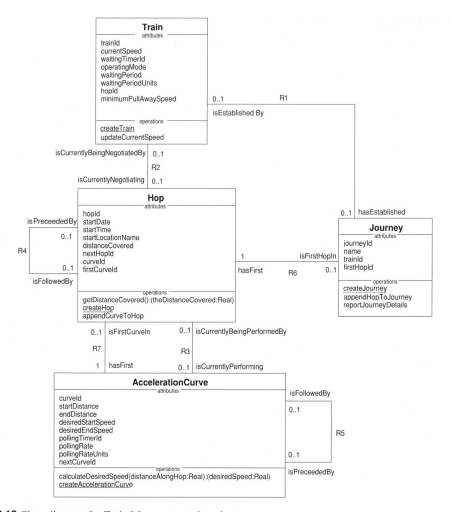

Fig. 10.10 Class diagram for Train Management domain

- The polling rate at which the actual speed is to be checked against the desired speed and adjusted accordingly (this must be high enough to give a smooth ride).

Once a train has been selected to make a journey and is in position at the first station on the route, association R1 is linked signifying that the journey has been established for that train.

Associations R2 and R3 identify which hop and acceleration curve that the respective train is currently executing.

On reaching a station on the journey, the train will wait at the platform until the next hop's start date and time (if they exist) before pulling away to negotiate that hop. Each train has a minimum speed that must be applied when pulling away from rest.

It can be assumed that the **distanceCovered** attribute of the HOP class is being kept up to date in real-time by the location tracking domain.

10.6.3.1 Specifying state actions using ASL

Some classes exhibit state-dependent behaviour, which is described using a state model. Figure 10.11 shows the statechart for the ACCELERATIONCURVE class – the entry actions for each state are specified using ASL.

Let's examine the action defined for the state `MaintainingAccelerationCurve` to illustrate how ASL can be used to specify state-dependent behaviour.

The desired speed of the train is dependent upon the distance covered along the current hop. Thus if you know this distance you can work out the speed. The current HOP object can be found by navigating association R3 from `this` instance of ACCELERATIONCURVE:

```
theCurrentHop = this -> R3
```

The `getDistanceCovered` operation can then be invoked on `theCurrentHop` to get the `distanceCovered`:

```
[ distanceCovered ] = H1:getDistanceCovered[] on theCurrentHop
```

If the `distanceCovered` is less than the `endDistance` of this ACCELERATIONCURVE then the curve is not yet complete and the train speed must be rechecked against the desired speed and adjusted accordingly:

```
if distanceCovered < this.endDistance then
  <ASL to adjust the speed of the train>
```

The desired speed is determined using the object-based operation `calculateDesiredSpeed` on this ACCELERATIONCURVE, which uses `distanceCovered` as an input parameter and returns `theDesiredSpeed`.

```
[theDesiredSpeed]=AC1:calculateDesiredSpeed[distanceCovered]on this
```

The speed of the train is adjusted to this `desiredSpeed` by invoking the `adjustSpeed` operation on `theTrain`, which is found by navigating association R2 from the `theCurrentHop`:

```
theTrain = theCurrentHop -> R2
[] = T2:adjustSpeed[theDesiredSpeed] on theTrain
```

The train speed must be adjusted at a rate that gives a smooth ride, so finally we set the polling timer to expire after a period defined by attributes `pollingRate` and `pollingRateUnits`. When this timer expires it will generate a signal AC2:TIMETOADJUSTSPEED directed at `this` instance of ACCELERATIONCURVE, such that a transition is made back into the `MaintainingAccelerationCurve` state and the action is re-executed:

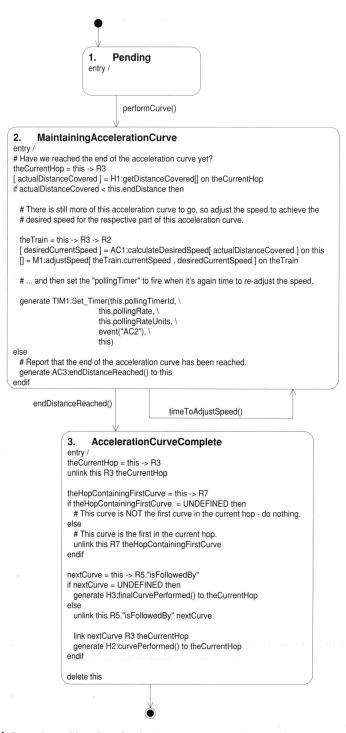

Fig. 10.11 State chart with actions for the ACCELERATIONCURVE class

Action specification

```
generate TIM1:Set_Timer(this.pollingTimerId, \
                       this.pollingRate, \
                       this.pollingRateUnits, \
                       event('AC2'), \
                       this)
```

If the `distanceCovered` was not less than the `endDistance` of this AccelerationCurve then the curve is deemed to have been completed, in which case a signal `AC3:endDistanceReached` is generated to `this` instance of AccelerationCurve – which causes a transition into the `AccelerationCurveComplete` state.

```
generate AC3:endDistanceReached() to this
```

You can probably work out most of the action defined in the state `AccelerationCurveComplete` yourself, but there are three points worth explaining:

- **Navigating reflexive associations** – When navigating reflexive associations like R5, you need to specify the **direction** of navigation explicitly. In ASL this is achieved by adding the role phrase as a suffix to the relationship specification R5 (where the role phrase is the one at the end of the association in the direction of the desired navigation). For example, the navigation...

```
nextCurve = this -> R5.'isFollowedBy'
```

...returns the next object of AccelerationCurve in the sequence of curves for the current Hop.

- **Testing the result of navigations** – Navigation of associations can result in no related object(s) and this can be determined using ASL. For example, when navigating the same association from the last object of `AccelerationCurve` in the sequence for the current Hop, the navigation will return an instance handle with a value of `UNDEFINED`. The required behaviour is different depending upon whether the `AccelerationCurve` is the last one or not...

```
if nextCurve = UNDEFINED then
  <ASL for behaviour when this curve is the last curve>
else
  <ASL for behaviour when this curve is not the last curve>
endif
```

...in this example the navigation will return `UNDEFINED` if there is no related object of AccelerationCurve in the specified direction of navigation.

- **Deletion of the 'current' object** – A state model is an abstraction of the state-dependent behaviour of each object of the respective class. The action defined for each state is thus defined in the context of a specified object which, in ASL, is referred to using `this`.

A 'final state' in xUML represents a state in which the respective object deletes itself at some point within its action. In ASL this is specified simply:

```
delete this
```

Typically this would be the last ASL statement within the action of a final state.

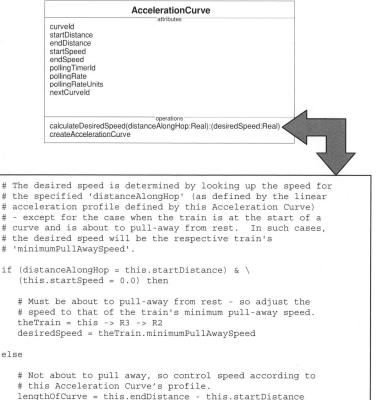

Fig. 10.12 Operation specification

10.6.3.2 Specifying operations using ASL

The state-independent operations for each class are also fully specified with ASL. The example in Figure 10.12 is an object-based operation, which therefore has access to the data item `this`, referring to the object executing the operation.

Note that the input and output parameters for the `calculateDesiredSpeed` operation have been made visible in the operations compartment for the ACCELERATION CURVE class.

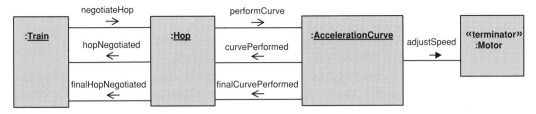

Fig. 10.13 Class collaboration model for the Train Management domain

Notice the use of the input parameter `distanceAlongHop` within the operation, and the assignment of the return parameter `desiredSpeed`.

Note also the use of ASL comments within this operation specification (all text after the # character is treated as a comment).

This example illustrates the use of ASL to specify the method for an object-based operation – and thus the respective object can be referred to using `this`. The methods for class-based operations and domain-based operations are also specified using ASL of course – however, `this` is meaningless in such a context given that such operations are not object-specific.

10.6.3.3 Inter-domain invocations

Each domain is analysed in a separate package. We wish to maintain a clean separation between the domains and thereby allow independent reuse. Therefore when a domain requires the use of a service provided by another domain, this is achieved using a `bridge operation` – it is this bridge that allows us to decouple the domains and maintain the separation of concerns. Bridges are covered in detail in Chapter 12.

For the moment, let us assume that the train is powered by an electric motor whose speed is proportional to the applied voltage, which is controlled by writing an integer value to a DAC.

The train speed is therefore adjusted by either increasing the integer value applied to the DAC (to speed it up) or decreasing it (to slow it down). The current motor voltage can be acquired by reading the value on an ADC.

You might recall from Chapter 7 that the TRAIN class delegates the responsibility for maintaining its desired speed to the ACCELERATIONCURVE – which can be observed on the CCM shown in Figure 10.13.

The ACCELERATIONCURVE adjusts the speed of the train by repeatedly invoking the `adjustSpeed` operation provided by the MOTOR terminator class. Let us also assume for the purposes of simplicity that the frequency of invocation is at a sufficiently high rate to get a smooth ride and, as such, that the integer value applied to the DAC only needs to incremented or decremented by one in order to accelerate or decelerate the train in accordance with the ACCELERATIONCURVE speed profile.

It is tempting to implement the ACCELERATIONCURVE class in such a way that it interacts directly with the DAC and ADC. However, if we build an ACCELERATIONCURVE class that 'knows':

- the train is powered by an electric motor;
- speed is directly proportional to applied voltage;
- there is a DAC to the control the speed of the electric motor, and;
- there is an ADC to monitor the currently applied voltage...

...then the class cannot be reused in a system where, for example, the motor is controlled by sending messages across an asynchronous communications network or indeed where the train is powered by a diesel engine.

If we had done it this way we would have effectively polluted the train management domain with knowledge of other subject matters – and such pollution is the primary barrier to reuse.

By partitioning systems up into domains and using bridge operations to map the required operation in one domain to provided operations in others, the problem of subject matter pollution can be contained.

If we revisit the state model for the ACCELERATIONCURVE shown in Figure 10.11 it is clear that the train's speed is controlled, or adjusted, from within state 2 (`Maintaining AccelerationCurve`), through the invocation of the bridge operation `M1:adjustSpeed` (where 'M' is the key letter of the MOTOR terminator class). This is graphically illustrated in Figure 10.14.

Firstly, note that the AccelerationCurve is unaware of what sort of motor powers the train – which of course is how it should be!

It also remains unpolluted with knowledge of how the speed adjustment is actually achieved – it delegates this responsibility to one or more other domains by invoking the bridge operation `M1:adjustSpeed` which (in this case) maps the requirement to adjust the train speed to applying an appropriate integer value to a specified DAC.

Each of the ASL statements within the bridge method are expressed in the context of the calling domain (the train management domain in the example), except for those contained between $USE/$ENDUSE directive pairs. These directives are used to change the domain context of the bridge from that of the calling domain to an alternative domain and back again.

For example, all the ASL statements prior to the first $USE HI directive in the `M1:adjustSpeed` bridge operation are executed within the context of the train management domain. The statements between the $USE HI and $ENDUSE directives are executed within the context of the hardware interface domain – we use a short form of the domain name (in this case HI) to identify which domain the execution context has changed to.

The hardware interface domain is responsible for providing operations that support reading and writing from hardware devices. Two such operations are the `ADC1:getADCValue` and `DAC1:setDACValue` operations, and this is why they are specified between the $USE HI and $ENDUSE directives.

The statements following the $ENDUSE directive are executed back in the context of the calling domain until either another $USE directive is encountered or the end of the bridge operation is reached.

Action specification

```
2.MaintainingAccelerationCurve
entry /
# Have we reached the end of the acceleration curve yet?
theCurrentHop = this-> R3
[ actualDistanceCovered ] = H1:getDistanceCovered[] on the CurrentHop
if actualDistanceCovered < this.endDistance then

    # There is still more of this acceleration curve to go, so adjust the speed to achieve the
    # desired speed for the respective part of this acceleration curve.

    theTrain = this-> R3-> R2
    [ desiredCurrentSpeed ] = AC1:calculateDesiredSpeed[ actualDistanceCovered ] on this
    [] = M1:adjustSpeed[ theTrain.currentSpeed, desiredCurrentSpeed ] on the Train

    # ...and then set the "pollingtimer" to fire when it's again time to re-adjust the speed.

    generate TIM1Set_Timer(this.pollingTimerId, \
                          this.pollingRate, \
                          this.pollingRateUnits, \
                          event("AC2")
                          this)
else
    # Report that the end of the acceleration curve has been reached.
    generate AC3:endDistanceReached() to this
endif
```

The invocation of the M1:adjustSpeed bridge operation results in the execution of this bridge method

```
voltageMonitoringADC = 3
voltageSettingDAC = 4

if (currentSpeed > desiredSpeed) then
    # Train needs to be slowed down - so decrement the voltage.
    $USE HI
    [motorVoltage] = ADC1:getADCValue[voltageMonitoringADC]
    newVoltage = motorVoltage - 1
    [] = DAC1:setDACValue[voltageSettingDAC , newVoltage]
    $ENDUSE
else
    if (currentSpeed < desiredSpeed) then
        # Train needs to be sped up - so increment the voltage.
        $USE HI
        [motorVoltage] = ADC1:getADCValue[voltageMonitoringADC]
        newVoltage = motorVoltage + 1
        [] = DAC1:setDACValue[voltageSettingDAC , newVoltage]
        $ENDUSE
    else
        # Train is travelling at desired speed - so do nothing!
    endif
endif
```

Fig. 10.14 Bridge method

Thus, the bridge operation M1:adjustSpeed encapsulates the pollution. OK, it is schizophrenic and therefore not reusable. But it is small. The pollution cannot be eliminated but it has been contained.

You might have noticed that the first two lines of ASL within the bridge method define the 'device' numbers for the ADC and DAC that monitor and set the voltage for the motor that controls the train. Accordingly you might argue that by doing this you have 'hard-wired' the relationships between the TRAIN in the train management domain and two counterpart DEVICES in the hardware interface domain – and that this is a bad thing. You would be right!

In xUML, of course, there is a better way of dealing with situations of this nature through the use of relationships that link classes in different domains – known as counterpart relationships. These are covered in more detail in Chapter 12.

10.6.3.4 Summary

The preceding UML model elements and action language segments provide a **complete agenda** that permits 100 per cent target code generation to any programming language onto any target environment.

10.6.4 Simulation and verification

> **OMG stated benefit**
>
> Such action specifications make possible high-fidelity model-based simulation and verification.

10.6.4.1 Use case specification

For many years now, users of ASL have been used to the idea that the analysis specifications should be fully exercised before any target code is generated. An executable model can be exercised to demonstrate that it realises each use case scenario. In order to exercise the model we need to be able to define the stimuli that will prompt the model into action and the initial conditions that apply to the use case scenario under test. We use ASL for both of these purposes: each stimulus is captured in a **test method** and each set of initial conditions is captured in an **initialisation segment**. This involves the following steps:

1. Use ASL to specify each use case in terms of one or more initialisation segments;
2. Use ASL to specify the external stimuli in test methods;
3. Execute the model either interactively for debugging or in batch mode for regression testing;
4. Compare the actual results with the expected results as defined by the sequence diagrams for each use case;
5. Iteratively refine the model until the expected results are produced for each use case;
6. Generate the target code.

Examples of an ASL initialisation segment and an ASL test method follow.

10.6.4.2 ASL initialisation segments

An ASL initialisation segment is used to capture the scenario upon which a use case is based. For example, the following initialisation segment specifies a special Christmas journey on

Action specification

the Jubilee Line of the London Underground for the year 2010. The train leaves Canary Wharf at 23:45 and makes one hop, with three acceleration curves, to North Greenwich.

```
# Create the 'Jubilee Xmas Special' journey
journeyNumber = 12345
journeyName = 'The Jubilee Xmas Special'
[theJubileeXmasSpecialJourney] = J1:createJourney [ journeyNumber, journeyName ]

# Create a train to make the 'Jubilee Xmas Special' journey.
idOfTrain = 65
waitingPeriod = 2
waitingPeriodUnits of TimeUnit = 'MINUTE'
[theJubileeSpecialTrain] = T1:createTrain[ idOfTrain, \
                                           waitingPeriod, \
                                           waitingPeriodUnits ]

# Create the single hop from Canary Wharf to North Greenwich for
  the 'Jubilee Xmas Special' journey
hopStartDate = 2010.12.24
hopStartTime = 23:45:00
[theCanaryToNorthGreewichHop]=J2:appendHopToJourney[hopStartDate,\
                    hopStartTime ] on theJubileeXmasSpecialJourney

# Create the three AccelerationCurves for this hop.
# Between 0 and 1000 metres, accelerate from 0 to 40 Km/h, with a
  polling rate of 500 milliseconds
[acceleration] = H3:appendCurveToHop[ 0.0, 1000.0, 0.0, 40.0, \
                 500, 'MILLISECOND' ] on theCanaryToNorthGreewichHop

# Between 1000 and 2500 metres, stay at a constant speed of
  40 Km/h, with a polling rate of 500 milliseconds
[cruise] = H3:appendCurveToHop[ 1000.0, 2500.0, 40.0, 40.0, \
                 500, 'MILLISECOND' ] on theCanaryToNorthGreewichHop

# Between 2500 and 4000 metres, decelerate from 40 to 0 Km/h, with
  a polling rate of 500 milliseconds
[deceleration] = H3:appendCurveToHop[ 2500.0, 4000.0, 40.0, 0.0, \
                 500, 'MILLISECOND' ] on theCanaryToNorthGreewichHop
```

10.6.4.3 ASL test methods

ASL test methods are used to stimulate the system and to formalise **extends** relationships. For example, the preceding use case might be **extended** to send an unsolicited doEmergencyStop signal to the TRAIN that is making the journey named 'The Jubilee

Xmas Special'. This can be specified as follows:

```
theJubileeXmasSpecialJourney = find-only Journey where \
                               name = 'The Jubilee Xmas Special'

theJubileeXmasSpecialTrain = theJubileeXmasSpecialJourney -> R1

if theJubileeXmasSpecialTrain != UNDEFINED then

    generate T8:doEmergencyStop() to theJubileeXmasSpecialTrain
else
    # Do nothing - there is no train established for this journey.
endif
```

Here, the analyst uses an ASL assertion, `find-only` to make it clear that there must be only one object of the JOURNEY class that meets that criterion. If zero or more than one objects are found at run time, an exception will be generated. The analyst can choose to specify an exception handler to deal with this.

10.7 Further benefits of an action language

There are further benefits derived from the use of a precise modelling formalism supported by an action language such as ASL. This section introduces some of these.

10.7.1 Reuse of domain models

> **OMG stated benefit**
>
> Such action specifications enable reuse of domain models, because each domain is completely specified, though not in a form embedded in code.

In development methods based upon domain models, the unit of reuse is the **Domain**. A domain does not correspond to a software unit because the choice of software units is affected by the degree of concurrency and distribution needed in a particular system.

Domain modelling is a process of acquiring expertise, capturing it in a form that is uncontaminated by other subject matters and making it accessible to others. It is fundamental that if a component is to be reusable it does not intertwine application-oriented behaviour with totally unrelated concerns, such as C++ structures, since it has proven historically impossible to untangle these to reuse them in a system based upon different technologies. This is why the xUML process is primarily based upon **reuse of expertise** expressed as domain models, rather than the more limited, less ambitious approaches based on reuse of implemented code.

In fact, representing behaviour using an implementation-oriented language such as C++ is just one way of compromising reuse potential. There is a more general principle at work here – that of keeping separate subject matters **separate**.

As we have already stated, a domain **represents a subject matter**. We deal with implementation language pollution by expressing required behaviour using the **implementation-free ASL**. We deal with subject matter pollution by containing the pollution within **bridge functions**. We deal with how to implement such domains and bridges by **separately** considering the subject matter of design. This is covered in more detail in Chapter 13.

10.7.2 Basis for design and coding

> **OMG stated benefit**
>
> Such action specifications provide a stronger basis for model design and eventual coding.

Use of configurable code generators allows generation of code to meet specific concerns, such as speed, space or reliability. Because the specification of behaviour makes no assumptions about class implementations and interactions, the code generator has not been robbed of its freedom to determine the most appropriate way to map the model onto the target environment.

10.7.2.1 Selection of data structures

Every software engineer knows that when first selecting a data structure for a class, the decision is based upon his or her current understanding of how the data are to be used. As more behaviour is specified, it often becomes clear that the selected data structure is no longer appropriate and that a significant amount of rework is needed. Use of an action language means that the choice of data structure can be left to a code generator, which will select the most suitable structure based upon the facts that it can derive from the model, such as:

- Is this a static class? That is, are there any `create` or `delete` statements pertaining to this class? If so, a predefined array might be most appropriate. Otherwise, some kind of dynamic structure – a list or a tree, would be better;
- Is this a read-only class? That is, are there any write accesses in the action language? If not, then the data can, if desired, be allocated to ROM.
- Is a relationship navigated in both directions? If so, and speed is more important than space, then references can be maintained at both ends of the relationship. Otherwise, a reference is only needed at the end from which navigation starts;
- Do objects need to be stored in sorted order? That is, does the action language contain

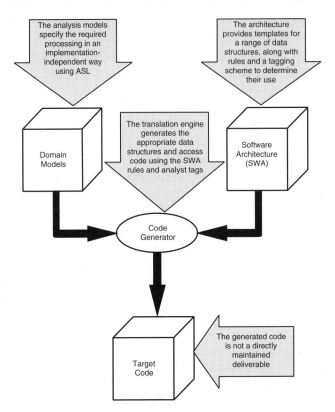

Fig. 10.15 Separation of concerns

any **ordered by** clauses? If so, then a sorted structure, such as a linked list or tree, might be more appropriate than an array.

Note that in normal circumstances, a change in the answer to any of these questions would require some data structure related code to be rewritten. By using an action language, we become immune to such changes. We just incorporate the changed behaviour and regenerate the code. The code generator will notice the new behaviour in the domain models and will generate new code with the more appropriate data structures.

This is possible because we have maintained a **clear separation** between the operations to be performed on data and the structure of the data (Figure 10.15).

The same benefits apply when moving from one technology to another, for example changing from memory-based storage to database storage.

10.7.3 System optimization

Many complex systems must be mapped to highly constrained environments. The developers' quality of life is seriously compromised if he or she is forced to deal with such concerns while producing a precise statement of required behaviour.

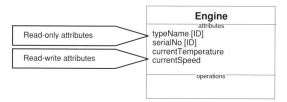

Fig. 10.16 A class specification

The deployment of an action language allows the developer to tackle the application requirements and the implementation constraints separately and independently. This avoids the risk that the system specification is somehow 'optimized' to address space concerns at the expense of performance. If the same specification must later be used to map onto a target with different characteristics, perhaps more onboard memory, but a smaller processor, then the optimization is the opposite of what is required. The specification may need to be extensively reworked, even though the application level behaviour has not changed. But life does not have to be that hard.

For example, an engine management system might have to be shoehorned into a target with 20kbyte of RAM and 32kbyte of ROM. The developer might identify classes like that shown in Figure 10.16 and can specify all processing in an implementation-free way using ASL.

The developer can also specify code generation rules that address the specific constraints of the target environment. Since this is an automotive system the headline constraints are:

- A level of safety criticality and reliability that requires **no dynamic memory management**. Therefore, the code generator will declare static arrays to hold all objects;
- Severely **limited amounts of RAM**. Therefore, the code generator will:
  - allocate all code and all static data to ROM;
  - hold relationships as bitmaps, with the position of the bit signifying the array index;
- A **very small processor**. Therefore, the code generator will evaluate as many expressions as possible at code generation time, which will make the system faster and smaller.

By considering the semantics of xUML itself we can see that in Figure 10.16 that the identifying attributes for the ENGINE class **must** be read-only attributes and so these can be deployed in ROM.

Note that code conforming to these rules would be very hard to write correctly and very hard to maintain. But we need no longer construct software whose performance and size is compromised by our desire to maintain it by hand. It is automatically generated and never changed by humans.

10.7.4 Code generation to multiple platforms

> **OMG stated benefit**
>
> Such action specifications support code generation to multiple software platforms.

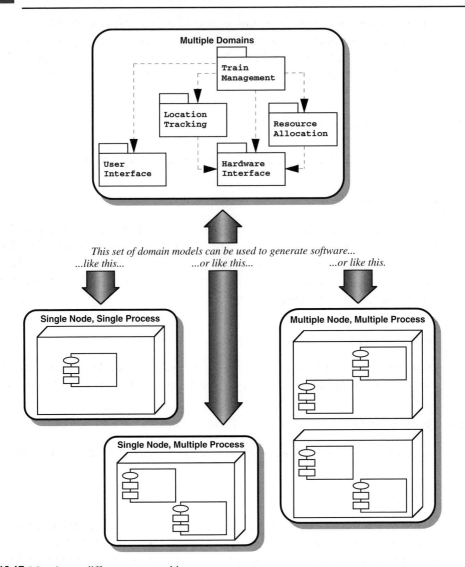

Fig. 10.17 Mapping to different target architectures

By keeping the specification of the application level behaviour totally uncontaminated by implementations concerns such as software and hardware boundaries, the domain models can be mapped to any software and hardware architecture.

So we can specify our system behaviour in terms of the domains making up the system. The domain structure is far more stable than the software structure and the domain models therefore have a much higher life expectancy. The domain models can then be used as the basis for generating code to a variety of target architectures, as illustrated in Figure 10.17.

Organizations are now achieving this by specifying their code generation and distribution rules within a domain model built with xUML. In other words, they are able to see **design as just another subject matter for analysis.** Details of the ways in which xUML is used to specify software architectures and code generators is covered in Chapter 13.

10.8 ASL Good practice guidelines

The following guidelines apply to the specification of state actions and operations.

10.8.1 Object creation

When an object is created all the identifying attributes must be assigned a valid value. Modellers must ensure that when specifying all the identifiers in an object creation that the identifying attribute values are unique. After an object creation consider if the newly created object needs to be linked to any other objects to comply with the multiplicity and conditionality defined in associations of the class diagram.

10.8.2 Write accessors

Modellers must not be allowed to write the value of any attribute that makes up part of any identifier for the object – since their assignment must only be done at creation time.

10.8.3 Object deletion

All of the associations to a particular object must be unlinked **prior** to its deletion.

10.8.4 Local variables

Modellers should ensure that the local variable names reflect the value being held.
In order to aid understanding when reading code it is important that modellers ensure that the local variables that they declare reflect the values that are being held. In the case of instance handles, use the convention 'the<qualifier><class name>', for example, `theNextTrain`.

10.8.5 Association navigation

Association navigation is the traversal of associations on the class diagram from one object, or a set of objects, in order to determine the related object or set of objects. The result of performing an association navigation is a single instance handle or a set of instance handles that are related to the object specified by the starting instance handle via the association specified.

In a completed model, check that all associations are navigated at least once. If they are not, why are they there?

Where an association is conditional in the direction you wish to navigate, that is it has a multiplicity of 0..1 or 0..*, then a navigation may not return any instance handles. Therefore it is wise to check before using the instance handles in subsequent ASL statements. For example:

```
thePatient = theBed -> R1
if thePatient != UNDEFINED then
  # A patient is using the bed
  ...
else
  # The bed is empty
  ...
endif
```

10.8.6 Finds

Modellers should ensure that the number of `finds` is kept to a minimum. Typically the modeller should only attempt to use a `find` on a class when absolutely necessary. This may be, for example, to obtain an object to be linked in an association.

Wherever possible `finds` should be performed on sets of instance handles that have been obtained via navigation rather than classes. This ensures that the `find` is across a subset of the objects rather than **all** the objects of the class.

When a `find` on a class is performed the system has to look at all objects of the class and evaluate the conditions specified on the `find`. With large object populations this can be very time and resource intensive.

The use of `finds` should be questioned as it may imply that an association is missing on the class diagram.

10.8.7 Instance handle verification

Modellers must ensure that instance handles are valid prior to attempting to use them.

Depending on the type of system being specified, the use of defensively written ASL should be considered. Therefore when an instance handle has been obtained via a conditional association navigation or a `find`, the modeller should ensure that the result of the operation provides a valid instance handle. An instance handle that does not refer to an object has the value UNDEFINED.

10.8.8 Typing

Modellers should ensure that all integer and real attribute values are constrained.

Modellers should avoid using the base types directly and should define their own 'user-defined types' based upon the existing types with the appropriate constraints for use within their domains. Similar arguments for the use of constrained types apply in xUML and ASL as they do in programming languages.

It is often the case that a domain corresponds to a type world. Therefore when we bridge operations between domains it is common to perform type mapping in the bridges.

10.9 Other action languages

A number of alternative action languages have emerged for xUML. These range from the rather too platform-specific C++ based variants to fully fledged platform-independent UML action languages. Here we simply list some of the possible alternatives to ASL – we do not make any attempt to cast judgement on them or to imply their relative merits, drawbacks, or maturity compared with ASL. So, in no particular order...
- **MoDAL** – a procedural action language incorporating both static and dynamic model specification developed by the UK's GCHQ (www.modeldrivenactionlanguage.org);
- **BridgePoint Object Action Language** – an action language developed by Project Technology Inc. (www.projtech.com);
- **SMALL** (Shlaer – Mellor Action Language, with one L being gratuitous) – a dataflow language developed prior to the emergence of the Precise Action Semantics for UML (Mellor and Balcer, 2002);
- **TALL** (That Action Language, with one L being gratuitous) – a functional language based on the Precise Action Semantics for UML (Mellor and Balcer, 2002).
- **Kabira Action Language** – an action language for the ObjectSwitch middle-tier server suite (www.kabira.com).

10.10 How to Build bad models

Although ASL is a well suited to specifying actions in models, there are plenty of opportunities to use it badly and build low-grade models. Here are just a few:
- Litter your ASL with constants. It is certain that they will change at regular intervals and this will allow you to soak up lots of time changing them and rebuilding the system. Capturing constants in specification classes is much more preferable!
- Use finds everywhere you can fit them in;
- Eliminate comments;
- Forget to validate instance handles that have been obtained through `finds` or navigation. After all, you might be lucky since there is a chance they will be valid;
- Don't bother linking related objects together through associations – related objects can often obtained by using suitably tortuous `finds`;
- Manipulate and navigate associations using referential attributes, rather than association primitives like `->` and `link`. This will have the dual effect of making much of the ASL almost impenetrable and hopelessly slow. It also means that the code generator cannot easily distinguish association manipulations from attribute manipulations;

- Use meaningless variable names. This is a tried and trusted technique, which many developers have honed to perfection over many years. Variable names like i and x ensure that your models provide a long-term source of mystery and intrigue to those who are exposed to them.

10.11 Summary

ASL is a small and simple language used to specify behaviour within the context of an xUML model. It complies with the OMG's UML Action Semantics, a part of the core UML standard. Most users learn ASL in a very short time: basic competence comes in a few days.

Models can be produced without using ASL but this chapter has highlighted the benefits that come from 'completing' the model with ASL. It is the ASL that makes the executable models do useful things, which in turn allows the required behaviour to be verified and then opens the door to code generation.

11 Modelling patterns

11.1 Introduction

Patterns are a powerful and intuitive concept. They are at the heart of, what is loosely termed in the engineering profession, 'experience'. If we start to reason about what lies at the core of the notion of experience, we find that it embodies the solutions for general classes of problem that an individual has repeatedly encountered. The study of patterns is an attempt to formalize and record such general solutions so that others can utilize these directly in their own work. This chapter explores some of the most useful patterns encountered by the authors in a wide variety of types of system development using the model-driven architecture (MDA).

There has been an ever-increasing interest in patterns in software engineering, stemming from the seminal work of the architect Christopher Alexander (Alexander, 1964, Alexander *et al.*, 1975, Alexander, Ishikawa and Silverstein, 1977). The popularity of patterns was secured after the success of the so-called 'Gang of Four' or 'GoF' book (Gamma *et al.*, 1994). The patterns community has found difficulty in agreeing on a single definition for 'pattern', so we will go back to Alexander who writes about patterns in the following words (1977):

Each pattern describes a problem which occurs over and over again in our environment, and then describes the core of the solution to that problem, in such a way that you can use this solution a million times over, without ever doing it the same way twice.

Alexander's interest, when he wrote this, was architecture for buildings and town planning, but the definition works equally well in object-oriented software. Our interest in this chapter is in analysis patterns in xUML (platform-independent patterns if you will). Design patterns, as espoused by the GoF book, are useful in our armoury for designing the software architecture, as discussed in Chapter 13.

Using Alexander's definition, we require an abstraction, that is a model, of an answer to a recurring problem, within the context of xUML analysis. In other words, our attention turns to frequently encountered problems within the business models that we produce and how we can generalize the solution to these. Such analysis patterns occur at varying levels of abstraction. Mowbray (1997) has produced a software design level model, given in Figure 11.1, which is equally applicable to analysis. The lowest level, objects and classes, comprises what are often called idioms, describing patterns applicable to one, or a small number of, xUML modelling elements. In the next level, microarchitectures, a few model

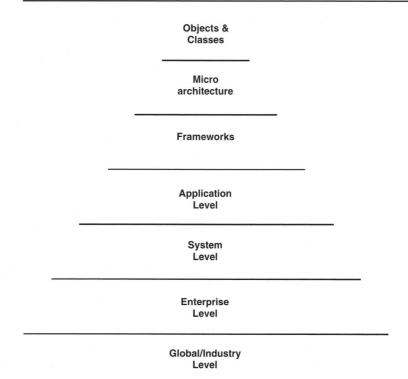

Fig. 11.1 A taxonomy of pattern levels

elements are combined that allow a small, partial, problem to be addressed. Frameworks are larger patterns that describe an approach to producing an entire domain, or part domain, for a particular class of problem, possibly combining microarchitectures to achieve this goal. Next, in the application level, domains and bridges are combined in ways that allow a single application to be built. Enterprise takes individual applications and combines them through the use of one or more enterprise level domains to deliver an enterprise-wide system. Finally, the global level addresses interoperability across enterprises to deliver global (multi-organizational) capabilities. This taxonomy is useful in helping us to reason about patterns at varying levels in the models we build or review, but it is not vital that we accurately ascribe a level to every pattern we encounter or document.

There are a number of languages and templates that are used to document patterns, a good starting point to explore these and other pattern-related literature is to visit the Hillside patterns' group website at http://hillside.net/patterns/. In this chapter we use a format adapted from the one employed in the GoF book. The subsection 'Motivation' for each pattern gives background and, if appropriate, an example of the problem and instantiated solution. In some of those subsections we present an incorrect, inappropriate, or naive approach that will subsequently be refined into the basis of the pattern. To give the reader a visual clue that this is not a valid way of modelling the problem, we use the symbol ☒. The subsection 'Structure' gives a generalized view of the pattern, where it is intended that the important UML elements are brought out and the user of the pattern will need to substitute appropriate elements such as attributes, association role phrases, association multiplicities, etc. to instantiate the pattern for their own problem. The subsection 'Applicability' describes where the pattern is useful and what major drawbacks it may have. Any significant variations are also described here.

The patterns that we describe are some of the most commonly encountered in xUML modelling. We do not have the space in this book to even attempt an exhaustive list – this would be futile anyway, since the catalogue of useful analysis patterns grows continually as we are exposed to new domains and consider new ways of exploiting xUML.

Patterns occur throughout all areas of xUML models and at all levels in Figure 11.1, ranging from those addressing fine-grained ASL constructs and how to manipulate single instances (Object and Classes level) to small assemblages of generic classes (Micro-architectures), reusable domains such as logging (Framework), and thence on to inter-domain patterns, delivering System and Enterprise level patterns. We shall concentrate on lower level patterns, ranging from the Object and Class level to Frameworks.

Patterns also come in semantic and presentational guises. A semantic pattern is one that embodies the meaning of a solution, whilst a presentational one prescribes a layout or notational convention usually in order to enhance or support a related semantic pattern. In general we shall focus on semantic patterns, which we define as patterns in the meaning of xUML models, however this class of pattern can often be enhanced by considering presentational patterns (patterns in the layout or appearance of the xUML model).

In some cases we present the pattern by showing an example of the problem and then its solution and finally generalizing that as a pattern. In other cases, where a readily understandable general form is apparent, we simply introduce the generalized form of the pattern and allow the reader to envisage where it may be useful in their own modelling.

Each of the following sections describes a single pattern, although we shall see that some patterns rely on the use of lower level patterns.

11.2 Specification

11.2.1 Intent

Factor out unchanging facts from their occurrences in actual instances.

11.2.2 Also known as

Type and actual.

11.2.3 Motivation

Many models become unnecessarily obscure because the modeller fails to properly distinguish 'types of thing' from 'actual things'. Use this pattern to separate these ideas. For example, in a command and control domain we may have Figure 11.2 as an initial attempt at the class diagram:

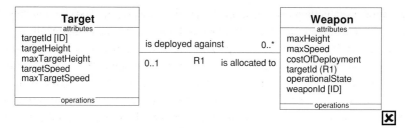

Fig. 11.2 Command and control model without specification classes

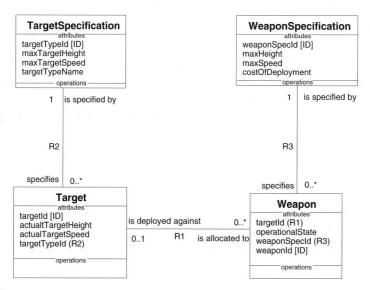

Fig. 11.3 Command and control model with specification classes

The WEAPON class is a model of an actual weapons system that is currently on the fighting platform or is deployed against a target. Once the weapon is spent then we delete the instance. Whilst the model in Figure 11.2 captures the idea that a target may be allocated to a weapons system, it exhibits some undesirable properties, since we have chosen to mix actual instances of weapon systems and targets with their specifications. If a commander wants to assess the suitability and cost of deployment for a particular weapon, he has to trawl through all the instances of WEAPON and then find one that is in a suitable state to deploy. Once he has fired all the weapons of a certain type then he looses the knowledge of the specification of that weapon type from the system. Update to the specification for a particular type of weapon involves updating all the redundant copies of the data held in each of the WEAPON instances. Things are even worse with targets, since he only has instances of a TARGET available when they have been detected and characterized as hostile, so there is no way to store information about and consider known types of hostile target without actually having one of them in the area.

The answer to these drawbacks is to factor out the specifications of the TARGET and WEAPON from the actual instances of these. This gives us the model in Figure 11.3.

This model employing specification classes improves on the first in a number of ways. There is no redundancy since we have factored out the specification facts from the actual instances that realize those facts. The operations and any life cycle of the specification classes are decoupled from the actual classes that realize them; in general such specification classes will display little or no behaviour (although we shall see an exception to this in the Assigner pattern, Section 11.11) and have static instance populations or will only be populated via system maintenance operations. We may also change the value of specification attributes over a period of time. This will lead to all actual instances of that type being updated rather than having to trawl through each actual instance to find the ones that require updating. The specification classes also allow us to retain knowledge of types of things even though we may not have any instance of that type in the system at present; for example, we know about target types without having any actual targets of that type. This is the reason that the multiplicity of the association from the *specification* class to the *actual* class is 0..*. Finally, the specification classes gives the analyst an agenda for thinking about other abstractions that may be important in the domain. In this case an association between the two specification classes would allow us to support queries about the effectiveness of a type of weapon against a target type.

11.2.4 Structure

Figure 11.4 gives the general form of the pattern. The role phrases on the association R3 are typical for this pattern. The actual class and specification class may have any number of attributes and operations as appropriate.

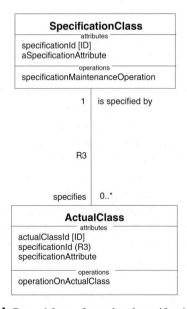

Fig. 11.4 General form of actual and specification classes

11.2.5 Applicability

This pattern will occur in many domains. Indeed, some organizations base their model layout on the principle that the classes on the left of a class diagram are the specification classes and those on the right are the actual classes. This is an example of a presentational pattern enhancing the efficacy of a semantic pattern.

11.3 Characteristic value

11.3.1 Intent

The characteristics of things are commonly known and fixed allowing them to be captured using the attributes in a class. This pattern is applicable when the number or type of the characteristics is unknown or subject to change.

11.3.2 Also known as

Colin's pattern (after the author who first characterized this pattern).

11.3.3 Motivation

Let us examine a model of a car manufacturer, which models the optional extras available on each of the types of car it produces. Each optional extra, called a Feature (such as electric windows, rain-sensing windscreen wipers, climate control or air-conditioning) has, in this case, a Boolean value that describes it on a particular model type of car. This gives us a model as in Figure 11.5. The model captures what features are offered on a particular type

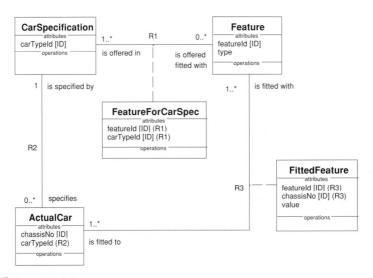

Fig. 11.5 A model of features on a car type

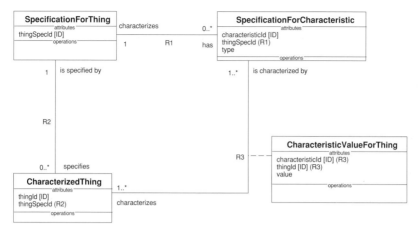

Fig. 11.6 The characteristic value pattern

of car and tells us what value for a particular feature was fitted to an actual car (for instance one actual car of type X may be built with electric windows, rain-sensing windscreen wipers and climate control, whilst another of the same type may be built with manual windows, manual wipers and no air-conditioning).

The model allows us to add new features to an existing car specification simply by creating instances of FEATURE (or reusing existing FEATURE instances) and linking them to the appropriate car type via the association R1.

An instance of the association class FITTEDFEATURE tells us the value of the feature as fitted to a particular car. The value type is given by the attribute type in the FEATURE class (in this case all features have the type Boolean) so we are able to say 'true or false' for all the features mandated by the association R1 for a CARSPECIFICATION.

11.3.4 Structure

It is simple to generalize the car model to give rise to a pattern that makes use of the specification pattern, as in Figure 11.6. An instance of the actual class in the specification pattern, in this case CHARACTERIZEDTHING, has its characterizing values held in the attribute **value** in the association class on R3. The structure is built upon the use of the Specification pattern (Section 11.2).

11.3.5 Applicability

This pattern is useful in a variety of situations but it is notable in models that capture meta-information about other models, such as the logging domain, discussed in Section 11.16.

One way of spotting where this pattern should be used is where a generalized class has a list of specialized classes under it, where the only specializations are in its attributes. This pattern is inappropriate if the specializations include different behaviour captured in state models or operations.

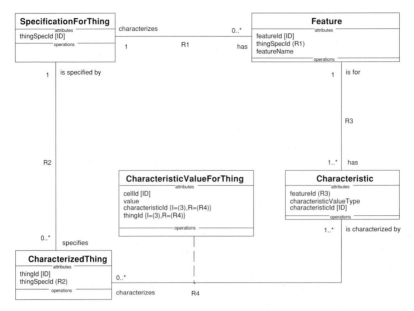

Fig. 11.7 Extended characteristic value pattern

A modification to this pattern that is sometimes encountered is that the association class is specialized for each of the types of value that have to be stored. This results in typical specialization classes to hold integers, reals, dates, etc.

A further refinement of the general pattern is to describe a feature for a type of thing to be characterized, which actually comprises many characteristics. This leads to the extended model in Figure 11.7.

11.4 Association timeframe

11.4.1 Intent

To precisely define the applicability of the timescale for an association.

11.4.2 Also known as

Association role phrase tense.

11.4.3 Motivation

Nebulous association names can present a serious obstacle to the readability of a model. In particular, failure to make explicit the period covered by an association tends to result in association names that simply 'state the obvious'. As an example consider Figure 11.8.

The model reveals very little about the requirement. Questions arise as to the meaning of the association R1, is it an abstraction of:

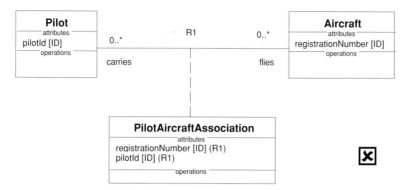

Fig. 11.8 Vague associations with temporal ambiguity

- The aircraft on which a pilot is scheduled to fly?
- The aircraft on which a pilot is currently flying?
- The aircraft on which a pilot has previously flown?
- The aircraft a pilot has travelled on as a passenger, as a captain or as a co-pilot?
- Some combination of the above? None of the above?

The role phrases of the association do not allow the reader to comprehend what the modeller intends and, in particular, over what timescale the association holds. By examining the description of the association and the association class, the reader could probably infer what is intended but a purpose of an analysis model is to enhance the readers' understanding, not their powers of deductive reasoning. We also note that the name of the association class is woefully inadequate in this example, conveying no additional meaning to its association.

It is therefore possible that several associations are needed, each of which has different characteristics and different rules; consider Figure 11.9.

This scheme brings out the critical issue, that we are only interested in the captains, and allows formalization of the different multiplicities and attributes for the associations that represent past, present and future associations.

11.4.4 Structure

It is important that the role phrases of associations convey:
1. Their precise meaning;
2. The period over which they apply.

These guidelines are equally applicable to naming association classes.

In general we must use precise language in association role phrases, paying particular attention to the lifespan of the association and using the appropriate tense in the role phrase to convey the association's timeframe.

11.4.5 Applicability

This pattern really lies at the lowest level of the taxonomy of patterns and is applicable to all associations wherever they are encountered.

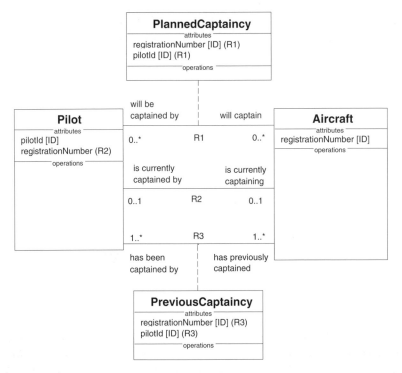

Fig. 11.9 Precise associations with no temporal ambiguity

11.5 Multivalued associative

11.5.1 Intent

To allow multiple occurrences of the same association link to be modelled.

11.5.2 Also known as

Many-to-many-to-many association.

11.5.3 Motivation

This pattern is closely related to the association timeframe pattern discussed in Section 11.4. We start by considering Figure 11.9, and in particular the history of the captaincy of each aircraft. The model, as it stands, simply records that a pilot has once captained a certain aeroplane in the past, but what if we wanted to record the time and date of *each* captaincy (as would be required in a logbook, for instance)? Our first attempt at this model might look like Figure 11.10.

Modelling patterns

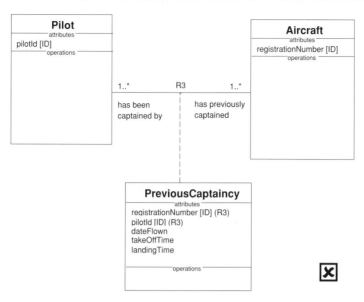

Fig. 11.10 An inadequate historic association

There is a problem with this model. This arises when we consider the same pilot captaining the same aircraft on different occasions. What values do we place in the attribute **dateFlown** in the association class for the second and subsequent flights in the same aeroplane? Is it the last flight date or the first? What may be intended is that we want a complete history of all flights (as in a logbook). We might be tempted to say that we simply link the same pilot with the same aircraft instance multiple times, giving rise to multiple instances of the association class, one for each link. This is erroneous, however; the UML states that the same two instances may not be linked more than once via the same association (except in reflexive associations, where we may link them once in either direction). We require a pattern here to model a many-to-many-to-many association (a primitive that is not supported in the current UML). The answer is to provide a class that is an abstraction of an occurrence of the link; thus our captaincy model becomes as in Figure 11.11.

The identifier for the CAPTAINCYOCCURRENCE class is worthy of further note. It comprises the identifier for the association class (which itself comprises the identifiers of the two participating classes in the association R3) and sufficient temporal attributes to ensure that each occurrence is characterized uniquely by the set of identifying attributes. In this case this relies upon the fact that captains cannot overlap flights in the same aircraft!

11.5.4 Structure

If we generalize this pattern, we get the model presented in Figure 11.12.

The extended identifier of the association link occurrence class does not have to be based upon time, although this is usual, so long as we ensure the identifier is guaranteed to be unique.

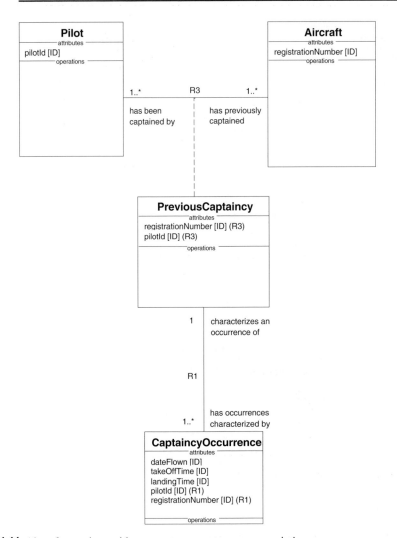

Fig. 11.11 Aircraft captaincy with a many-to-many-to-many association

11.5.5 Applicability

This pattern is suitable for any occasion where we require the details of multiple occurrences of a link between two instances, although it most often arises in the modelling of an historic relationship, as shown in the example.

11.6 Compatibility

11.6.1 Intent

To model the allowable combinations of two sets of class instances in an extensible fashion.

261 Modelling patterns

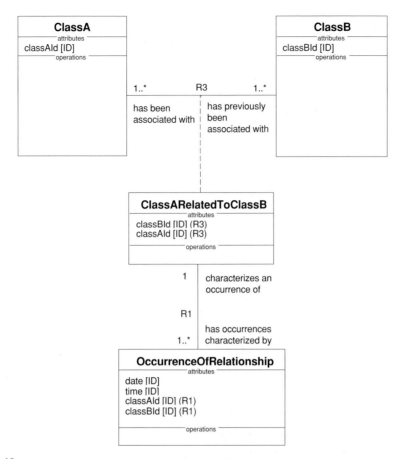

Fig. 11.12 Generalized many-to-many-to-many association

11.6.2 Also known as

Feasible combinations or legal combinations.

11.6.3 Motivation

In many domains there will be a need to capture rules regarding legal combinations. Table 11.1 shows a typical example from a command and control domain.

In the example, there are a number of different environments, such as Air, Surface and Subsurface, in which targets may reside. Note that some types of target can move from one environment to another. For example, a submarine can move between the surface and subsurface environment. The environment in which a target currently resides constrains the types of weapon that can be deployed against that target.

It is often tempting to formalize the different types of things using subclasses and then capture the legal combinations using associations. This would result in a model as in Figure 11.13.

Table 11.1 Example legal combinations in a Command and Control domain

Type of Vehicle	Effective Weapon Type
Air Vehicle	Gun, Sea Wolf SAM
Surface Vehicle	Gun, Sea Wolf SAM, Exocet SSM, Torpedo
Subsurface Vehicle	Torpedo

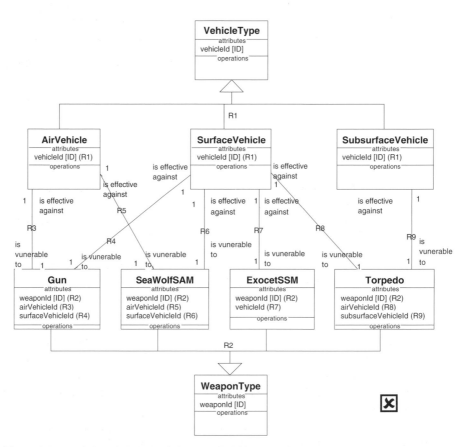

Fig. 11.13 Explicit use of classes and associations to model legal combinations

This model has many shortcomings, although it displays the dubious advantage that all the different types of things and all the combinations between them are explicitly formalized as classes and associations. However, when new types of thing appear (new associations or new types), the analysis model has to be changed to reflect the new things as subclasses and their legal combinations as associations. From a maintenance point of view, this is clearly undesirable. It would be preferable to use a scheme that does not require iteration back to

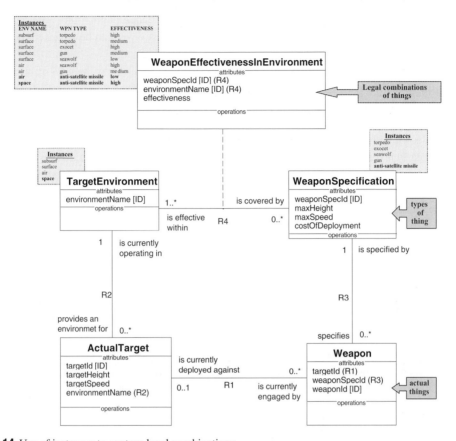

Fig. 11.14 Use of instances to capture legal combinations

the analysis model when such new requirements materialize. The lack of aesthetic appeal concerning the layout and repeated association role phrases (a presentational anti-pattern) should also give a clue that this model is, at the very least, suboptimal.

This can be achieved by adopting an extension to the specification pattern (Section 11.2), as shown in Figure 11.14.

In this scheme, the things that were previously abstracted as subclasses of VEHICLE-TYPE and WEAPONTYPE have been abstracted as **instances** of TARGETENVIRONMENT and WEAPONSPECIFICATION. The associations between types of weapons and environment, formerly modelled by individual associations, are now captured as **instances** of an association class WEAPONEFFECTIVENESSINENVIRONMENT.

This technique has the advantages that new environments or weapon types can be accommodated with no change to the analysis model. They are just added as new instances, so capturing new combinations. The addition of both specification instances and links between them can be achieved at run time, if desired, through maintenance operations on the specification classes.

Figure 11.14 has annotations indicating typical instance populations. The instances in bold show how easily new environments and weapons types can be added to the scheme, purely by changing the data.

This level of maintainability (known as 'run-time binding') is the goal of many xUML models. Often, though, problems like this are solved using a generalization – specialization technique similar to the one first presented. Unfortunately, such techniques deliver inextensible analysis, since the addition of new types of thing requires the explicit addition of classes and associations. Once the model has been translated into the target system, this clearly is hard to achieve whilst the system is executing and it is almost always necessary to return to the analysis models, add the appropriate classes and associations and reimplement.

One minor drawback of this pattern is that we must consider instance populations to fully understand the model; in some sense the model is less clear than the first explicit approach. If the problem we are modelling turns out to yield a small and completely stable hierarchy, then the explicit approach may be desirable, since we shall never have to add specification classes and associations.

11.6.4 Structure

In the general form of the pattern we encounter any multiplicity and conditionality for the association R2 between the specification classes. The association class is most often present, although is not always necessary, since it holds attributes particular to the combination between two specification classes involved in the association (Figure 11.15).

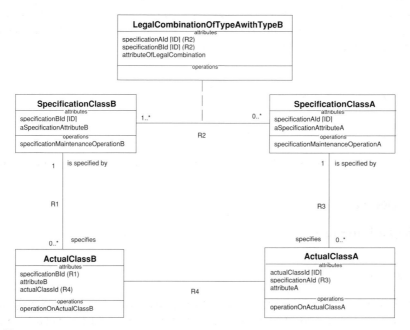

Fig. 11.15 General form of legal combinations with specification classes

The association R4 is the resultant combination of actual classes, formed by consideration of the legal combinations of their specifications modelled by R2. The actual role phrases and multiplicity of R4 are determined by the particular application of this pattern.

11.6.5 Applicability

There are many occasions when an analyst has to make a choice between the use of specialization and the use of specification. As we saw in the Vehicle and Weapons example, it is possible to represent a given problem using either technique. The use of specification classes produces a more general-purpose model. It is extensible simply by creating new specification instances and no extra classes are necessary. However, such a generalized model is harder to read and understand. It is more meta in nature.

Consider the following when assessing the suitability of this pattern. If the different types of things to be modelled, that is different types of weapons or vehicle in the example, are unknown or unstable then choose the specification technique. If the different types of things have specific behaviour that needs to be modelled, for example the difference between the behaviour of a gun and a torpedo matters, then use the specialization approach. Modelling situations are not always clear cut. If you choose the specification technique, then you can help your reader understand the model by adding example instances, as in Figure 11.14.

It is important to understand the difference between superclasses and specification classes: a specification class captures attribute **values** common to a set of instances, for example the maximum altitude for each type of aircraft; whilst a superclass captures attributes common to a set of instances, for example the current altitude for each aircraft, where the values of the attribute in the superclass vary with each instance.

11.7 Multiple classification

11.7.1 Intent

To model independent properties of a class as independent specialization hierarchies.

11.7.2 Also known as

Orthogonal properties.

11.7.3 Motivation

The example in Figure 11.16 is a model of aeroplanes. These are heavier or lighter than air and may be powered or unpowered; all combinations are permissible (from left to right on the subtypes, these represent powered aircraft (such as Concordes), gliders, airships and

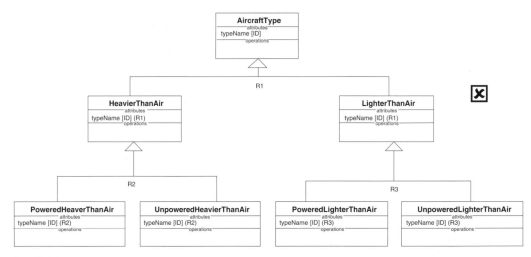

Fig. 11.16 Orthogonal properties modelled with a single hierarchy

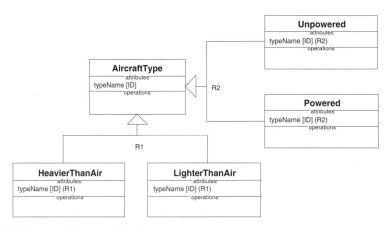

Fig. 11.17 Use of multiple classification

balloons). The problem that is being modelled reflects the differing behaviour of aircraft that can be heavier or lighter than air and also powered or unpowered. The model fails to separate the abstraction of whether or not the vehicle has a propulsion unit from its relative density, therefore we have to repeat the model of behaviour for, say, powered aircraft in both the POWEREDHEAVIERTHANAIR and the POWEREDLIGHTERTHANAIR subtypes.

An alternative to this is to use the multiple-classification pattern shown in Figure 11.17.

This model states that an Aircraft is either heavier or lighter than air and, **at the same time**, is powered or unpowered. This allows us to model the behaviour of each specialization in each distinct specialization hierarchy, independent from those in a separate hierarchy and so avoiding redundancy in the model.

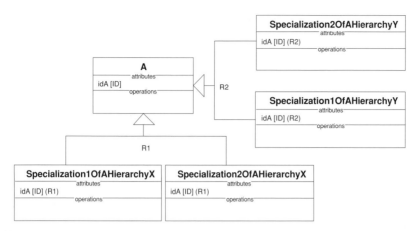

Fig. 11.18 General form of multiple classification

11.7.4 Structure

The general form of the model is given in Figure 11.18. It should be noted that we may have as many separate specialization hierarchies as necessary. It is common in this pattern for the separate subtypes to have their own state models, modelling the behaviour of the general class A, from the viewpoint of a particular specialization.

11.7.5 Applicability

Use this pattern where the behaviour and properties of an entity to be modelled are independent of another set. The pattern may be (rarely) extended to three or more separate hierarchies. Differing behaviours of the specializations can be modelled in state models and this can be combined with polymorphic signal events (see Chapter 9) to provide a flexible and sophisticated model. Differing behaviour can also be modelled by deferring the implementation of supertype operations down one of the generalizations (R1 or R2 in this case, see Chapter 8).

The modeller is, of course, not limited to two specializations in each hierarchy. In fact, it is cases where there are many specialization classes that this pattern proves its utility, since the number of classes required is considerably less than trying to express the problem with a single hierarchy. For example, if one hierarchy had four specialization classes and the other had five, then the multiple classification pattern would require ten classes in total whereas the single hierarchy model would require 25 as a minimum.

This pattern is one of the ways that the construct of generalization – specialization in UML is shown to be more general and powerful than the programming language construct of inheritance. We are unable to show such an elegant model if we restrict ourselves to the inheritance view, since inheritance blends both supertype and subtype in one entity, not allowing any additional branches of the hierarchy to be added. This does not mean, of course, that this construct cannot be implemented in the mapping to a realization. Such a mapping inheritance may optionally be used, but it is not sufficient; in fact it is possible to implement generalization – specialization in languages that do not support inheritance (see Chapter 13).

11.8 Dynamic classification

11.8.1 Intent

To use specialization classes to model significant and distinct states for an entity where especially different or critical associations, and/or attributes, apply to distinct states.

11.8.2 Also known as

Subtype-migration.

11.8.3 Motivation

We often encounter modelling problems where the subject matter under investigation leads us to model classes that have distinct modes of behaviour or, under some circumstances, exhibit additional attributes or associations. As an example of this, let us consider a model of an Aircraft from the viewpoint of an airport-scheduling domain. An aircraft that is in-service is always allocated to exactly one future route and an aircraft that is out-of-service, for repair, has exactly one technical repair request. An initial model might look like Figure 11.19. This model is imprecise in that the reader, without resorting to reading the association

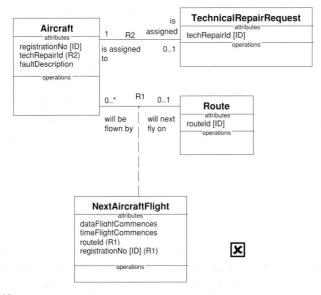

Fig. 11.19 Airport scheduling without dynamic classification

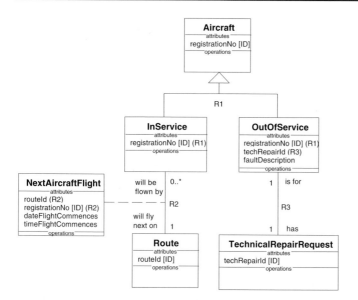

Fig. 11.20 Airport scheduling with dynamic classification

descriptions, cannot tell that an aircraft that is in-service cannot have a repair request and must have a next flight and, perhaps more importantly, an aircraft that is out-of-service must not have a next route assigned to it. In addition, what value should the attribute **faultDescription** take for an aircraft that is in-service? The AIRCRAFT class may also have a state model that reflects its behaviour in service and throughout repair. These two aspects of its behaviour do not overlap, modelling both in a single class serves only to obfuscate the essential behaviour in each mode; in other words the model is weak.

We can improve the declarative properties of the model by introducing the idea of dynamic classification, as in Figure 11.20. This model is stronger since it allows us to assert that in-service aircraft must be assigned to a route and that out-of-service aircraft must have a repair request and have a fault description. Each specialization will encapsulate the behaviour for its mode (in service or under repair). An individual instance of AIRCRAFT will transition from one specialization to another and back again as it goes out-of-service and is then repaired. It is often the case that the in-service specialization handles the transition to out-of-service and then once repair is effected the out-of-service one handles the return to in-service. The subtype instances will be created and deleted as required and linked to the single, long-lived instances of AIRCRAFT as it goes through its differing modes. A variation on this pattern is to allow the supertype, AIRCRAFT in this example, to handle the creation, transition and deletion of its subtypes.

11.8.4 Structure

We can use dynamic classification with any number of subtypes and combine it with multiple classification in both static and dynamic hierarchies. Such a general form is shown in Figure 11.21.

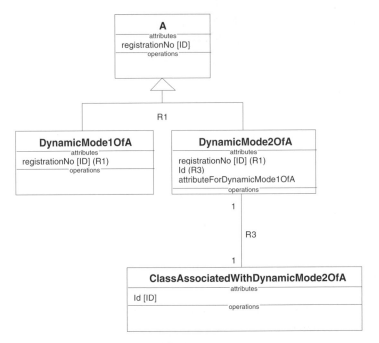

Fig. 11.21 General form of dynamic classification

11.8.5 Applicability

This is a useful pattern where an entity displays significant and non-overlapping modes of operation that have distinct requirements that exhibit themselves as attributes or associations on the specializing subtypes. Care should be taken not to apply this pattern inappropriately where the expedient of straightforward state modelling will suffice.

11.9 Ordered items

11.9.1 Intent

To maintain the knowledge of the order of the instances of a class.

11.9.2 Also known as

Lists.

11.9.3 Motivation

On many occasions we want to maintain an ordered list of instances. There are two basic patterns that allow us to achieve this goal. This is such a common and wide-ranging problem

Fig. 11.22 Attribute-based ordering

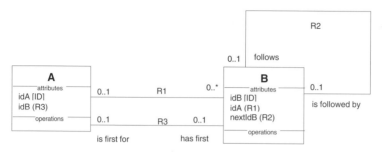

Fig. 11.23 Association-based ordering

that we shall use the generalized models straightaway. We shall assume that the instances are obtained from association navigation.

11.9.4 Structure

11.9.4.1 Attribute form

The first form of the pattern, as in Figure 11.22, relies mainly upon the processing defined in ASL to achieve an ordered set if navigating from class A, thus:

```
{unorderedBs} = myInstanceOfA -> R1
{anOrderedSetOfBs} = {unorderedBs} ordered-by orderingAttribute
for anOrderedB in {anOrderedSetOfBs} do
  # Do something with each B in turn
endfor
```

The ASL processing is supported by just one attribute within the class whose instances have to be sorted.

This approach has the benefit of simplicity, particularly when maintaining the instance population of B. The disadvantage is that we have to manipulate the whole set of Bs, order them and then iterate over that second, ordered set. Our architecture and the number of instances of B in the original set will determine the efficiency of this scheme.

11.9.4.2 Reflexive form

An alternative approach is to model the ordering of B in the class diagram, explicitly modelling the order by a reflexive association and also maintaining an association to the first member of the ordered set, as in Figure 11.23. In this scheme we have the overhead of maintaining the reflexive association, R2, whenever we delete or create an instance of B, but it does allow us to traverse over the ordered set of Bs, for a given A, simply by navigating R3 and then R2; in fact, we may be able to dispense with R1 altogether if we never have to retrieve the set of Bs for a given A in one operation. As R2 is conditional at both ends, this models a linear chain of Bs, if we make R2 unconditional at both ends then the resultant model is of a circular structure.

11.9.5 Applicability

Both forms of the pattern are generally applicable to situations where an ordered set of instances needs to be maintained. The modeller needs to weigh up the additional cost of instance creation and deletion in the reflexive form against the benefits of an increase in declarative precision and the potential to not have to handle the entire ordered set of instances in one go. The attribute form can offer a lower overhead when creating and deleting instances; however, this benefit can be eliminated if the ordering attribute forms a contiguous value range and we have to renumber the instances if we insert or delete an instance somewhere inside the existing ordered set. The attribute form should generally be avoided if the analyst has to invent the ordering attribute, rather than it naturally occurring from the business requirements.

11.10 Resource requester

11.10.1 Intent

To model competing client requests for a resource so as to handle contention and priority issues.

11.10.2 Motivation

For many domains there may be several clients that asynchronously request services of the domain. Sometimes we service such requests as they are delivered by the architecture, although often it is important to assess the requests, priorities and action them according to the rules and policies dictated by the viewpoint of the domain being invoked.

11.10.3 Structure

In this case, we model the requester and request as classes, as in Figure 11.24. A requester makes a request for one or many resources. Note that we make use of the multivalued associative pattern (see Section 11.5), if a requester can request the same resource on many occasions. In this example we have added the time and date of the occurrence of the request and also a priority that is assigned by the requester. These attributes allow us to action the requests in an appropriate way, for example on a first-come-first-served basis.

Figure 11.25 gives an outline state model for REQUESTEDRESOURCEOCCURRENCE, which shows that these are created in the `requestReceived` state and that another

Modelling patterns

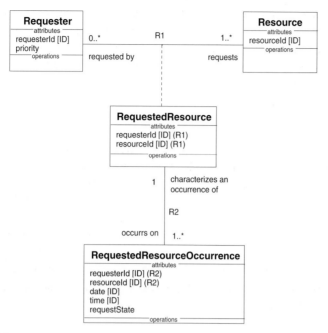

Fig. 11.24 Resource requester

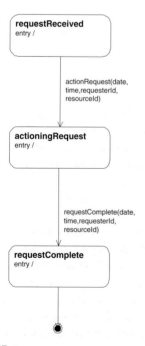

Fig. 11.25 Resource occurrence outline state model

class in the domain, acting in the role of an assigner (see Section 11.11), can drive the REQUESTEDRESOURCEOCCURRENCE through its life cycle to completion. It is this assigner class that implements the scheduling and prioritization policies supported by the attributes of priority and date and time received. There are many variations on this basic theme, where requests may be actioned in either a single or multithreaded fashion for a single resource instance.

11.10.4 Applicability

It is common to combine this pattern with the specification pattern (see Section 11.2), so that requesters make requests upon a type of resource, and then these are met from a pool of actual resources. When the specification pattern is used, the RESOURCE specification class is often chosen to play the part of the assigner, since it can control all RESOURCE instances that it defines.

An example of the difference between requesting a specific thing and a type of thing can be seen when purchasing a car. When a new car is purchased the buyer typically orders a type of car and a car with the appropriate specification is found from stock or built. When a second-hand (previously-owned) car is purchased, the buyer typically selects a particular car.

11.11 Assigner

11.11.1 Intent

To provide a single point of control to allow resolution of contention for shared resources and so prevent race conditions.

11.11.2 Also known as

Resource allocator, contention resolver.

11.11.3 Motivation

Modellers are frequently faced with the situation in which client requests arrive for resources in an unordered, asynchronous fashion. It is necessary to ensure that resources are allocated in a way that prevents race conditions arising and that allocation is performed obeying the rules and policies of the domain containing the requested resource.

11.11.4 Structure

The assigner pattern is built upon the auspices of the resource requester pattern. The principle is that a single instance of a class should resolve the requests for a pool of resource instances that clients are competing for, so acting as a single point of control. It is common to find

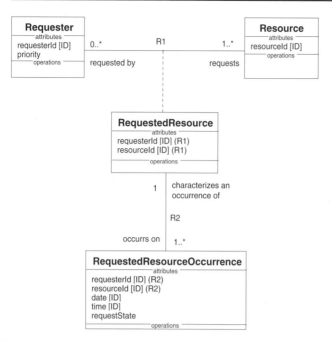

Fig. 11.26 The assigner

that such an assigner role can be allocated to a specification class for the actual RESOURCE instances that are being requested, since all instances of a particular type will have only one instance of their corresponding specification class. If such a suitable specification class cannot be used then another class must be employed that either splits the allocatable instances into distinct sets or is a singleton (Gamma *et al.*, 1994) (Figure 11.26). As an aside, the Shlaer–Mellor OOA method (Shlaer and Mellor, 1992) had this idea built into the method as a primitive, but we do not have that luxury in UML.

11.11.5 Applicability

This pattern can be used in any situation where there are competing requests for a pool of resources that must be assigned to a client.

11.12 Hierarchy

11.12.1 Intent

To model hierarchy or tree like structures.

11.12.2 Also known as

Tree or Composite (Gamma *et al.*, 1994).

Fig. 11.27 A non-hierarchical management structure

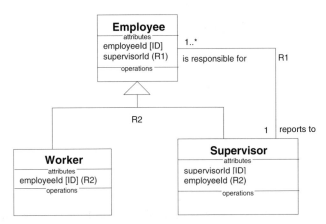

Fig. 11.28 A hierarchical management model

11.12.3 Motivation

Let us suppose we want to model a management hierarchy where employees report to managers. Our first naive attempt might look like Figure 11.27.

Such a model exhibits the inability to model supervisors reporting to their own supervisors, that is a hierarchy management structure is not supported by this model. A better way to model this is as a tree structure, where all members of the organization are employees and are specialized into either workers, who have no supervisory role, or supervisors, who must have at least one person reporting to them. Figure 11.28 is such a hierarchical model. Each person in the organization is an EMPLOYEE, in this case every EMPLOYEE reports to exactly one SUPERVISOR. If the company had a policy whereby the managing director and other board members were unsupervised then we would make R1 conditional at the SUPERVISOR end.

11.12.4 Structure

The general form is obvious from Figure 11.28 and is given in Figure 11.29.

11.12.5 Applicability

The hierarchy pattern can be used whenever a tree like structure is required. By changing the multiplicity of the association R1 to '0..*' at the COMPOUNDNODE end, it can be used to model a matrix like structure.

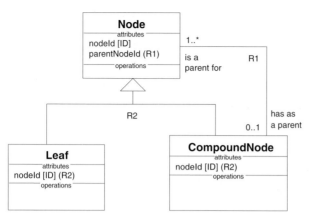

Fig. 11.29 A general tree structure model

11.13 Instance deletion

11.13.1 Intent

To form an agenda of the essential actions that takes place when deleting an instance.

11.13.2 Motivation

Let us consider the closure of a bank account. Figure 11.30 presents a simple model for a typical current account. In this simple model we assume there is one branch of a bank that maintains the account. What steps must we go through before we can delete the instance of account? Each account instance is linked to one or many customers, so we must unlink ourselves from these customers. However, the association R1 has an association class, so we must first unassociate the association class and delete it (remembering that

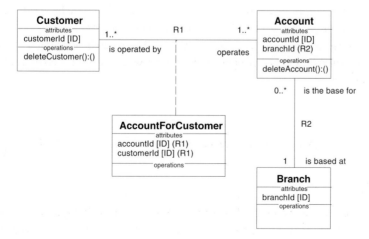

Fig. 11.30 A bank account model

association classes only exist because of an extant association link). We must also consider the CUSTOMER class when deleting R1; the class model states that a Customer must have at least one account, so what if the account we are about to delete is the last one for a particular customer? There are a number of options, we could always inform the respective CUSTOMER instance that an account is about to be deleted and it could decide whether or not to delete itself (presumably calling upon appropriate logging services before deletion), or the ACCOUNT could assess whether or not it is the last instance outstanding for one or more of its CUSTOMER instances and call an appropriate deletion service on these. The association R2 is easier to deal with since it is conditional at the ACCOUNT end and BRANCH is a static class; we simply can unlink the ACCOUNT instance from its associated BRANCH.

Let us consider the processing that takes place within the ACCOUNT instance operation deleteAccount, called when deleting an instance of ACCOUNT. In this case the ACCOUNT does the work as to whether or not this is the last Account for a Customer. If it is, then we call a deletion service on the CUSTOMER. The association class has no deletion service and is simply deleted directly by the ACCOUNT.

```
# first navigate to the Customer class
{theCustomers} = this -> R1.Customer

# now we iterate over the Customer instances
for aCustomer in {theCustomers} do

  # navigate to the association class
  anAccountForCustomer = this AND aCustomer->R1.AccountForCustomer

  # now "clean up" R1 by unassociating the association class,
  # unlinking R1 and deleting the association class instance
  unassociate aCustomer R1 this from anAccountForCustomer
  unlink this R1 aCustomer
  delete anAccountForCustomer

  # now see how many further accounts the customer has
  {remainingAccountsForTheCustomer} = aCustomer -> R1.Account
  if countof {remainingAccountsForTheCustomer} = 0 then
    # there are no further accounts for this Customer, therefore
    # we call the deletion operation on the Customer
    [] = CUST1:deleteCustomer [] on aCustomer
  endif

endfor

# now unlink Account from the related Branch
theBranchForThisAccount = this -> R2
unlink this R2 theBranchForThisAccount

# and finally it is now safe to delete the instance of Account
delete this
```

11.13.3 Structure

In general, we have to perform the following steps for any deletion of an instance.
A deletion action must:
1. Find all currently related instances, including association classes;
2. Unlink currently related instances;
3. Check for unconditional association violations, so that model integrity is maintained;
4. Delete unconditionally related instances, either directly or by calling an operation or generating a signal event;
5. Delete this instance.

These steps ensure that we maintain an instance population that has integrity with respect to the class diagram. It should be noted that on a microscopic scale, at each step in the operation outlined above, the instance population might be out of step with the class diagram. In fact, if we choose to send a signal event to a class, in order to signify that it should delete itself, then the instance population will not be in step with the class diagram until the thread of control that we initiated from the first deletion has completed; such a thread may be of considerable complexity if a cascade of deletions ensues due to following a mesh of unconditionally related instances. It is the duty of the analysts to ensure that any deletion service will leave the associations and instances in step with the class diagram, once the thread of control, initiated by the originating deletion, has terminated. This is often best achieved by ensuring that classes have their own deletion services that 'clean up' the associations in their immediate vicinity and call any further deletion services as appropriate (these in turn may cause further deletions in order to maintain model integrity, resulting in a cascade of deletions). However, deletion services are not necessary for purely passive cases, such as the association class in the example, where the originating class can simply synchronously delete the appropriate passive class instance.

11.14 Instance creation

11.14.1 Intent

To form an agenda of the essential actions that takes place when creating an instance.

11.14.2 Motivation

This is very closely allied to instance deletion, so much so that we shall simply state the generalized form of the rules here for a creation operation. The aim of these, as with instance deletion, is to ensure that model integrity is maintained after the instance has been created and any ensuing thread of control has completed.

11.14.3 Structure

Therefore, a creation operation must:

1. Create the new object;
2. Assign attribute values, all attributes of every identifier must be set when the instance is created;
3. Find unconditionally related objects;
4. Link new object to related objects.

11.15 Unordered operations

11.15.1 Intent

To allow a set of operations to occur for a client instance in any order and then to notify that client that the set has completed.

11.15.2 Also known as

The 'to do' list or the Hanrahan pattern (after the memorable phrase made by the BBC news correspondent Brian Hanrahan during the Falklands war: 'I counted them all out, and I counted them all back').

11.15.3 Motivation

In this pattern, an object needs to wait for a set of operations to complete before it can move on through its life cycle. These operations can be performed concurrently and in any order. It is the job of this pattern to prevent pernicious race conditions from developing.

Consider a hire car. There are a number of types of operation that have to be performed to ready it for an initial or subsequent hire, such as cleaning the windscreen, checking the oil, checking the tyre pressures, etc. These operations may be performed in any order but we must ensure that all operations are completed before we hire out the car (Figure 11.31).

Both the HIRECAR and PENDINGOPERATIONFORHIRECAR classes have state models, as in Figures 11.32 and 11.33, respectively.

The CCD for the Hire Car Preparation domain is given in Figure 11.34. When an on-hire car is returned, the client, represented by the BOOKINGCLERK terminator class, generates the signal 'carReturned' to the respective instance of HIRECAR which results in it making a transition into the RETURNED state. The entry action associated with this state invokes the object-based operation `getCarReadyFor-Hire` which is responsible for:
1. determining which operation types need to be actioned in order to ready the car for subsequent rehire (we shall see how this might achieved in practice when we take a look at the general form for this pattern);
2. setting up the appropriate instances of the association R1 to be consistent with this result, and creating the respective objects of the association class PENDINGOPERATIONFORHIRECAR in the AWAITINGACTION state (by invoking the class-based operation `createOpPendingForCar` (passing in the instance handles of the HIRECAR and the OPERATIONTYPE);

Modelling patterns

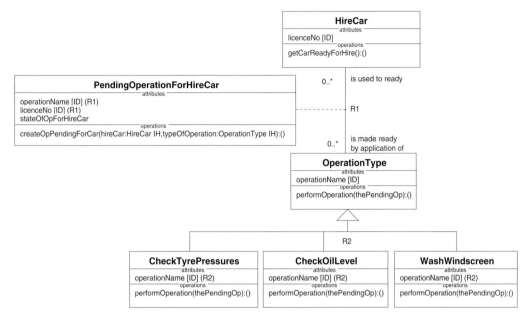

Fig. 11.31 Hire Car preparation

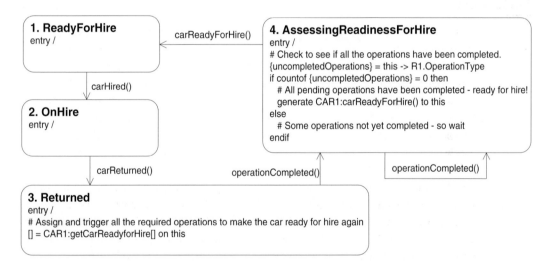

Fig. 11.32 Statechart for the SERVICE OPERATION class

3. triggering the execution of each of the selected OPERATION TYPES (by generating the signal 'actionOperation' to the respective instances of PENDING OPERATION FOR HIRE CAR; the order of generation not being significant).

On receipt of this signal, each instance of PENDING OPERATION FOR HIRE CAR makes the transition from the AWAITING ACTION state to the IN PROGRESS state and executes the entry action which, in this case, is to invoke the polymorphic operation PERFORM OPERATION,

282 Model Driven Architecture with Executable UML

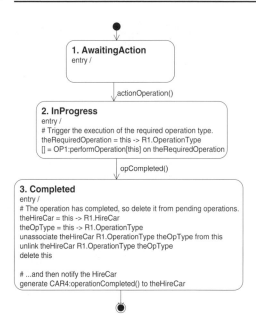

Fig. 11.33 Statechart for the PENDING OPERATION HIRECAR class

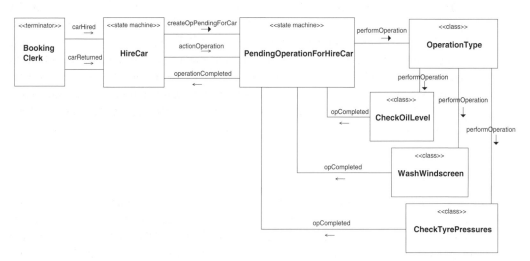

Fig. 11.34 Class collaboration diagram for hire car preparation

which propagates down to the appropriate subtype. The instance handle of the PENDINGOPERATIONFORHIRECAR is passed as an input parameter in this call such that when each operation completes it can send back the signal 'opCompleted' to indicate completion of the operation. On receipt of this signal the PENDINGOPERATIONFORHIRECAR object transitions into the COMPLETED state where it deletes itself and the respective instance of R1, and then notifies the HIRECAR object of completion by generating the signal

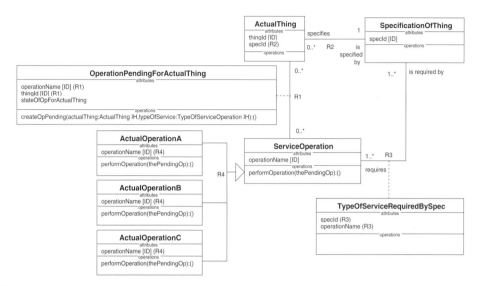

Fig. 11.35 General form of independent actions

'operationCompleted'. In fact, this is the key to the success of this pattern – the association instance must be deleted **before** the signal is sent.

On receipt of such a signal, the HIRECAR object makes a transition into the ASSESSINGREADINESSFORHIRE state and checks to see if all the operations have been completed or not by testing to see if there are any instances of association R1 still in place. If so, it simply waits for the next similar signal notification, otherwise it generates the self-directed signal 'carReadyForHire' which returns the HIRECAR object to the READYFORHIRE state.

The operations can be called in any order and they can complete in any order – this pattern will guarantee that the HIRECAR object does not return to the READYFORHIRE state until all of the operations have been completed.

11.15.4 Structure

The general form of this pattern is directly derived from the preceding example. However, it is augmented with a specification class, SPECIFICATIONOFTHING, which allows us to model the required operations for any actual instance of the specification. The class diagram for this general form is given in Figure 11.35. The state models and CCD are of a very similar form to that given in the hire car example and therefore are not repeated here. The SERVICEOPERATION class may have as many subtypes as required.

11.15.5 Applicability

This pattern is only useful when the operations are unordered and can execute independently. If this is not the case then a more sophisticated scheme is necessary, where both ordering and communication between jobs is modelled.

11.16 Logging

11.16.1 Intent

To provide a generic capability of logging occurrences from other domains in an xUML model.

11.16.2 Also known as

Audit.

11.16.3 Motivation

There are numerous situations where we require to 'remember' that something has happened and be able to retrieve reports based upon certain criteria later on. Let us consider the customer account model that we examined earlier (Figure 11.36).

As we have seen, when the last ACCOUNT for a particular CUSTOMER is deleted, the relevant instance of CUSTOMER will also be deleted. In this case we are required to log the fact that a CUSTOMER has been deleted, along with certain facts about the customer, such as the customerId and, say, the date the deletion occurred (in reality, of course, there would be many more facts we would want to log about this occurrence). Later on, in order to produce a monthly management report, we may want to retrieve details of all customers that have been deleted this month. In order to support this requirement, we shall make use of a logging domain, that is capable of storing logs in a log trail, in this case a trail called 'deleted customers'.

There are other types of incidents that we shall want to log; these fall into the general category of 'errors'. Imagine that when we navigate R1, from the closing ACCOUNT to a CUSTOMER, we do not get back a valid CUSTOMER, that is somehow the integrity of the class diagram has been broken at this point as the ACCOUNT no longer is linked to

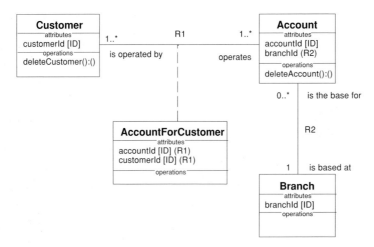

Fig. 11.36 An example domain that requires a logging service

Modelling patterns

any CUSTOMERS via the association R1. There are two possibilities here, one is that the analyst checks for an empty set of CUSTOMERS after performing the navigation, in which case we would want to call the logging domain and explicitly add an instance to an error log; however, this would probably be overkill since we know from the class diagram that there should be at least one customer, so what usually happens is that the responsibility for logging such a incident falls to the architecture as part of an exception handling mechanism. The architecture would log the fact in an error log trail that an empty set was returned from the navigation of an association, R1, in the banking domain, with other relevant information, such as the identifier for the starting instance of the navigation. This information can then be used by analysts to track down such logical errors in their models.

In addition, we may have a security requirement that states that we log the identity and time of queries on the CUSTOMER class, so that we can check that privacy rules are not being abused. A logging domain can be used to support this form of audit requirement.

11.16.4 Structure

This pattern can be considered to be at the framework level in the pattern taxonomy given in Section 11.1. In Figure 11.37, there is the outline for a generic logging domain that

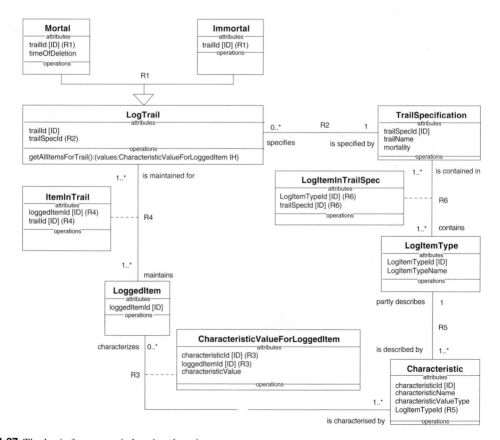

Fig. 11.37 The basis for a generic logging domain

can be tailored to meet individual requirements. It makes use of two previously explored patterns, Specification and Characteristic Value. The right-hand side of the model comprises specification classes for the log trail types that we require, such as 'Deleted Customers', the items that are to be logged in that trail, such as 'Customer' and 'Last Held Account', and finally, for each of those item types, we specify which characteristics we want, such as the 'name' and the 'date deleted' for the Customer LogItemType.

We maintain zero to many trails against its specification, for example we may have a separate instance of LOGTRAIL for deleted customers every month. In this example trails will either be maintained forever, that is they are **immortal**, or they timeout after a while, and so are **mortal**. The characteristic value pattern is used between LOGGEDITEM and CHARACTERISTIC on the association R3. The association class stores the value of the logged characteristic for an item in its attribute **characteristicValue**, the type of which is given in the class CHARACTERISTIC.

Any such logging domain will have operations to retrieve logged items. We give an example of such an operation in the class LOGTRAIL, which has a simple operation that returns all the characteristics for that trail instance. In reality, we would probably want more sophisticated operations that allowed all instances of LOGGEDITEMS to be retrieved for a LOGTRAIL and then allow further detailed queries that returned the characteristics for one or more of the items returned.

11.16.5 Applicability

The logging domain, often supported by appropriate persistence mechanisms provided by the architecture, occurs in many different guises in a wide range of differing types of application. The ability for the architecture to log warnings and errors allows an xUML model to be debugged by analysts when deployed on a target, in a situation where a simulation environment for the models may not be available or appropriate. There are many sophisticated services that may be added to the basic framework to manage both log and log specification instances at run time, as well as providing comprehensive reporting facilities.

The logging domain itself may add data to the logged items, such as placing a time stamp on every item as it is received.

11.17 Plant control

11.17.1 Intent

To form a basis of an application that requires real-time control of plant based upon sensor readings and effecting change to the state of the plant by means of one or more actuators.

11.17.2 Motivation

There are general classes of application type whose analysis can proceed from a starting point of a standard, or pattern-based, domain chart. This represents an application level pattern in Mowbray's taxonomy, see Figure 11.1.

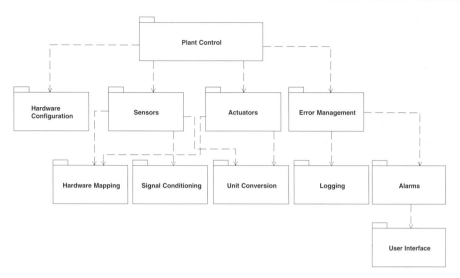

Fig. 11.38 Domain chart for the plant control pattern

11.17.3 Structure

The general structure of the domain chart for this pattern is given in Figure 11.38.

Such a model contains a variety of subject matters and actually has much inherent complexity. The following discussion gives just a flavour of the nature of the domain chart and the way the domains interact.

The plant control domain is the application domain that contains a model of the actual plant that we are controlling (e.g. a car engine controller or vehicle stability system), it relies upon hardware configuration to provide the actual hardware that is present for a particular version of the application. Hardware configuration itself will rely upon the specification pattern (Section 11.2) to capture the actual types of hardware (both sensor and actuators) fitted to a particular type of application, thus we can reconfigure the entire application for another version of the application by updating the instance populations in hardware configuration. The sensors domain models the sensors and their attributes that we have present in the application. It relies upon the services of hardware mapping to get the raw sensor values from an actual device, therefore hardware mapping provides an isolation layer to the actual hardware present in the application, so if we change the pin upon which we read water temperature, say, then sensors will not have to change. Sensors gets raw data from hardware mapping (e.g. in volts or counts) and so uses the services of unit conversion to convert these into engineering units (say temperature, pressure, etc.). Sensors also use the services of signal conditioning to clean up the raw data and eliminate transient errors due to noise. Error management is used to keep track of recurring errors, raising appropriate alarms or logging when critical situations occur. Alarms knows when to display a critical condition to the user through the user interface (if present).

11.17.4 Applicability

This is really the bare bones of such a domain chart for real-time plant control. In actuality, many more domains, specialized to a greater or lesser extent for the actual application type, would be brought into this general framework to produce the end application. Almost certainly we would require the use of a real-time scheduler which is often modelled as a domain in its own right.

11.18 Anti-patterns

A successful way to learn about patterns is to study other peoples' mistakes. There are many examples that have been encountered over the years and this short section gives a flavour of some of the current 'most popular' blunders.

11.18.1 The manager

In this anti-pattern we assign the intelligence of the domain to a single class. Often part of the class name is 'manager' or 'controller' and it is usually divorced from the real subject matter of the domain. The 'real' classes within the domain end up as data repositories and are acted upon by the manager. This form of localization of control is the antithesis of a good executable model where responsibility should be layered and so shared between classes. The reason this anti-pattern is so pernicious is that any change to the business logic or data will usually greatly affect the model in the manager; since there is no layering it must handle every change throughout the domain. Another insidious effect of employing such an overarching manager class is that it usually becomes very complicated and fiendishly difficult to maintain. Such manager classes are usually single instance classes and often have very complex statecharts to describe them. UML supports an abundance of state modelling techniques including hierarchical decomposition of states, history and concurrent state actions. These tend to be needed to describe manager classes. In xUML the statechart represents the behaviour of an instance, therefore decomposition of states is not needed because the dynamic behaviour is distributed between classes and separate mechanisms to express concurrency are not required because of the dispersal of state machines for each instance.

11.18.1.1 Signs of the anti-pattern
There are three main clues to recognizing the use of the manager class. First, the CCD will contain a class that is connected to many other classes that simply respond to its signals with little or no further layering, as in Figure 11.39. Secondly, another tell-tale sign is that classes appear with 'manager', or 'controller' in their names. Thirdly, the offending manager class usually has a single instance (is a singleton), which in itself is an abuse of a design pattern in analysis (see Section 11.18.4).

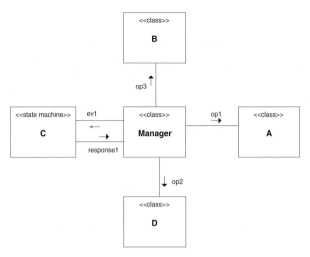

Fig. 11.39 A manager class revealed on the class collaboration diagram

11.18.1.2 Antidote

We require each class to encompass both its data and appropriate responsibilities, leading to a layered CCD, with classes delegating tasks to other classes, down a hierarchy, in order to complete a task. The original manager class can often be eliminated since it frequently does not play a part in the real subject matter of the domain and simply acts as a bucket for the business logic. If a class changes then the effects are often minimized to its nearest neighbours. Layering of the CCDs is discussed in Chapter 7.

11.18.2 The implicit association

Here the analyst uses an identifier of one class as an attribute in another and part of the operations of the second class relies upon a `find` on the first class, employing its identifier `read` from the attribute of the second class. The analyst is actually hiding the requirement to maintain an association between the two classes. This pattern is harmful in at least three ways. First, it hides the fact that there really is a relationship between instances of the related classes, so obscuring the model and making it harder to comprehend and maintain. Secondly, the `find` operation, on most architectures, will be considerably more expensive to perform than a simple relationship navigation. Thirdly, there is no way of specifying the conditionality that should pertain to this implied relationship.

11.18.2.1 Signs of the anti-pattern

This anti-pattern manifests itself in the appearance of attributes in one class that are an identifier for other classes within the domain, but are not referential, that is they do not formalize a relationship. For instance, if we consider a simple bank account and branch model, as in Figure 11.40.

The ASL in the operation `findAllOverdrawnAccounts` might look like:

Fig. 11.40 An implicit association anti-pattern

```
{accountsForThisBranch} = find Account where branchId = this.branchId
{theOverdrawnAccounts} = find {accountsForThisBranch} \
                              where balance < 0.0
```

The first find is an indication of the implicit association, where we use the identifier of the BRANCH to locate the instances of ACCOUNT that belong to this particular BRANCH instance. Of course, in this simple case we are simply missing an association between the BRANCH and the ACCOUNTS that are based at it. Therefore, the first line of ASL in the operation in BRANCH would become a straightforward association navigation and the reader of the class model would then understand the important relationship that holds between accounts and branches.

11.18.2.2 Antidote

This is an easy one to solve. Use a relationship instead.

11.18.3 Spider state models

State modelling is an intuitive and powerful way of capturing state-dependent behaviour. It can, however, be used inappropriately as a modelling technique for state-independent behaviour. If this anti-pattern is applied then we tend to end up with a state model that has a central idle state (or something similar), surrounded by other states that simply perform an action and then immediately cause us to go back to the idle state and await the next thing to do.

11.18.3.1 Signs of the anti-pattern

State models will resemble deformed arachnids, as in Figure 11.41.

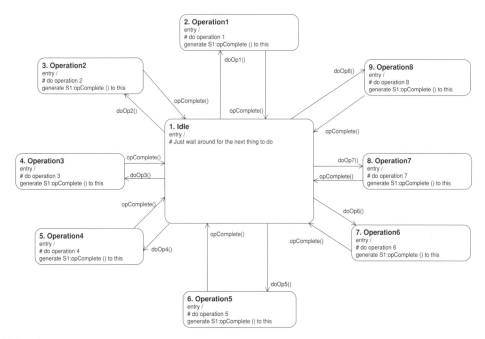

Fig. 11.41 A spider state model

11.18.3.2 Antidote

The antidote to this anti-pattern, once it has been diagnosed, is also straightforward. Use operations, rather than the response to signal events, to model state-independent behaviour.

11.18.4 The design pattern

This is, of course, not really an anti-pattern but rather the inappropriate use of a design-oriented pattern in analysis. With the popularity of the GoF book, patterns such as 'State' (Gamma *et al.*, 1994) are used in analysis where the simple expedient of using a statechart would be much more appropriate. This is not to say that all design patterns are inappropriate in analysis, far from it, but it should not be assumed that just because a design or implementation idea is well regarded that it will be appropriate in analysis where our toolkit of the xUML can be very different from the implementation language paradigms that underpin many design patterns. Another example of such a design pattern is the 'Bridge', which is a way of implementing multiple classification.

11.18.4.1 Signs of the anti-pattern

Symptoms of this problem often manifest themselves in the documentation for classes and associations, with frequent references to the names of well-known design-oriented patterns.

11.18.4.2 Antidote
Question the validity of the use of design patterns in analysis and see if there are simpler ways of using xUML to model the business requirements.

11.19 Conclusion

Patterns occur at every level of abstraction in our xUML models. Recognizing and reusing patterns can be a way of greatly increasing the efficacy of an analysis team. However, like all good things, they can be overused and oversold. Some situations require novel and even lateral thinking, and not just the unthinking application of a patterns' catalogue that will yield poor analysis models. Where novel solutions have been arrived at, bear them in mind as candidate patterns for the next time a similar situation is encountered. If they prove useful a second or third time, possibly in a modified form, then you have the makings of a proven pattern that will be of utility to other analysts. It should be emphasized that a pattern arises out of the repeated utility of a set of constructs, not just a single application of them. The authors have witnessed and used all of the patterns, and many more, discussed in this chapter, as well as lamentably seeing many examples of the occurrence of anti-patterns.

We have seen one example of a pattern at the application level, the Plant Control (Section 11.17). There are many other domain charts that capture application level patterns for differing application types, such as e-commerce, insurance, military command and control systems, air traffic control, etc.

This section has only scratched the surface of the fascinating cornucopia of patterns that arise in executable modelling. A starting point for further reading on analysis patterns is Martin Fowler's book (1997), which places the emphasis on the reusable properties of patterns in analysis.

12 Integrating domains

12.1 Domain interfaces

In the earlier chapters we have seen how to model domains using xUML. Each of these domain models is executable and can be tested. We have also emphasized the importance of domain separation. We avoid building domains with explicit knowledge of other domains to minimize the coupling between them. However, real systems are made of many domains put together. A small system might contain five or six domains whilst a large system may well have in excess of 20 domains. In this chapter we will focus on how systems are built from sets of domains and how we marry up the requirements of one domain to the provisions of another.

The domain is the primary unit of reuse. It can also be thought of as a logical component. To facilitate the assembly of systems by plugging together a compatible collection of domains, each domain has a well-defined interface, comprising:

- **Provided services** – the set of services provided by a domain for other domains to use;
- **Required services** – the set of services that a domain declares it requires other domains to fulfil.

The reason for defining interfaces is so that we can connect sets of domains together to realize whole systems. To integrate a set of domains requires us to match up each required service of one domain to one or more provided services of other domains. To help us match up interfaces it is important to think about each service and what it expects from its matching half.

It is insufficient to state a service's name and set of parameters. This gives us little more than the syntax of the service. We also need to describe:

- what the service means, that is its semantics;
- how it behaves;
- its relationship to other services.

We use the term **contract** to describe some of these important properties of interfaces. In Section 2.6.6, we made the analogy between domains and integrated circuits. The contract is analogous to the definition of the pins provided by an integrated circuit. We shall examine this aspect first, then move on to see how we 'wire' our domains together to form a system.

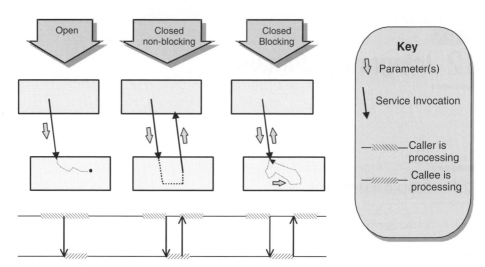

Fig. 12.1 Contract types

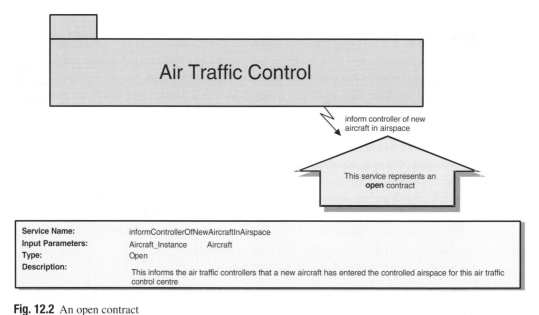

Fig. 12.2 An open contract

12.2 Contract types

For both provided services and required services there are three types of contract, reflecting the run-time behaviour of each service (Figure 12.1).

Integrating domains

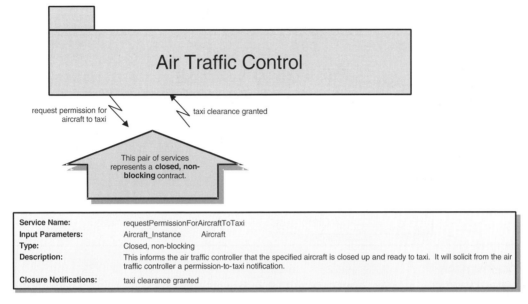

Fig. 12.3 A closed, non-blocking contract

12.2.1 Open contracts

Use an open contract when the caller does not wish to be informed when the requested service has been executed. The caller assumes that the service will **eventually** be executed but must not make any assumptions about when the service will be carried out. No data can be returned to the caller (Figure 12.2).

12.2.2 Closed, non-blocking contracts

Use a closed, non-blocking contract when the caller needs to be informed when the requested service has been executed but does not need to, or want to, suspend its activity until that confirmation is received. The term used to describe the acknowledgement or confirmation is a closure notification. The caller can continue while awaiting the closure notification because the closure notification is delivered asynchronously to the caller (Figure 12.3).

A closed, non-blocking contract can have many closure notifications. In the example in Figure 12.3 taxi clearance granted is one closure notification, taxi clearance rejected might be an alternative closure notification.

12.2.3 Closed, blocking contracts

Use a closed, blocking contract where execution of the caller needs to be suspended until the contract has been completed. Note that this is the only type of the service that can have input **and** output parameters (Figure 12.4).

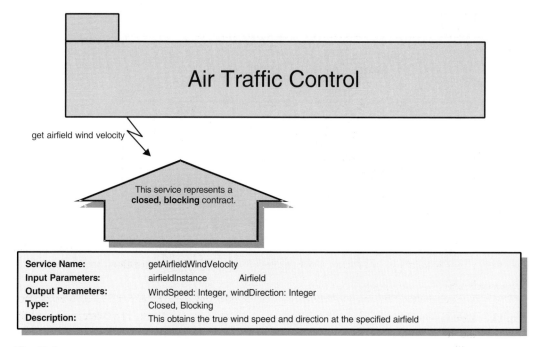

Fig. 12.4 A closed, blocking contract

12.3   Required services

Let us consider an implementation view for a moment. Most programming languages have libraries of software available to implementers: the Standard Template Library (STL) in C++; the Swing libraries in Java; and the predefined packages in Ada are examples. It is common practice for the user of these libraries to call the services directly from their code. The client, therefore, becomes directly dependent upon the library interface. The server is unaware, and independent, of its client but the client is aware of and dependent upon the server. This becomes a problem when there is a need to switch to a different library since the client code has to be changed everywhere that the library interface is referenced.

The same problem can occur when modelling. A client package can become tightly coupled to the interface of the server package. It is perhaps even more of a concern in models than in code because there are more opportunities for selecting or changing the package that provides the services.

A solution to this problem is for the client to define an interface of the services it requires. The client domain is then only dependent on that interface and not on the server domain that provides the services. The server is unaware, and independent, of its client and the client is unaware of and independent of the server. This is called **anonymous coupling** (Yourdon, 1978).

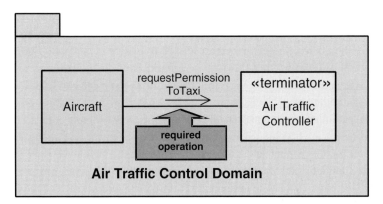

Fig. 12.5 A required operation

The domain that defines the required service simply defines an interface. The domain is not responsible for the realization of that interface. The realization of the interface is achieved in a **bridge**, which we shall come to shortly.

Each domain version publishes a number of **required services**, each of which is associated with a «terminator» class. The set of required services defines the **required interface** for the «terminator» class.

The «terminator» class is named according to the perspective the requesting domain has; it is not named after the domain that the modeller suspects will provide the service. In Figure 12.5 the air traffic control domain declares a required service, 'requestPermissionToTaxi', on its «terminator» class named AIR TRAFFIC CONTROLLER. The domain that will provide this service is not specified. The «terminator» class is named AIR TRAFFIC CONTROLLER because from the perspective of the air traffic control domain that is the destination of that request. In the final system of integrated domains, this required service may map to a provided service on the ATC GUI domain. This naming style for terminators is intended to make each domain comprehensible on its own preserving the isolation of individual subject matters.

Each domain can have many «terminator» classes like this and each of these can have a required interface with many required operations upon it.

12.3.1 Required service – open contract

The calling domain invokes the required service and then immediately carries on with its business. A typical use of this would be where an occurrence in the calling domain needs to be recorded for logging or audit purposes. The calling domain does not need to wait for confirmation that the occurrence has been successfully logged but can just get on with its job. The required service can carry input parameters only.

12.3.2 Required service – closed non-blocking contract

The calling domain invokes the **required s**ervice and immediately carries on with its business. However, it expects to get a response at some time in the future. The response, or **closure notification**, will take the form of a **provided service**, typically on the same «terminator» class. There can be many closure notifications for a single required service. So in this case the contract links a required service to one or more of the domain's own provided services. A typical use of this type of contract is where the required service may take some indeterminate amount of time and the calling domain can carry on doing useful things in the meantime.

The hospital patient management domain provides a suitable example of this. The domain has a required service, 'findSuitableBed', which takes parameters including the patient identity, the sex of the patient and the type of ward he/she should be in. Two closure notifications are defined: 'bedAllocated' and 'noBedAvailable'. These are provided services on the hospital patient management domain. One important issue with closed, non-blocking contracts is that the closure notifications must carry enough data to provide a suitable context for the domain to work out what the closures relate to. In this example the patient's identity would be suitable. The level of concurrency assumed in the model means that any number of 'findSuitableBed' requests could be made and each is awaiting a response and so providing sufficient context is essential.

This example suits a closed, non-blocking contract because the length of time to get a response may be difficult to determine and the domain can carry on with other duties, such as dealing with other patient's needs whilst the response is being obtained.

The required service can carry input parameters only. The closure notifications can carry input parameters only although, of course, from the perspective of the domain that called the initiating required service, the parameters passed with the closure notifications are effectively return parameters.

12.3.3 Required service – closed blocking contract

An object in the calling domain invokes the required service and waits for the service to be completed. This is used wherever the calling domain needs a result to be able to carry on. A typical use of a closed blocking contract would be where the calling domain depends upon another domain for some algorithmic calculation. The required service can carry input parameters and output parameters. Note that although some common programming languages restrict the number of output parameters on a method call to one, no such restriction applies in xUML.

In a closed blocking contract only the object that invokes the service blocks (its thread of processing blocks at the point of the call). Any other threads of control within the domain carry on unhindered by this blocking invocation.

12.4 Provided services

Each domain publishes zero to many **provided services**. These are services that the domain makes available for other domains to use. Of course, with **required services**, as described

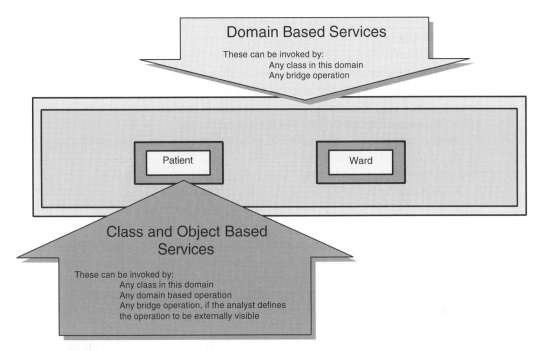

Fig. 12.6 Provided services

in the previous section, the domains don't use these services directly but through **bridges**. The provided services are domain-based, class-based or object-based (Figure 12.6).

Provided services are also associated with a «terminator» class and the set of provided services on a «terminator» class define the **provided interface**. The «terminator» class is named from the perspective of the domain providing the service How does this domain perceive the users of its services? This ensures that all names in the domain relate to the same subject matter viewpoint and avoids inappropriate assumptions about the technology employed by the other domains.

Method definitions

The set of provided services made available to an entity external to the domain is called a provided interface.

The set of required services expected of an entity external to the domain is called a required interface.

A «terminator» class is an abstraction of the provided interface and the required interface of an entity outside of the domain. A «terminator» class may not have any attributes.

A domain may have zero or more «terminator» classes.

12.5 Simple bridges

The domain level sequence diagram introduced in Chapter 7 provides a driver for the process of integrating domains and specifying bridges. The interactions on the domain level sequence diagram show where domains require services of other domains and also provide clues as to the appropriate types of contract to use.

The analysis patterns in Chapter 11 illustrated a trivial banking domain that requires a logging service and what the logging domain might look like. Let us develop that example further to show how a simple bridge works. First some more definitions:

> **Method definition**
>
> A bridge operation is a realization of a required operation that uses one or more provided operations of one or more domains.
> A bridge is a class containing a set of bridge operations for a single required interface.

12.5.1 Example client domain

The greatly simplified banking domain features two classes CUSTOMER and ACCOUNT as shown in the class diagram in Figure 12.7.

This domain requires that a historical record be kept whenever an account is closed. However, it is not the purpose of the banking domain to perform that function itself – it

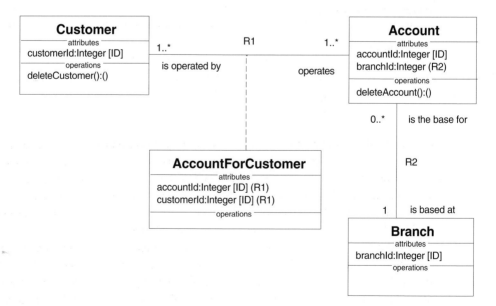

Fig. 12.7 Class diagram for banking domain

Integrating domains

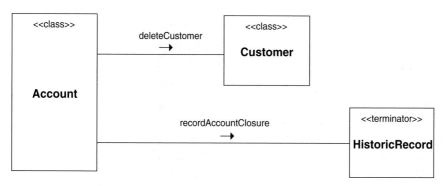

Fig. 12.8 Class collaboration diagram for banking domain

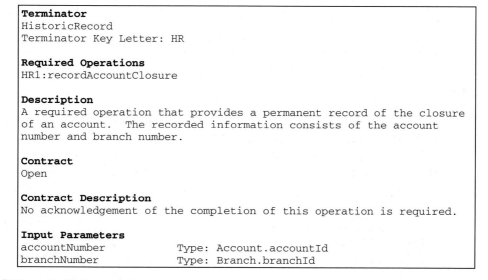

Fig. 12.9 HISTORICRECORD terminator

chooses to publish a required interface. This required interface is very simple and has only one operation. It states that the domain requires a service from some other domain to save this historical record. The banking domain, which will be the client in this example, does not state which domain will provide a suitable service and so has no knowledge or dependence on any other domain. The CCD shows, in Figure 12.8, the interactions between the classes and from the class to HISTORICRECORD «terminator» class.

The «terminator» class, the name given to the collection of required and provided interfaces for an external entity, is called HISTORICRECORD. HISTORICRECORD is not the name of the domain providing the service, since that would pollute one domain with explicit knowledge of another, but it is the banking domain's view of the entity it needs to satisfy its required interface. The terminator description is given in Figure 12.9.

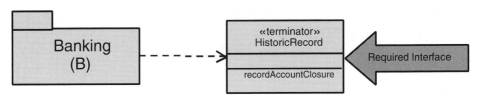

Fig. 12.10 Banking domain and required interface

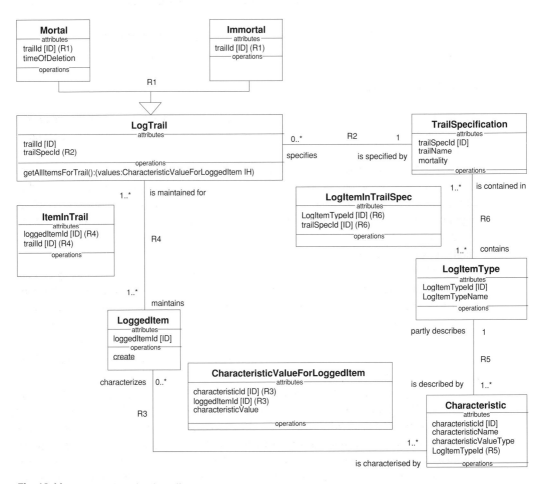

Fig. 12.11 Logging domain class diagram

The single required operation has no return parameters and the banking domain does not need any acknowledgment to carry on with its task, so an **open** contract is specified. The required operation takes two input parameters, the accountNumber and the branchNumber. The domain and its «terminator» class with the single required operation are shown

Fig. 12.12 Logging domain class collaboration diagram

in Figure 12.10. There is a dependency relationship from the domain to its terminator class.

The HISTORICRECORD «terminator» class has no provided interface.

12.5.2 Example server domain

So far we have only considered the client view with the banking domain. Now let us look at the domain that might be able to provide this service and then specify a bridge that connects the two.

Our example logging domain from Section 11.16 provides an operation called `logItem`. When this is called, an object of a 'LoggedItem' is created and the parameters passed to `logItem` are held as characteristics of the Log. We don't need to know the details of the logging domain to be able to bridge to it. We only need to understand the services it provides and how to use them. However, for completeness we will look at the logging domain class diagram (Figure 12.11).

The CCD (Figure 12.12) shows a CLIENT «terminator» class that represents any domain wishing to use a logging service. The CLIENT «terminator» class has a provided interface but no required interface. The `logItem` operation is provided by the domain and is realized by calling the `create` operation on LOGGEDITEM. This in turn creates and links the necessary objects.

This provided operation has an open contract since the client is not informed of the outcome of the operation.

The logging domain provides a generic service and so knows nothing about the clients that use it. The `logItem` operation provides flexibility by taking a single input parameter called 'loggableCharacteristics' that is a set of tag–value pairs. The tags correspond to the **characteristicName** attribute of the prepopulated CHARACTERISTIC class. This allows the logging domain to be completely generic but it can be instantiated for a particular purpose by populating the specification classes, TRAILSPECIFICATION, LOGITEMTYPE, LOGITEMINTRAILSPEC and CHARACTERISTIC. In this example the value parts are assumed to be integers, although it could just as easily support text values (Figure 12.13). Mixed type values could be supported but it would make the interface a little more complex.

The logging domain and its «terminator» class with the single provided operation are shown in Figure 12.14, linked with a realization relationship.

```
Domain Scoped Provided Operations

LOG1::logItem - externally visible

Description
This provided operation causes a LoggedItem to be created along with
the associated Characteristics.  The input parameter takes the form
of a set of tag and value pairs.  The tag representing the name of
the characteristic and the value the corresponding value of the
characteristic.  This allows any number of characteristics to be
logged in a single loggedItem.

Contract
Open

Contract Description
This operation does not provide any response to the client and so is
classified as open.

Input Parameters
loggableCharacteristics         Type: logCharacteristicType


Method
[] = LI1:create [{loggableCharacteristics}]
```

Fig. 12.13 Domain operations provided by the logging domain

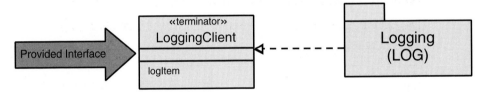

Fig. 12.14 Logging domain with its provided interface

12.5.3 Example bridge

So, we have a banking domain with a single required operation, `recordAccountClosure`, and a logging domain with a single provided operation, `logItem`. Let us simply assert that `logItem` is a suitable operation to satisfy the requirements of the required operation `recordAccountClosure` so we can build a **bridge**. The bridge realizes the required interface of the banking domain and depends upon the provided interface of the logging domain as shown in Figure 12.15.

A bridge operation is a realization of a required operation, so in our example the bridge operation we need to define will be the method that realizes the `recordAccountClosure` operation. The bridge operation will need to do the following:
- Select the server domain that has the requisite provided operation;
- Map the parameters of the required operation to the parameters of the provided operation;
- Invoke the provided operation.

These are shown in Figure 12.16.

Integrating domains

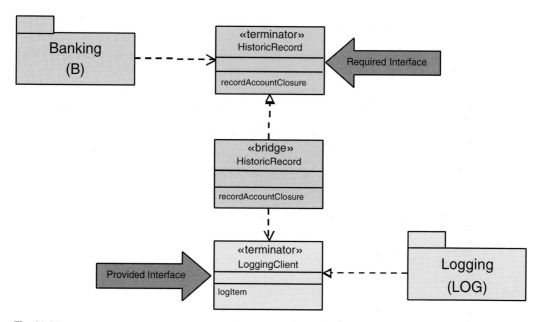

Fig. 12.15 A simple bridge

```
Bridges in Build Set for Domain: Banking (B)

Required Operations for Terminator: HistoricRecord (HR)

HR1:recordAccountClosure

Contract
Open

Input Parameters
accountNumber          Type: Integer
branchNumber           Type: Integer

Multi Domain Bridge Code

$USE LOG
{loggableCharacteristics} is logCharacteristicType

characteristicName1 = "Account Number"
characteristicName2 = "Branch Number"

append [ characteristicName1, accountNumber] to \
  {loggableCharacteristics}
append [ characteristicName2, branchNumber] to \
  {loggableCharacteristics}

[] = LOG1::logItem[ {loggableCharacteristics} ]

$ENDUSE
```

Fig. 12.16 A Bridge for the recordAccountClosure required operation

The first part of the bridge definition is a repeat of the definition of the required operation `recordAccountClosure`. The new part here is the multi-domain bridge code section. The ASL which defines the method for a bridge operation has a special property not found in ASL anywhere else in a model, namely it will typically refer to elements in two or more domains. As a domain is a namespace we need to be able to resolve names unambiguously within the bridge operation. It is the `$USE` clause that allows us to specify the domain context. The use clause is a bracketing construct, `$USE...$ENDUSE`, and all statements between the `$USE` and the `$ENDUSE` are in the context of the specified domain.

In our example all of the ASL statements within the bridge are in the context of the logging domain. Note that the `$USE` clause refers to the short name of the logging domain (LOG). The ASL in a bridge operation has as a default context – the context of the calling domain – therefore any ASL statements in a bridge operation that are not within the bracketed `$USE` construct are considered to be in the context of the calling domain. Whilst multiple `$USE` clauses are allowed (which gives the ability to express **forking** bridges (see Section 12.5.5), they are not permitted to be nested.

The penultimate line of the bridge operation

```
[] = LOG1::logItem[ {loggableCharacteristics} ]
```

is a call to the `logItem` operation provided by the logging domain. This has to be within the `$USE` clause because it is an explicit reference to a model element from the logging domain. All other statements in the bridge operation are required to convert the input parameters of `accountNumber` and `branchNumber` into `loggableCharacteristics`. The first line

```
{loggableCharacteristics} is logCharacteristicType
```

declares the tag and value pair structure by reference to the `logCharacteristicType`, which is a user-defined type in the logging domain and so needs to be inside the `$USE` clause. The assignments

```
characteristicName1 = "Account Number"
characteristicName2 = "Branch Number"
```

set the tag values and then the append statements

```
append [ characteristicName1, accountNumber] to \
{loggableCharacteristics}
append [ characteristicName2, branchNumber] to \
{loggableCharacteristics}
```

populate the tag and value structure before making the call to `logItem`.

Strictly speaking the middle four statements don't need to be inside the `$USE` clause since they only refer to items local to the bridge itself. However, there is little point having two use clauses in this case.

12.5.4 Characteristics of a simple bridge

The job of the bridge is simply to map a required operation to a provided operation or operations (for a forking bridge, see Section 12.5.5) and therefore it should do little else. On many occasions the parameters map very easily and no explicit parameter conversion is required. Such bridges have a $USE clause containing a single line of ASL invoking the provided operation – undoubtedly the best kind of bridge! Occasionally, complex parameter conversions are needed. The preceding example was chosen to illustrate the sort of thing bridges can do and to emphasize the point that the parameters on the provided and required operations don't need to match: they must be mappable, but they don't need to match.

In simple bridges the provided services invoked are usually domain operations. These are attractive because the bridge needs to know nothing about the domain other than its domain operations. In fact many projects make a rule that the only services that can be invoked from a simple bridge are domain operations. A bridge can invoke a class operation but this means the bridge knows about a class within the domain. For a simple bridge to invoke an object operation or send a signal event it must have an instance handle. To obtain an instance handle it must perform a `find` operation, which immediately implies that the bridge has knowledge of classes and attributes. Whilst this isn't wrong, it does 'fatten' up the bridge and increase the coupling between the bridge and the server domain and therefore this approach should be used with caution.

The most common mistake made by modellers specifying bridges is that the bridges get too fat. Functionality that belongs in a domain creeps into the bridge. This is easy to stop if you are aware it is happening. The ASL within a bridge operation should minimize its dependence on model elements within a domain. The following checklists may help in this regard:

Simple bridge ASL will preferably refer to these elements within a domain:
- Domain operations;
- Class operations;
- User-defined types.

Simple bridge ASL may, with caution:
- Perform `find` operations;
- Invoke `object` operations;
- Generate signal events.

Simple bridge ASL should NOT:
- Read or write attributes of classes;
- Link, unlink or navigate associations within a domain;
- Create or delete objects;
- Perform complex set operations unless justified by type conversion;
- Perform logical or arithmetic operations unless justified by type conversion;
- Perform timer operations.

When we look at more advanced bridge structures later in this chapter we shall see that these guidelines change a little. However, for simple bridges they provide a good checklist.

```
Bridges in Build Set for Domain: Banking (B)

Required Operations for Terminator: HistoricRecord (HR)

HR1:recordAccountClosure

Contract
Open

Input Parameters
accountNumber           Type: Integer
branchNumber            Type: Integer

Multi Domain Bridge Code

$USE LOG
{loggableCharacteristics} is logCharacteristicType

characteristicName1 = "Account Number"
characteristicName2 = "Branch Number"

append [ characteristicName1, accountNumber] to \
   {loggableCharacteristics}
append [ characteristicName2, branchNumber] to \
   {loggableCharacteristics}

[] = LOG1::logItem [ {loggableCharacteristics} ]

$ENDUSE

$USE STATS

[] = STATS1::noteClosureEvent[branchNumber, accountNumber, "Acct"]

$ENDUSE
```

Fig. 12.17 A forking bridge

12.5.5 Forking bridge

A bridge operation could invoke provided operations of more than one domain. This is known as a **forking bridge**, although this name has greater mirth value than methodological merit!

As an example of such a construct let use assume that our bank Section 12.5.1 maintains statistics so that the performance of branches can be monitored to aid the decision-making process in the next round of branch closures! Each time an account is closed the statistics domain notes such a closure, associating the date and time along with the account number and branch number. The `noteClosureEvent` operation takes the parameters 'businessUnit', 'closureIdentifier', 'typeOfClosure'. The `branchNumber` will map to `businessUnit`, the `accountNumber` will map to `closureIdentifier` and we set `typeOfClosure` to "Acct". The parameter mappings on this occasion are very simple and our bridge ASL might then look like Figure 12.17.

This example is a suitable use for a forking bridge. Be cautious, however, of putting any business workflow decisions in the bridge as these properly live in the client domain. In fact, all forking bridges should be examined closely to see if business processes and decisions are being modelled within the bridge operations – a situation that should always be avoided.

12.5.6 What do required interfaces and bridges achieve?

The idea of a provided interface is common. Most programming languages have libraries that publish operations for others to use (Section 12.3). Most UML books will show examples of interface classes and they are almost invariably provided interfaces. Users of these provided interfaces depend on them directly and as a result knowledge of the Swing library becomes embedded in your Java or knowledge of the STL becomes embedded in your C++. If you want to change to use a different library then you have a problem. References to the existing library are typically scattered throughout the client code.

The required interface is a solution to this. It states explicitly the services that a domain requires. It makes no assumptions about the domain or domains that will provide that service. This means that a domain model with its set of provided and required interfaces is a self-contained specification – domains fit neatly into a component view. Domains are logical components. With explicit required and provided interfaces domains are more reusable since there are no implicit dependencies.

However, sets of self-contained domains don't define a system. We need to select and integrate domains to build systems. Integration is achieved by mapping required services to provided services; such mappings are specified in bridge operations. Even in the integrated model there is no dependence from one domain directly to another. The only connection between domains is captured in the bridges. This supports the fundamental quality criteria of software engineering, **cohesion** and **coupling** (Yourdon, 1978). Domains are highly cohesive and loosely coupled. So, coupling between domains is explicit and contained in the bridges.

Using a bridge achieves loose coupling by indirection. At this point in the story, implementation-minded members of the audience get fidgety and raise the issue of performance: 'Surely these bridges will make it run very slowly!', they cry. The answer to this is that when we translate the models into the target language, as we will see in Chapter 13, we have the opportunity to optimize away the indirection if necessary.

12.5.7 Matching contracts

We defined the contract types for our required and provided operations. To match required to provided operations we need to consider the contract types. If a required service is the same contract type as a provided service then the contract types are obviously compatible. However, it is not essential that the types map exactly. For example, consider a required operation that has an open contract. This states it does not expect a return parameter and does not require any acknowledgement that the operation has completed. Assume we find a candidate provided operation that seems to do the right sort of job but it has a closed, blocking

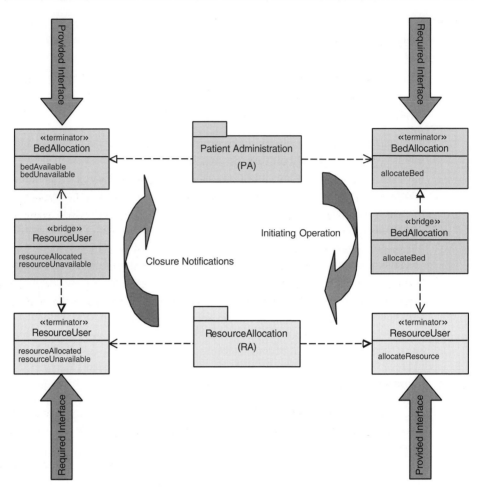

Fig. 12.18 Bridges in a closed, non-blocking contract

contract. It returns a status parameter to indicate the success or otherwise of the operation. At first sight these are not compatible. However, we can map the required operation with its open contract to the closed, blocking provided operation. The bridge simply throws away the return parameter. This works because the provided operation has greater capability than the required.

The reverse mapping of a closed, blocking required operation to an open, provided operation will not work since the server domain will not provide the response the client expects.

A closed, non-blocking required contract on a domain A will map to a closed, non-blocking provided contract on a domain B. However, more than one bridge is required to match these types of contract. There is a bridge operation for the initiating operation which is a required operation on domain A mapping to a provided operation on domain B. Each closure notification, of which there may be many, maps a required operation on domain B to a provide operation on domain A. This is illustrated in Figure 12.18.

12.6 Advanced bridges

The first part of this chapter has focused on simple bridges. Entire large systems have been successfully specified and built using simple bridges alone. They are sufficient to do the job but they don't necessarily produce the most elegant models. If this is your first reading then you may wish pass over the rest of this chapter and return to it when you have had some experience of using simple bridges. On the other hand, if you want to know the details of the advanced 'power' features of bridges, then read on.

It is not uncommon when modelling to become convinced that the same thing exists in more than one domain. This is a troublesome thought for a modeller since, by definition in xUML, a class can exist in only one domain. In the air traffic control example it is easy to think that AIRCRAFT exists in multiple domains. This is a hint that some abstraction is required. The same real-world thing, an aircraft, may be instantiated in multiple domains but as objects of different classes. But this isn't the whole story. If there are many objects in different domains how do you link them together, after all they are all instances of the same real-world thing? This introduces the concept of **counterpart**s that will be explored in the following sections.

12.6.1 Counterpart classes

The concept of **counterparting** is based upon the view that a real thing manifests itself at a number of different levels of abstraction. For example, a multiplexer in a telecommunications network would manifest itself as:

- A *node* in the network management domain, representing a network element with traffic carrying capability;
- An *Add Drop Multplexer* in the ADMUX domain, representing a device that is capable of cross-connecting multiplexed signals between SDH ports;
- A *Fallible Component* in the alarms domain, representing a component that can suffer faults and generate alarms;
- A *Hardware Unit* in the equipment domain, representing a set of slots configured with a legal combination of cards;
- A *Comms Entity* in the communications domain, representing a system component that is the source and sink of messages to configure and control the network;
- An *Actual Resource* in the resource allocation domain, representing a resource for which competition exists;
- An *Icon* in the user interface domain, representing a shape on a display screen.

The counterparts are specified as part of the system specification process (Table 12.1). Each real-world thing is represented at different abstractions in various domains – the **counterpart classes**. The objects of counterpart classes in the various domains are related by special kinds of relationships that span domains – the **counterpart relationships**.

Table 12.1 Counterpart relationships for the 'network management' system

Counterpart relationship number	Domain 1	Class 1	Domain 2	Class 2
CPR1	ADMUX	Add drop	Network	Node
CPR2	ADMUX	Add drop	Alarms	Fallible
CPR3	ADMUX	Add drop	Equipment	Hardware unit
CPR4	ADMUX	Add drop	Communications	Comms entity
CPR5	ADMUX	Add drop	Resource	Actual
CPR6	ADMUX	Add drop	User interface	Icon

Consider our air traffic control system. This system clearly needs to deal with aircraft, but an aircraft manifests itself at a number of different abstraction levels, in different domains:

- In the air traffic control domain, it is an aircraft subject to the rules and policies established by the aviation authorities;
- In the radar data processing domain, it manifests itself as a large metallic object that reflects electromagnetic radiation;
- In the GUI domain, it is a colour-coded shape with associated pop-up menu options displayed on a screen;
- In the software architecture domain, it is an instance of an executable software class with associated attribute values and operations.

We clearly need to formalize the idea that each of these classes, modelled in their separate domains, have counterparts in other domains that together represent all aspects of the actual aircraft. We achieve this by specifying a set of counterpart relationships. This allows the run-time system to create the set of counterpart class objects whenever a new Aircraft is detected, and delete them when it leaves our controlled airspace. It also allows us to synchronise the counterpart objects, for example by changing the colour of an 'Icon' object when the 'Aircraft' object changes to the 'Taxiing' state (Figure 12.19).

There are two types of counterpart relationship:

- Counterpart associations, like those illustrated so far;
- Counterpart generalizations, which are less common and have consequently been relegated to a later section.

12.6.2 Counterpart associations

Consider the air traffic control domain, which has a class AIRCRAFT. In the user interface domain there is a counterpart class called ICON. The ICON class specifies the graphical representation of the aircraft on the air traffic controller's display. A particular object of an ICON corresponds to a particular object of an AIRCRAFT. They are counterpart classes. It is, of course, important to the correct operation of the air traffic control system that we know

Integrating domains

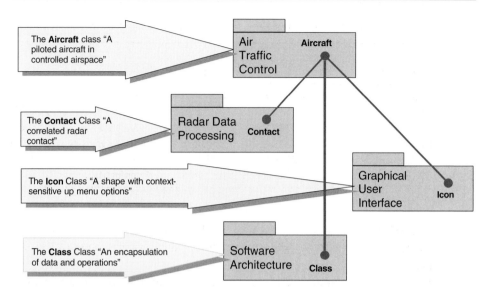

Fig. 12.19 Counterpart classes

which AIRCRAFT corresponds to which ICON! Within a domain we use associations to link related objects together; we want the same type of mechanism to work between domains but we must preserve the idea that each domain does not need to know about the other. A counterpart relationship is the special type of association we use to achieve this.

A counterpart association is a counterpart relationship that links objects of classes in different domains. You may be wondering why this is any different to a normal binary association within a domain. The answer is that we want to achieve the effect of an association but without polluting either domain with knowledge of the other. Each domain uses a special type of «association terminator» class to represent its view of its peer, namely the counterpart class. Like conventional associations, counterpart associations can be linked, navigated and unlinked. The link operation allows us to associate two counterpart objects together. The navigate operation allows us to find a particular object from its counterpart object and it can be used in either direction. The unlink operation allows us to disconnect the counterpart objects. Each counterpart association (CPR) is allocated a number of the form 'CPR<n>' which is comparable with the style used within a domain 'R<n>' (Figure 12.20).

12.6.2.1 Counterpart association «association terminator» classes

In our air traffic control example let us assume that the air traffic controller has to make the final decision for an aircraft to taxi, take off or land. The air traffic control domain therefore sends a 'requestPermissionToTaxi' signal to the air traffic controller when clearance is needed. However, this is actually achieved by making the icon, which represents the aircraft, flash on the controller's display. The air traffic control domain does not, and should not, care how the request is delivered to the air traffic controller. Similarly, the user interface domain does not care which application requested it to make an icon flash. So, although AIRCRAFT

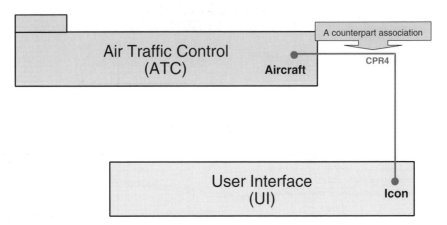

Fig. 12.20 Counterpart classes with counterpart association

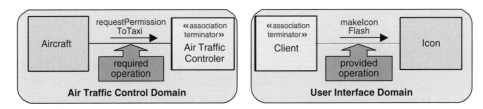

Fig. 12.21 A class collaboration diagram with «association terminator» classes

and ICON are counterparts, we want to preserve the anonymity between domains. This is achieved by using an «association terminator» class in the air traffic control domain to represent the controller who will receive the message, and another «association terminator» class in the user interface domain to represent the client that requested the icon to flash. In this way, the terminators are just like the terminators we have already seen and the beneficial **anonymous coupling** between the domains is retained. The key difference is that we can attach counterpart associations and define counterpart object-based operations on these special terminators. An «association terminator» class is associated with one, and only one, class in the domain in which it is defined. So, in our example, the air traffic controller «association terminator» class is associated with only the AIRCRAFT class in the air traffic control domain.

The «association terminator» class is depicted as the source of incoming signals, destination of outgoing signals and operations on the CCD. In Figure 12.21 we see fragments of the CCD from both domains with the «association terminator» classes represented.

Counterpart associations exist where we want to link counterpart objects in participating domains and therefore an «association terminator» class exists at both ends of the counterpart association.

Integrating domains

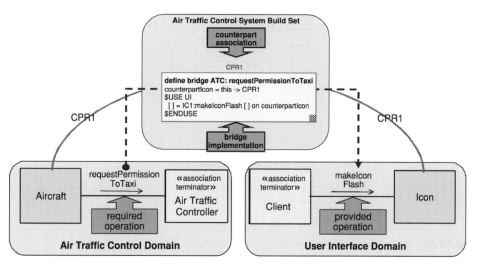

Fig. 12.22 A bridge implementation with a counterpart association

12.6.2.2 Bridges using counterpart associations

As with simple bridges, the required services of one domain are mapped to the provided services of another domain using **bridges**. The analyst must define the **bridge operation** that maps each required operation to each provided operation as in the example in Figure 12.22.

The bridges for each system specify how the required services are mapped to the provided services. Each bridge is a realization of the required interface part of a single «association terminator». Each bridge operation is a realization of a single required operation.

The form of the bridge operation is the same as for simple bridges but we can now do more in the ASL of the bridge operation. Additionally, we can:

- link counterpart objects together:

```
link-counterpart theAircraft CPR1 theIcon
```

- unlink counterpart objects:

```
unlink-counterpart theAircraft CPR1 theIcon
```

- navigate counterpart association:

```
theCorrespondingIcon = theAircraft -> CPR1
```

Note that these ASL constructs can only be used in bridge operations.

In this example the bridge is as simple as a counterpart bridge can get in that it only navigates the counterpart relationship to obtain the instance handle for the counterpart Icon and then invokes an object-based operation on it.

As before, the $USE directive is not an executable statement; it indicates the domain whose namespace should be used until the next $ENDUSE directive. This allows the context

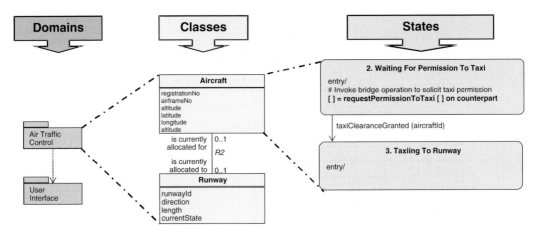

Fig. 12.23 An example bridge invocation

of the bridge execution to be switched between the different domains in the system for the purpose of invoking provided operations on more than one domain.

12.6.2.3 Invoking a bridge operation

Bridge operations can be invoked from any operation or state machine action. They look just like normal operation calls. In our simple bridge example the invocation looked like a call to a class-based operation. A counterpart operation behaves like an object-based operation except that the keywords 'on counterpart' can be used to make it clear that the operation is actually to be mapped to an object-based operation on a counterpart class. Note, however, that at this point, no indication is provided regarding which counterpart class (or even domain) is to be used, as this would represent serious domain pollution, the consequences of which should now be fully appreciated. It is equivalent to using the keyword 'on this' – the use of 'on counterpart' is simply to aid readability (Figure 12.23).

The ASL in the bridge operation, as well as mapping from one counterpart object to another, can do all the things listed for simple bridges including mapping parameter values from one domain to another.

12.6.3 Counterpart generalizations

Service domains are commonly used to capture generic abstractions that apply to classes in many other domains. For example, whereas an inheritance-oriented model might look like Figure 12.24 an executable UML model would look like Figure 12.25.

In this example we have abstracted the shared properties that make an item serviceable into a generalized class. Again we use domain separation to distinguish subject matters. We can describe a lot about the maintenance of serviceable items without caring what they actually are. In this case, the serviceable items correspond to things in a maintenance domain, but they could also be aircraft or vehicles in some other domains. The separation will simplify the development and maintenance of these domains and increase the chances of being able to reuse the generic domain.

Integrating domains

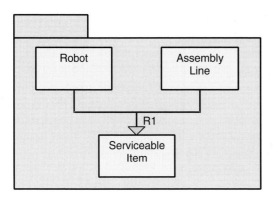

Fig. 12.24 An inheritance relationship

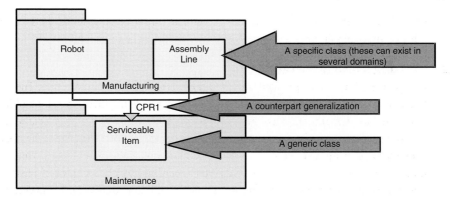

Fig. 12.25 A counterpart generalization

12.6.3.1 Specific counterpart «specialization terminator» classes

Comparable with the terminator for a counterpart association, «specialization terminator» classes deal with counterpart generalizations (Figure 12.26). They have associated outgoing specific counterpart «specialization terminator» class operations.

The «specialization terminator» class is depicted as the source of incoming signals and specific/generic services on the CCD. This type of «specialization terminator» class appears only at the 'generic' end of a counterpart generalization. This is a key difference compared to counterpart associations and has the benefit that the specialized domains have no explicit link to the generic domain within them.

The «specialization terminator» class exists only for the **generic** class. It is named '<generic class name> Counterpart'.

A «specialization terminator» has associated services:
- **Generic required operations** – representing generic operations that must have specific implementations in each of the specific classes participating in this specific – generic relationship.

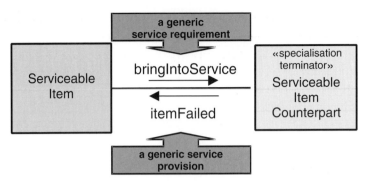

Fig. 12.26 A «specialization terminator» class

- **Generic provided operations** – representing operations provided by the generic class that may be invoked by the specific classes.

Whenever a specific object is created/deleted, exactly one corresponding generic object is created/deleted. Once the counterpart generalization has been instantiated, behaviour can be propagated:
1. from the generic object to the specific object;
2. from the specific object to the generic object.

Clearly, the analyst must exercise care to avoid loops!

12.6.4 Summary of bridge operations

The primary differences between bridge operations and other types of operation are:
- The $USE directive in ASL, which indicates the key letter of the domain whose namespace is to be used until the $ENDUSE directive. A bridge operation can contain as many $USE directives as required.
- No direct class or association references are allowed. This limits the risk of the bridge becoming tightly coupled to the structure of the domains that it bridges, and from performing work that should really be performed within the domains;
- All user-defined types revert to their base types to facilitate mappings from one domain's type to another;
- Only externally visible operations can be invoked. This includes all domain-based operations and those class-based and object-based operations that are specified as part of a provided interface.

12.6.5 Summary of terminators

All «terminator» classes have:
- An analyst-specified name (e.g. Air Traffic Controller);
- An analyst-specified key letter (e.g. ATC);
- A description.

The «terminator» class is an abstraction of a real **external entity** and is named using the language of the domain it is in and not the domain that might realize it.

The «terminator» class, comprising signals and operations but no attributes, is depicted on the CCD. There are three types of «terminator» class:
- Non-counterpart «terminator» class;
- Counterpart association «association terminator» class;
- Counterpart specialization «specialization terminator» class.

12.6.6 How not to specify bridges

Examine the bridge operations in Figure 12.27. Notice that they are infested with constants. The experienced reader will realize that the main problem with constants is that they keep on changing. For example, this approach will cause problems when adding new drives to the model.

Therefore, the model in Figure 12.28 shows an IO domain that allows construction of bridge operations that are free from constants, and permits run-time reconfiguring of the hardware. Note that all constants have now been captured as objects in the server IO domain (Figure 12.29).

12.6.7 Using specific–generic relationships

Specific–generic relationships (SGRs) provide some very powerful capabilities that go beyond the scope of this book. Further details can be found at www.kc.com.

12.7 How to manage domain integration – build sets

We have seen that we can represent the set of domains that make up a project on a domain chart utilizing the UML package diagram. However, projects evolve, domains are developed independently and we may want to start integrating domains before all of the domains are ready. In fact this incremental development is to be encouraged since it avoids the big bang integration that is rarely an enjoyable experience. How then can we manage the evolving set of domains that we want to integrate gradually? Specifically, we must address the realities of:
1. Incremental development of domains;
2. The need to include 'simulated' domains if hardware is unavailable (e.g. the 'Simulated PIO' domain);
3. The need to build simulations comprising a subset of the set of domains on the domain chart;
4. The need to specify a series of **system versions**, each comprising a set of **domain versions**.

To specify a system, we specify a **build set**. This is comparable to a 'parts list', which specifies:

```
Build Set for Domain: WindTunnelModelMotionControl (WTMMC)

Required Operations for Terminator: DriveHardware (DH)

DH2:applyBrake

Contract
Open

Input Parameters
driveId                    Type: Integer

Multi Domain Bridge Code
switch driveId
case 1 # port canard
  $USE IO
    ioCardNo = 1
    regNo = 0
    bitNo = 2
    bitValue = 1
    []=BIT2:setBit[ioCardNo, regNo, bitNo, bitValue]
  $ENDUSE
case 2 # stbd canard
  $USE IO
    ioCardNo = 1
    regNo = 0
    bitNo = 3
    bitValue = 1
    []=BIT2:setBit[ioCardNo, regNo, bitNo, bitValue]
  $ENDUSE
...
case 7 # inboard stbd taileron
  $USE IO
    ioCardNo = 1
    regNo = 1
    bitNo = 0
    bitValue = 1
    []=BIT2:setBit[ioCardNo, regNo, bitNo, bitValue]
  $ENDUSE
endswitch
------------------------------------------------------------------
DH3:applyVelocityVoltage

Contract
Open

Input Parameters
driveId                    Type: Integer
velocityVoltage            Type: Integer

Multi Domain Bridge Code
switch driveId
case 1 # port canard
  $USE IO
    ioCardNo = 1
    regNo = 4
    regValue = velocityVoltage * 3276.75
    []=REG2:setReg[ioCardNo, regNo, regValue]
  $ENDUSE
...
endswitch
```

Fig. 12.27 Examples of how not to specify bridges

Integrating domains

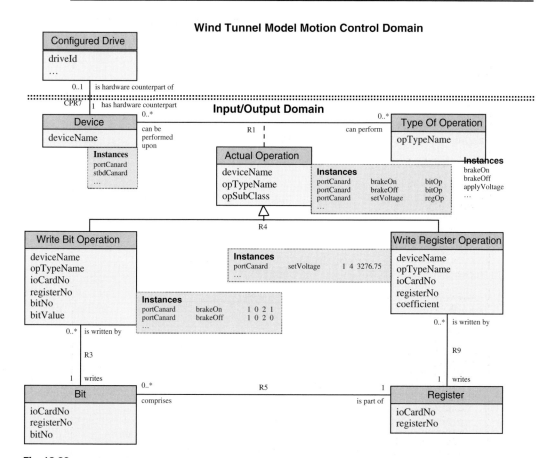

Fig. 12.28 Mapping between domains

1. The component versions (domain versions) for this system;
2. The pin-to-pin connections (counterparts and bridge mappings) between the components.
   The principles of build sets are described as follows.

Each domain chart summarizes a system, as in Figure 12.30.

Each domain has one or more versions (Figure 12.31).

The build set associates each domain reference on the domain chart with an actual domain version (Figure 12.32).

The build set specifies the counterpart relationships and bridge operations (Figure 12.33).

A build set specifies:
1. The set of domain versions;
2. The set of bridge mapping versions;
3. The set of scenario specification versions.

Each domain version will have a complete set of **stubbed bridge implementations** (for single domain simulation) that can be overridden by **build-set specific bridge implementations** for each build set to which the domain version is assigned.

```
Build Set for Domain: WindTunnelModelMotionControl (WTMMC)

Required Operations for Terminator: DriveHardware (DH)

DH2:applyBrake

Contract
Open

Input Parameters

Multi Domain Bridge Code

$USE IO
  counterpartDevice = this -> CPR7
  opName = "brakeOn"
  []=DEV2:performBitOperation[opName] on counterpartDevice
$ENDUSE

-----------------------------------------------------------------------
DH3:applyVelocityVoltage

Contract
Open

Input Parameters
velocityVoltage         Type: Integer

Multi Domain Bridge Code

$USE IO
  counterpartDevice = this -> CPR7
  op_name = "applyVoltage"
  []=DEV3:performRegOperation[opName, velocityVoltage] on
  counterpartDevice
$ENDUSE
```

Fig. 12.29 Example of mapping between domains

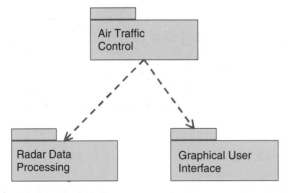

Fig. 12.30 Simple domain chart

323 Integrating domains

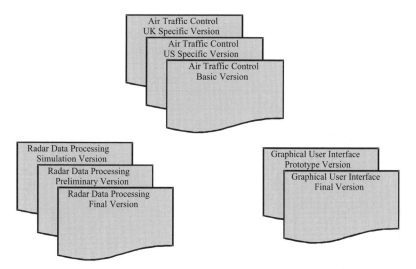

Fig. 12.31 Domain versions

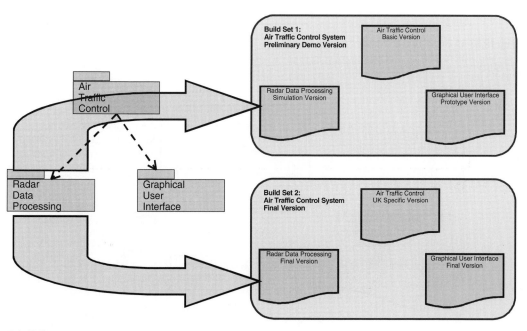

Fig. 12.32 Build sets

12.7.1 Why not use 'foreign classes'?

Of course, we could honour the tradition of using 'foreign classes' within each domain package to show which classes in other domains we are to interact with (Figure 12.34). The folly of this approach is obvious. It instantly renders the packages system-specific and

324 Model Driven Architecture with Executable UML

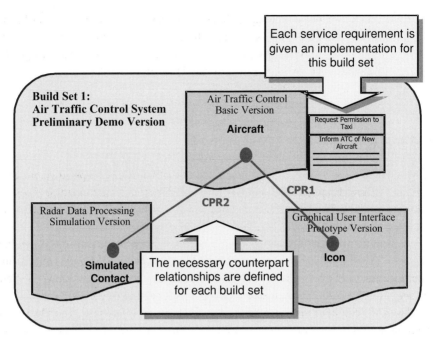

Fig. 12.33 Contents of a build set

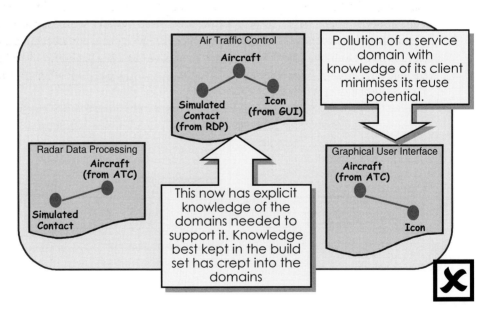

Fig. 12.34 Foreign classes

increases their complexity. The separation of concerns, that we worked so hard to achieve, is abandoned, and we are back on the road to hell. It will take time, but the domain will become progressively more schizophrenic, embedding more and more knowledge of subject matters unrelated to the one that it was intended to formalize.

It's easy to do, but don't do it.

12.8 Summary

In this chapter, we have considered how a set of domains can be put together to achieve a whole system. In may be worthwhile revisiting Chapter 5 on domain partitioning since knowledge of how domains can be integrated helps with the understanding of how to identify the different types of domain found in a system.

The important thing to remember about contracts is that an interface is more than a list of operations and the parameters they take. Our experience has shown that this simplistic approach frequently leads to systems that will not integrate. It is important to express the meaning or semantics of the interface and contracts can help with that.

The specification of required and provided interfaces and the use of bridges to connect them promotes low coupling between the PIMs that describe the domains. This increases the chances that the domains can be reused and can be treated as valuable assets.

A simple bridge was developed and illustrated using ASL. Entire systems can be built using simple bridges alone. In fact, bridges should, wherever possible, be simple. Don't let domain functionality creep into the bridges. The concepts of counterpart associations and generalizations spanning domains were introduced, allowing powerful and subtle concepts to be utilized in bridges. We have chosen to provide an introduction to them but it is recommended that you become familiar with the simple bridges first. Further details on advanced bridges can be found at www.kc.com.

13 System generation

13.1 Introduction

In previous chapters, we have seen that xUML provides a rigorous, executable formalism that allows complete model verification. The benefits of such formalism, applied within an MDA process, do not stop at early model testing; they allow complete system generation from the models produced during analysis. This is only possible because of the complete, self-consistent and executable nature of the formalism, encompassing static, dynamic and process modelling.

There is a wide range of possibilities when it comes to generation of the final system, ranging from legacy hand-elaborated design and implementation, through pattern driven development and manual 'code-generation', to sophisticated formalism-centric auto-code generation.

In this chapter, we shall touch on a number of these possibilities and show how we can exploit the expressive and powerful nature of xUML to simplify the process of transforming PIMs, that are a verified expression of a system's business requirements, into the final target code. We shall concentrate on a code generation approach based upon the translation of xUML models. This methodology is founded in the xUML metamodel, which will also be briefly considered.

This approach of the use of metamodels to capture the mapping rules from the executable expression of the analysis models (PIM) to the implementation (PSM) is at the heart of the OMG's MDA, as discussed in Chapter 2 (www.omg.com/mda). To recap, consider Figure 13.1 showing how a single mapping can be reused to translate PIMs expressing differing subject matters into their PSMs. This chapter explores how we can build and exploit such reusable mappings, expressed in the, by now familiar, formalism, the xUML.

13.2 Conventional approaches to system implementation

There are many ways we may realize a system given a set of requirements. This section explores some of that spectrum of possibilities.

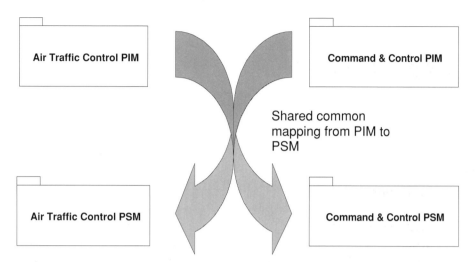

Fig. 13.1 An example mapping from PIM to PSM in MDA

13.2.1 The traditional elaborative approach (the legacy system development method)

This is the approach that is still taken by the majority of system developers to this day. We shall not take much time examining this method of development since it exhibits many undesirable traits. That said, if we choose to use xUML modelling in elaborative development then this provides us with some benefits since the models are rigorous and have been tested, thus ensuring that we have tested models that meet the functional requirements. However, the models usually built in an elaborative development are without such executable properties and so can only be assessed as 'correct' and complete by highly subjective human review processes.

The basic approach that most elaborative developers take to producing a system is given in Figure 13.2, assuming that non-executable modelling is employed.

The diagram assumes that a traditional, non-rigorous, non-executable form of analysis is employed. This, combined with the waterfall nature of the development life cycle, leads to a number of undesirable qualities, discussed as follows.

This approach may be summarized by saying that analysis models are built that are vague and incomplete. These are then elaborated into design, where we add elements of our design, along with further functional requirements (shown by the solid black arrows on the left). This increases the 'fidelity' of the models, adding detail that was deemed too 'low level' for the analysis models. Finally, we elaborate the design into implementation, adding further user requirements that were not captured in either the analysis or design. The boundary between analysis, design and implementation is moot. Arbitrary decisions on what is reflected in the analysis, design or just in the code are made on an ad hoc basis by the developers. We do not have any strong completion criteria for the analysis and design models, except to say that they are complete when they are 'good enough' (or when the budget is exhausted)!

Model Driven Architecture with Executable UML

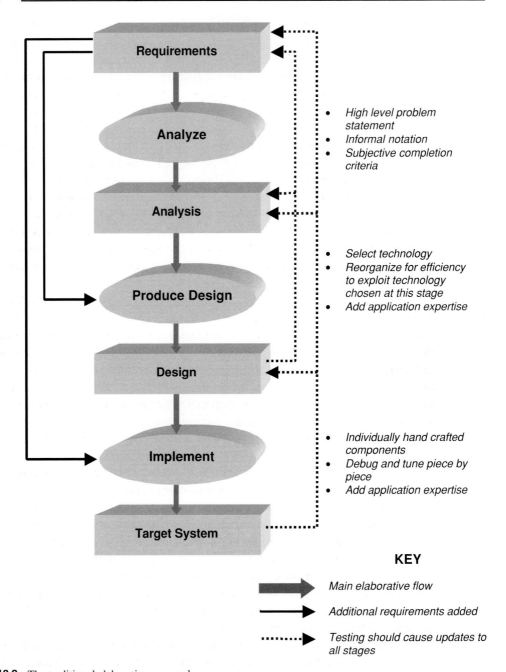

Fig. 13.2 The traditional elaborative approach

If a requirement is captured in analysis then it is repeated in design and implementation. Such redundancy (the same fact in three places) is a source of many errors. It also adds to the maintenance overhead of the models.

The code is the usual place to implement changes to fix problems with the original business requirements (thus requiring changes to the requirements and the analysis and design models) or non-functional requirements (requiring update to the design models) or in our understanding of the business (thus requiring update to the analysis and design models). Because the boundaries are blurred between the models, it is not always apparent what artefacts need updating when the code is changed. Thus, the statement of requirements and the design and analysis models must be updated to prevent early obsolescence of the modelling artefacts (the updates are shown by the dotted arrows on the right-hand side of Figure 13.2). The design may be updated from code using reverse engineering tools, but this helps little in the maintenance of analysis models, which invariably has to be done manually. This is error prone and subject to time and budget pressures. As a result, analysis models often fail to reflect the real system, even after only a few product iterations, thus becoming so much 'shelfware'.

Each element of the analysis model is turned into a bespoke design and then code element on an ad hoc basis leading to many undesirable properties for the finished code. It is non-uniform, there are no rules specified on how an analysis element is transformed into a design element and then onto code. Since each developer produces design and code artefacts in their own way, the resulting code is difficult to understand and traceability back through the analysis model to requirements is difficult at best. Rigorous coding standards can help to alleviate such symptoms of bodging a design but are incapable of elevating coding from the realms of a craft to an engineering discipline.

Testing of the target code has to cover two distinct areas. First, it obviously has to address the functional correctness of the developed system; that is have the requirements been satisfactorily addressed? The analysis models are incapable of fully undertaking this problem since they are not executable, so running the final target code is the first opportunity to test the system for functional correctness. Secondly, a significant proportion of the code that is written for an application is not concerned with directly implementing the requirements, it is infrastructural (examples include management of linked-lists or collections, allocation and deletion of memory, contention for shared resources, persisting or retrieving data, pooling of database connections, etc.). The quantity of the 'support' code tends to increase when middleware is used, further obscuring the business requirements realizations with the platform code that supports it. Testing of this type of software is done alongside the main functional testing, the two interfering with each other.

Because each software element is handcrafted there will be a distinct lack of uniformity in the code, allowing for the introduction of non-systematic errors which are notoriously difficult to find and resolve.

Reverse engineering, if used, gives us a 'picture' of the code. This view incorporates the business constructs modelled in the analysis, along with all the design artefacts and programming constructs from our chosen implementation. Therefore, there is an enormous gap between the view that we require in analysis (business-centric) and the reverse-engineered view (code-centric). This view is further (and possibly irrevocably) broken from the analysis view if a non-OO paradigm is chosen for the implementation.

It is vital to maintain the analysis models as a current and valid representation of the requirements. The maintenance links (shown as dotted arrows in Figure 13.2) have to be implemented by hand because of the limitations of reverse engineering, so, at best, the analysis model maintenance is costly and error prone and, at worst, does not occur at all, resulting in models that become little better than 'shelfware'.

The process is non-repeatable; once an individual has created the design and code artefacts, there is no guarantee that another (or the same person after a few weeks) could reproduce, or fully justify, the design and implementation.

The traditional approach assumes that a developer can develop (understand, characterize and handle changes to) both the business requirements for a system and the technical architecture. The two become inextricably linked. The rate of change of both these aspects in a modern system seems to be ever increasing. This entangled approach to analysis and design places an almost impossible demand on the breadth of knowledge of a single developer. We must separate the concerns of technology that implements a system from the business requirements that are implemented by it. The traditional approach is rooted in old-fashioned systems that assume technology that is based upon a single executable with simple scheduling and simple persistence. In the case of modern enterprise systems there are a myriad of concerns, comprising complex behaviour to do with distribution, failover, scheduling, security, heterogeneous systems and so forth. In the embedded world complexity is growing at an equal pace, with advanced RTOSs, distribution technologies, etc. The intertwining of the technology and business concerns makes any change to either the implementation technology or the business requirements considerably more difficult than when we keep these concerns separate. This is why the elaborative approach is contributing to the enormous rate of project failures in modern software developments.

13.2.2 The use of xUML in a traditional elaborative approach

If xUML is used then the bleak picture painted in the preceding section is at least improved. The completion criterion for analysis is clear and straightforward, namely that the models execute and demonstrate compliance against the requirements. This has a subsidiary effect in preventing business requirements from creeping into the design and code, since such additional requirements should have been demonstrated by the execution of the xUML analysis. Testing is also simplified, with the basic compliance of the analysis models against the requirements having already been demonstrated.

13.2.3 Design pattern based development

The software development life cycle is essentially unchanged from the basic waterfall form given in Figure 13.2, however, here we use a design pattern catalogue to formalize the lessons learnt in previous developments. This is clearly a better approach than reinventing the wheel every time developers come across a similar problem. We can use patterns easily with an elaborative approach to increase the quality of the artefacts in both analysis and design. The quality and uniformity of the code can be further improved by the systematic application of coding standards.

The main drawback of this approach is spotting that the problem matches one in the catalogue. Even if this fit is correctly identified, it is often not an easy task to suitably modify the generic pattern to its particular instantiation in the system. This leads to local modifications of the pattern, which are not reused or catalogued.

Over-exuberance by developers, who have realized the power of patterns, often leads to over use of a favoured pattern or patterns; for instance, it is common to see the factory pattern applied indiscriminately throughout a design.

The use of patterns by themselves ameliorates, rather than eliminates, the problems of the traditional approach. It is necessary to deal with the shortcomings of the elaborative approach in a much more radical way by use of translation in the MDA process.

Of course, the misapplication of patterns is all too frequently observed at all stages and in all types of system development approaches – see Chapter 11 for some of the most popular anti-patterns.

In summary, beware of any advice that suggests that the application of patterns in themselves will solve all software woes.

13.3 Translation-driven development

This approach differs from the traditional and pattern-driven approaches by separating the development of the analysis models and the solutions that will realize the system. We link the two distinct subject matters by defining a systematic and complete mapping from all xUML analysis constructs to their 'solution'. Because xUML is underpinned by precise executable semantics, we are able to specify an **abstract** implementation of those semantics, free from application bias (i.e. we do not define this mapping in an ad hoc case-by-case basis).

At the heart of an xUML architecture is a set of mapping rules that take elements of the populated xUML metamodel and transform these ultimately into a PSI.

13.3.1 Example of a translated model

Before we examine the theoretical underpinning of this approach, let us explore a typical analysis fragment and its implementation. For our small example, we take an air traffic control domain (ATC), in which we have modelled classes such as AIRCRAFT and AEROPLANETYPE (Figure 13.3) and we use Java as our implementation language. A very simple approach to implementing this analysis model may lead us, at first, to the following code for the AIRCRAFT analysis class:

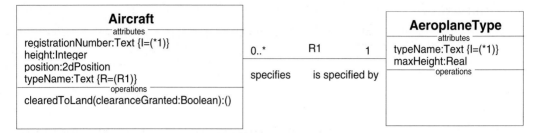

Fig. 13.3 A simplified ATC model

```
class Aircraft {

  // Attributes
  private int registrationNumber;
  private int height;
  private 2dPosition position;
  private String typeName;

  // Attribute read accessors
  public int readRegistrationNumber () {
    return registrationNumber;
  }

  public int readHeight () {
    return height;
  }
  public 2dPosition readPosition () {
    return position;
  }
  public String readTypeName () {
    return typeName;
  }

  // Attribute write accessors
  public void writeRegistrationNumber (int theNewRegistrationNumber) {
    registrationNumber = theNewRegistrationNumber;
  }
  public void writeHeight (int theNewHeight) {
    height = theNewHeight;
  }
  public void writePosition (2dPosition theNewPosition) {
    position = theNewPosition;
  }
  public void writeTypeName (String theNewTypeName) {
    typeName = theNewTypeName;
```

}

```
// UML operations
public boolean clearanceGranted clearedToLand () {
// Translated ASL goes here!
}
}
```

Even on casual inspection of the code, it is obvious that such an implementation is derived from the analysis model. If we look at the attributes for the analysis class AIRCRAFT, then these have systematically become private members of the implementation class and accessor methods have been provided to read and write each of them; the design therefore employs the principle of encapsulation. Note, we make no judgement here as to whether or not this is an appropriate design paradigm to employ in this situation!

After we have hand-coded a number of our analysis classes it will become apparent that all of our classes in this system will have the **same** structure, with the appropriate changes to attribute names and types. That's because the **same** rules are being applied. In fact, it would be possible to write a template, or **archetype**, for a typical, but unspecified, analysis class. If we use the convention of placing substitution text in angle brackets thus:

```
<substitutionGoesHere>
```

then the archetype would look something like:

```
Class <theClassName> {
  // Repeat for as many attributes as we
  //   have got in the analysis class
  private <attributeType1> <attributeName1>
  private <attributeType2> <attributeName2>

  ...............

  // Attribute read accessors,
  //   repeat for as many attributes as we have got
  public <attributeType1> read<attributeName1> () {
    return <attributeName1>;
  }
  public <attributeType2> read<attributeName2> () {
    return <attributeName2>;
  }

  ...............

  // Attribute write accessors,
  //   repeat for as many attributes as we have got in the analysis class
  public void write<attributeName1> (<attributeType1>
                                     theNew<attributeName1>) {
    <attributeName1> = theNew<attributeName1>;
```

```
    }
    public void write<attributeName2> (<attributeType2>
                                       theNew<attributeName2>) {
      <attributeName2> = theNew<attributeName2>;
    }

    ..............

}
```

What we have started to do is to capture the translation rules that take analysis entities and transform them into their implementation. This example is intended to show, in the simplest of cases, that such a proposition is practicable. We should soon realize that such an implementation has many shortcomings. Some of these deficiencies arise due to the semantics of xUML; for instance, the implementation has both a read and write accessor for `registrationNumber` (the preferred identifier of Aircraft) and yet we should never write this value outside the context of creating an instance of Aircraft. In order to address this we could define a constructor for the class that sets the value of all identifiers at instance creation time; in addition, we would only provide a read accessor for such identifying attributes. This one, simple modification has already complicated our archetypal rules (we would need to express conditional substitution to formalize this rule). Next, consider the ASL statements that need to be supported in connection with the class, for instance the `find` statement. In order to find an instance we must place the class instances in a container of some sort, further expanding the design. We shall see in the following sections how we can easily capture more advanced translation rules such as these by using the xUML itself to express the rules of substitution.

It is common in legacy approaches to consider the requirements for a class (or any other model element) in an ad hoc fashion, approaching each class as if it were unique. In order to undertake the definition of translation rules in a systematic way we must first examine the underlying principles in the xUML that makes translation possible. We will then have a basis upon which we can write a translator for any analysis model.

13.3.2 The basis of the translational approach

In order to understand the steps in implementing a system in a translational approach, we must first understand the theoretical basis of analysis. The reader should now be familiar with the idea of modelling as being a process of formalizing abstractions within the context of a domain; what the analyst is doing, from an architectural viewpoint, is instantiating the xUML metamodel. Figure 13.4 shows that during the process of modelling, the analyst is, in essence, instantiating elements of the xUML formalism, which we shall express as an xUML metamodel.

Figure 13.5 shows a small and simplified part of the xUML metamodel. It is sufficient to start the discussion of how we apply the mapping indicated by the arrow in Figure 13.4. We do not have space in this book to consider the full model, but we shall discuss some of the major elements of the excerpt given in Figure 13.5.

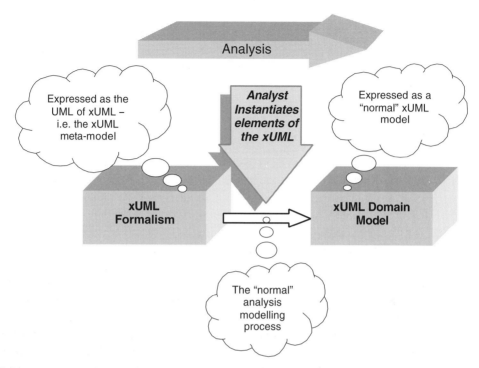

Fig. 13.4 The xUML modelling process

In the following discussion we frequently refer to the xUML metamodel so we shall call this the xUML^2 (the xUML of xUML). We shall also use the convention that classes, and other xUML^2 model elements, will be *italicized*, whilst those from 'standard' (xUML^1) domain models will not.

A good way of interpreting any UML model is to reason about an instantiation of it for one or more typical scenarios. The xUML^2 model is no exception. In order to think about an instantiation we require an 'ordinary' analysis model. Figures 13.6 and 13.7 show part of a simplified model for an air traffic control (ATC) system. The abstractions in this domain are unsurprising, with the viewpoint of air traffic control in mind. However, with a view to the architecture we consider this, and any other analyst created models, in a completely different light. Such analyst created models form the elements that allow us to instantiate the xUML^2 model.

Figure 13.7 has a number of xUML tags added that aids both the translation process and executing the models. Referential attributes are shown by tags such as {R=(R1)}, indicating that these attributes formalize the association R1. The tags, such as {kl=A, no=2}, are shown on classes, giving the class key letter and number, so providing shorthand identifiers for classes. Such tags are always present in an xUML model but the analyst may choose to elide them on the class diagram.

We consider such models as an instantiation of the xUML^2 model. If we instantiate the model in Figure 13.5, with the domain model in Figures 13.6 and 13.7, then we obtain the following instance tables derived from the ATC domain model.

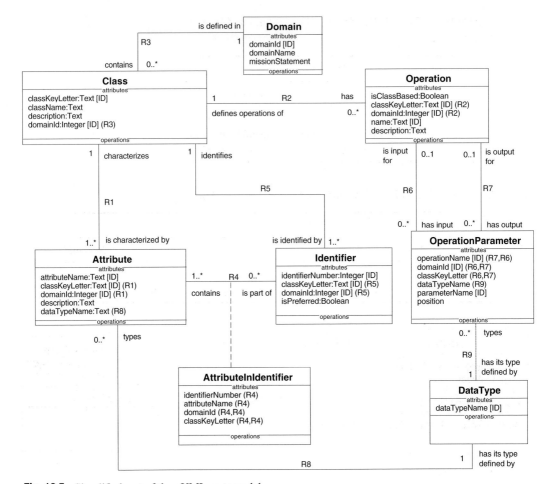

Fig. 13.5 Simplified part of the xUML metamodel

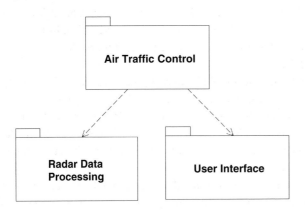

Fig. 13.6 A simplified ATC domain chart

Table 13.1 The xUML^2 *domain* class instantiated for the ATC system

	Domain	
domainId	domainName	missionStatement
1	Air Traffic Control	to boldly control …
2	Radar Data Processing	to boldly process…
3	User Interface	to boldly display…

Table 13.2 The xUML^2 *class* class instantiated for the ATC domain

		Class	
classKeyLetter	className	description	domainId
A	Aircraft	An aircraft under ATC control	1
AT	AeroplaneType	An aircraft type designated by the aviation authority	1

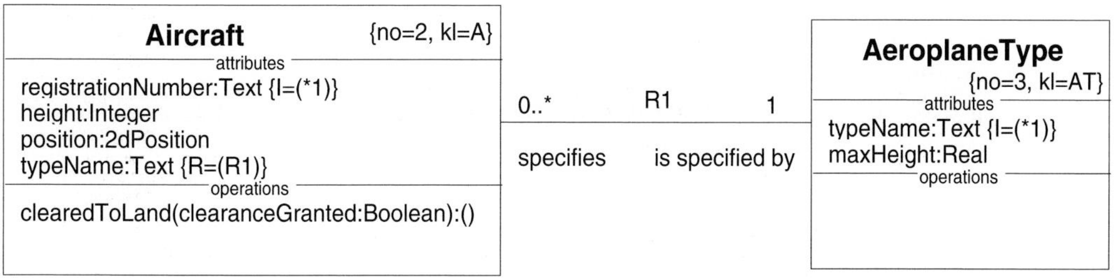

Aircraft {no=2, kl=A}
attributes
registrationNumber:Text {I=(*1)}
height:Integer
position:2dPosition
typeName:Text {R=(R1)}
operations
clearedToLand(clearanceGranted:Boolean):()

0..* R1 1
specifies   is specified by

AeroplaneType {no=3, kl=AT}
attributes
typeName:Text {I=(*1)}
maxHeight:Real
operations

Fig. 13.7 A tagged ATC model

First, consider our air traffic control system. It has three domains. We instantiate the xUML^2 class *Domain* with these instances, giving us the instance table in Table 13.1.

Secondly, consider the instances of *Class*, (again from the xUML^2), instantiated for the ATC domain (*domainId* = 1). *Class*es are defined in exactly one *Domain*, given by the association *R3* (Table 13.2).

Now, for each of the *Class*es in the ATC domain, we have the following 'Attribute' instances, which characterize a *Class* via the association *R1* (Table 13.3).

Finally, consider the instances of *Operation* in the ATC domain (Table 13.4).

There are other classes in the xUML^2 model, given in Figure 13.5 (such as *Identifier*, *DataType*, *OperationParameter*, etc.), which will also be instantiated in a similar manner to the classes explored here.

Given these instance tables, we can see that the xUML^2 model is like any other UML model. We have seen in previous chapters that analysis classes always belong to a domain (in fact this policy is enforced by the xUML^2 model in the cardinality expressed by *R3*). The xUML itself defines the domain for the xUML^2 model given in Figure 13.5. We shall

Table 13.3 The xUML^2 *Attribute* class instantiated for the ATC domain

			Attribute	
attributeName	classKeyLetter	domainId	description	dataTypeName
registrationNo	A	1	the A/C tail number	Text
height	A	1	Height above sea level	Integer
position	A	1	Lat/long	2dPosition
typeName	A	1	referential	text
typeName	AT	1	An aviation authority recognized type	text

Table 13.4 The xUML^2 *operation* class instantiated for the ATC domain

		Operation		
isClassBased	classKeyLetter	domainId	name	description
FALSE	A	1	clearedToLand	Allows ATC to inform the ...

see later that a full code generator comprises more domains than the xUML^2 model, such as tagging, build sets and error handling (see Figure 13.11).

It would be a daunting task if we had to produce the xUML^2 model for every development. In fact, the model may be obtained with commercial code generators and populated automatically from xUML tools.

13.4 The design process

We have considered the instantiation of xUML^2. The next step is to move from the world of analysis into the solution space for the problem, that is we shall now consider how to realize the business requirements that have been expressed in xUML models.

Before we can start considering the application of solutions based in computer technology, we must yet again consider the requirements of the system that we intend to build, although the agenda is different on this occasion. Throughout the xUML process, there has been an emphasis on separation of concerns; this step is no different. So, in starting the design, we consider the requirements that must be met by the realization. There are a number of prime considerations and constraints that lead us to consider a number of differing design viewpoints. These primary characteristics of any design that must be considered include:
- **Performance** – what are the requirements for throughput, number of users (human and non-human), responsiveness, liveliness, latency and so forth? What typical usage patterns are there – does data arrive in a predictable manner or is it subject to highly irregular behaviour? What are the requirements for these factors in the future? Does the system have to cope with ever-increasing demands over its lifetime? Are the performance limits

hard (i.e. must cope with the worst case scenario) or soft, in which case we can engineer against an average environment?
- **Robustness** – what availability is required of the system? What are the consequences of partial or total failure? Can part of the system fail whilst the rest carries on? Do we require any form of rollback (i.e. transactional) support? What response do we require to partial or total failure? Do we require any resilience to hardware failure?
- **Distribution** – is the system intended to be distributed locally (across a LAN), geographically (across a WAN), or simply multi-processed within one machine? In what ways can we change system distribution? What will be the units of distribution (domain instance, class, class instance state, etc.)? Are the units of distribution statically allocated or dynamically, in order to provide load balancing?
- **Utilization** – do we have to achieve, or avoid, certain utilization levels for the hardware resources that we have available?
- **Persistence** – what level and type of persistence is required? Does the system exhibit a uniform or a complex heterogeneous mix of persistence types?
- **Safety** – is any, or all, of the system to be considered safety-related or critical? If so, what standards and prohibitions must we meet?
- **Through-life maintenance** – is there any requirement to replace some, or all, of the system whilst it is still running? If not, what level and frequency of downtime is allowable?
- What is the acceptable **defect density**?
- What are the **security** considerations? Who is authorized to use or maintain the system? Are there heterogeneous security levels?
- What **scheduling** requirements are present?
- What special requirements must be met for **startup** and **shutdown**?

The answers to these questions (and others) will lead to a set of design characteristics that the system must exhibit. We observe that the characteristics listed interact with each other in complex ways. For instance, the effect of distribution will profoundly influence performance, reliability and throughput (and other characteristics), whilst persistence and transactional support often go hand-in-hand.

Next we consider the constraints that may be imposed on the development:
- Are there any constraints on the hardware or resources to be used (is this an embedded application)? Have the machines and/or networks been chosen prior to the development?
- Are there any legacy systems that must be incorporated? If so, do they have well-defined APIs or do we require xUML wrappers for them?
- Have any of the implementation domains been procured (database, middleware, etc.)? Can we utilize already existing licences for databases, development tools, compilers, etc?
- Do we have any externally defined standards to meet (e.g. coding standards)?
- Have particular development languages or operating systems been mandated?

It is the job of the architect to resolve these competing design characteristics whilst considering the constraints. It should be borne in mind, however, that the principle of Ockham's razor should be applied, that is we should strive to define the simplest architecture that will meet these characteristics today and for the foreseeable future of the system, rather than the cleverest or most fashionable creation that the architects' imagination can conceive.

340 Model Driven Architecture with Executable UML

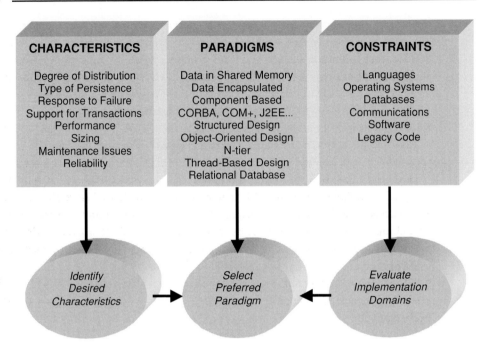

Fig. 13.8 The design process

Finally, we consider the design paradigm and language (or languages) that we shall employ to implement the system. Because xUML is object-oriented, it is tempting to jump to the conclusion that the design and implementation must also be OO. This is not necessarily the case. With the characteristics of the architecture and constraints identified we may choose to use structured design and a third generation language such as C (or a myriad of other styles). Translation of the models into their implementation allows us great flexibility in the design approach we use. We do not have to be wedded to one design paradigm or language type.

Pictorially we can think of these stages as in Figure 13.8.

We observe that some of the design paradigms are mutually exclusive (for instance object-oriented design and structured design) whilst others may be used in conjunction to produce the overall design paradigm (such as an object-oriented, component-based, N-tier architecture).

13.5 Transformation of instantiated xUML^2 models – design

At this stage, we have a number of elements of the puzzle that will eventually lead to the final system. We have the instantiated xUML metamodel (xUML^2), we have the chosen design approach containing our architectural characteristics, constraints and embodying our chosen paradigm and we also have the definition of the xUML virtual machine (i.e. the semantics of xUML model behaviour).

System generation

We shall see that the process of design proceeds in parallel, and largely independent, to the analysis effort (see Section 13.11). So actually, our xUML^2 may not be fully populated yet (i.e. the analysts have not finished). This presents no difficulties since what we shall define is a mapping from **any** consistently instantiated xUML^2 model, not just the particular one that applies at the moment to our particular business problem.

Now the stage is set for us to define the transformations that take an xUML^2 model and produce the code. In many ways we shall treat the xUML^2 model like any other xUML model. There are two main ways of exploiting this instantiated model. The first, which we explore here, is to write processing for the model that transforms the model elements into the code directly (termed the **formalism centric approach**). In effect, we capture the design and coding steps in the ASL specified process models of the xUML^2. The second approach is to produce a separate design model, populate this from the instantiated xUML^2 model and then write the ASL to generate code from this (the **explicit design model approach**). We shall briefly examine the pros and cons of this latter approach in Section 13.5.2.

We can think of these approaches as counterparting elements of the xUML^2 model into design and finally code. Figure 13.9 expands Figure 13.4 to include design and code of the architecture.

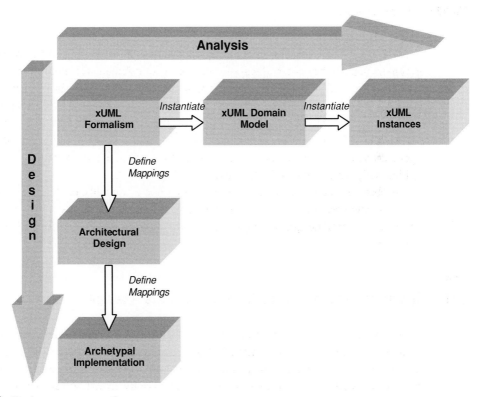

Fig. 13.9 Design as counterparting

We are now in a position to define what is meant by the software architecture. It comprises the design process in Figure 13.8 and the transformation rules that define the counterparting in Figure 13.9, given the operational semantics of the xUML. In other words, it is the way that we transform analysis models into an implementation that meets our non-functional requirements and the requirements of a well-formed xUML model.

This sounds more complicated than it actually is, so let us look at a simple example. Given the formalism-centric approach (short circuiting the explicit architectural design step) the processing we require on the xUML^2 model is given in principle by the following ASL:

```
allDomains = find-all Domains
for aDomain in {allDomains} do
  # generate code for aDomain
  {classesForDomain} = aDomain -> R3.Class
  for aClass in {classesForDomain} do
    # generate code for aClass
    {attributesForClass} -> R1.Attribute
    for anAttribute in {attributesForClass} do
      # generate code for the attribute
    endfor
  endfor
endfor
```

This scheme iterates over the set of domains, in turn iterating over the classes within each domain and so forth. It is apparent that this ASL is completely general, in the sense that it will produce code for any xUML-specified system. The counterparts, from analysis to implementation, are produced by the 'magic' of the comment lines `# generate code`. We shall see below how we actually can achieve this step.

We do not have space in this book to explore a full architecture, so we shall look at the loop to generate code for each class in more detail. We can take Figure 13.9 and see how it looks concentrating on the xUML^2 *Class*.

Figure 13.10 implies that we must define the mapping from an xUML *class*, right the way into (in this case) a Java class. The ASL that produces the code for a *class* might look like:

```
1.  for aClass in {classesForDomain} do
2.    [classFileName] = FM20001:constructClassFileName [aClass]
3.    [classFileNameHandle] = FM20002:openFileForWriting [class
         FileName]
4.    $FORMAT classFileNameHandle
5.      class [T:aClass.className] {
6.    $ENDFORMAT
7.    {attributesForClass} -> R1.Attribute
8.    for anAttribute in {attributesForClass} do
9.      attrType = anAttribute -> R8.DataType
10.     $FORMAT classFileNameHandle
```

System generation

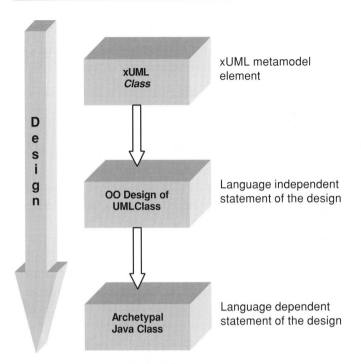

Fig. 13.10 Counterparting the xUML *class* in design

```
11.      private [T:attrType.dataTypeName] [T:anAttribute.
         attributeName];
12.      public [T:attrType.dataTypeName] read[T:anAttribute.
         attributeName] () {
13.        return [T:anAttribute.attributeName];
14.      }
15.      public void write [T:anAttribute.attributeName]
16.        ([T:attrType.dataTypeName] theNew[T:anAttribute.
         attributeName]){
17.        [T:anAttribute.attributeName] = theNew[T:anAttribute.
         attributeName];
18.      }
19.    $ENDFORMAT
20.    endfor
21.    $FORMAT classFileNameHandle
22.    }
23.    $ENDFORMAT
24.  endfor
```

Taking a closer look at this simple code generator, reveals a typical (if simplified) structure. On lines 2 and 3 we have a couple of utility operations (details are not shown). The first:

```
[classFileName] = FM20001:constructClassFileName [aClass]
```

takes an instance of *Class* from the xUML metamodel and returns a filename (which will be fully qualified with the path to the generated filename, for example

```
/myPath/myGeneratedCode/AirTrafficControlDomain/Aircraft.java
```

It is common to have a small number of such operations that control the whereabouts of the output of the code generator.

Line 3 shows an operation that, given this generated filename, opens the file for writing and returns the file handle, so we can subsequently write to this file. It is obvious that this operation could fail (for instance because of the file system being full), however we do not show any exceptions or error handling here, which would be necessary in a real, robust code generator.

Next comes a mysterious construct, on line 4:

```
$FORMAT classFileNameHandle
```

This is a directive in ASL (available for use with the iCCG code generation suite, see www.kc.com) which allows us to write all of the following text, up to the $ENDFORMAT directive, to the given file (or in fact to a local ASL text variable, useful for intermediate formatting). The text within the $FORMAT directive is output as literals (including white space and new lines), with the exception of constructs such as:

```
[T:anAttribute.attributeName]
```

which defines a substitution rule, in this case indicating that the text resulting from the evaluation of the expression introduced by the [T: should be inserted in this position. In our case, we have the attribute read accessor:

```
anAttribute.attributeName
```

The metamodel processing rules, expressed in standard ASL, combined with the literal output and textual substitution rules of the special $FORMAT directive, allow us to complete the counterpart mapping of xUML metamodel elements to their implementation; in other words, we have specified a code generator! It is possible to perform code generation without the special $FORMAT directive, using native language constructs (such as printf in the language C), however these tend to be much more cumbersome to use.

The code generator specified above will give us a simple coded class close to the hand-coded one presented at the start of the chapter. It still has many deficiencies, for instance we create a write accessor even for attributes that form part of an identifier (which are not permitted to be written outside the scope of object creation). We solve this by extending our use of the metamodel (see Figure 13.5), navigating R4 to see if the attribute is part of an identifier, only if it is not do we create the write accessor. Thus we get the following modification to the relevant part of our code generator:

```
1.  for anAttribute in {attributesForClass} do
2.    attrType = anAttribute -> R8.DataType
```

```
3.    $FORMAT classFileNameHandle
4.      private [T:attrType.dataTypeName] [T:anAttribute.
        attributeName];
5.      public [T:attrType.dataTypeName] read[T:anAttribute.
        attributeName] () {
6.        return [T:anAttribute.attributeName];
7.      }
8.    $ENDFORMAT
9.    {identifiersForThisAttribute} = anAttribute -> R4.Identifier
10.     if countof {identifiersForThisAttribute} = 0
11.       # i.e. this attribute is NOT part of an identifier
12.       $FORMAT classFileNameHandle
13.         public void write[T:anAttribute.attributeName]
14.           ([T:attrType.dataTypeName] theNew[T:anAttribute.
           attributeName]){
15.           [T:anAttribute.attributeName] = theNew[T:anAttribute.
           attributeName];
16.         }
17.       $ENDFORMAT
18.     else
19.       # No output — this attribute forms part of at least one
          identifier,
20.       # therefore we do not want to generate a write accessor
          for it.
21.     endif
22.   endfor
```

This shows how we can start to think about tailoring the rules of the code generator depending upon the instances (and attribute values) of the populated xUML^2.

Of course, this is just the first of many improvements we can make to the generator. For instance, we could further improve the read and write accessor generation by examining the process models (ASL written by the analysts) and only generating those accessors that are actually used. It is important that the architects concentrate on generating correct code at first and then turn their attention to such optimizations.

There are also improvements to be made, if we wish, in the layout and appearance of the output code. For instance, in the aircraft example, the write accessor for the new value of the position attribute will be generated as `theNewpositon`. We could pass this through a prettify function within the code generator prior to output to produce 'camel case', that is `theNewPositon`. The architect must consider whether such niceties are worth the effort, since it is the models, not the code, that will be inspected and maintained.

If we were to include the optimization of only including attribute accessors that are actually used by analysts in their ASL, where should we look to find instances of these attribute operations? The answer to this probably comes as no surprise. Such ASL constructs are represented as instances of metaclasses in the metamodel of ASL. Again, we shall follow the convention of calling this model ASL^2. Therefore, if we consider the xUML^2

model as providing a view of the macroscopic structure of an xUML model (*classes*, *attributes*, *operations*, etc.), whilst the ASL^2 model provides the microscopic process view, with metaclasses such as *specified process* specialized into classes such as *create*, *generate*, *delete*, *if*, *switch* and so on; in fact we meet all the classes that a parser of ASL would understand. We can use these classes, in a similar way to the classes in the $xUML^2$ model, to generate code for our ASL. Since ASL represents a distinct subject matter, the ASL^2 model is a separate domain.

This formalism-centric approach has been successfully employed for a number of years for systems that are now operational. The architectures produced range from highly constrained safety critical embedded applications to massively distributed enterprise-wide applications. Target languages include C, C++ and Ada.

13.5.1 Complete code generator

We have seen that the fundamental part of the architecture is based upon the xUML metamodel. This, however, does not cover all the subject matters present in a code generator. Figure 13.11 gives a typical domain chart covering the needs of a code generation system for xUML.

The application domain is Code Generator; in the formalism-centric code generator this is a small domain that orchestrates the production of code. It calls upon other domains to generate code. The xUML domain (the $xUML^2$ that we have already explored) uses a population domain to populate its instances, this in turn relies upon the API of the model repository, usually the API of the analysts' xUML CASE tool. We have seen that the xUML domain contains classes such as *operation*. The processing contained in these, and the state actions, etc., are held in the Action Specification Language domain, which is a metamodel of ASL. This is populated from an ASL parser that in turn relies upon Lexer and Format Services. Error Handling is a generic service that allows reporting of such things as ASL syntax errors, sub-optimal use of find statements, redundant model elements and in fact any report that can be generated by writing processing against any of the models in the other domains. The xUML domain also makes use of a Tagging domain; this is a metamodel of the way the xUML models may be tagged, sometimes known as colouring. This is further discussed in Section 13.8. Finally, the Build Management domain handles the configuration management view of the system build, modelling the versions of each xUML domain used in a particular build and the bridges mapping required to provided services, whose assemblage makes up the complete executable system.

A number of these domains are common to all code generators and therefore can be provided 'off-the-shelf'. Such standard domains and the bridges that link them to make up the framework of the code generator can be purchased as part of a commercial offering. Therefore, it is not necessary to develop all of these domains but only those that characterize the specific architecture that is to be built. This may still be a non-trivial task but it is considerably simpler than the alternative proposition of developing the whole code generator infrastructure from first principles.

System generation

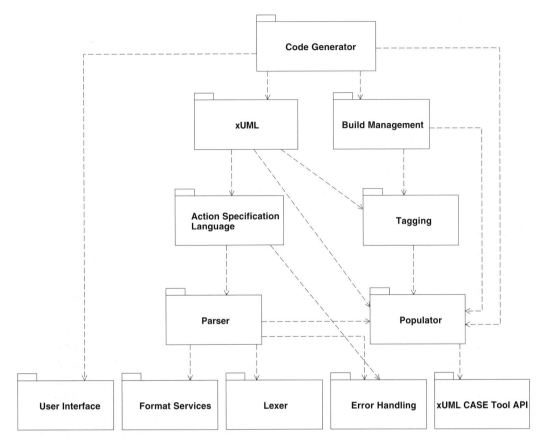

Fig. 13.11 A code generator domain chart

13.5.2 Explicit design model approach

In the previous section, we concentrated on a code generation strategy that centred on the xUML formalism itself. The design is expressed **implicitly** in the ASL that implements the code generation rules against the xUML metamodel (and other metamodels such as ASL and tagging). As an alternative to this approach, we could choose to build an explicit UML model of the design, populate this from the xUML[2] (the population rules would of course be expressed in ASL) and then write an ASL code generator that operates against the instantiated design model.

An example domain chart for a component-based architecture is given in Figure 13.12. Here the application domain captures the rules and polices of a component-based architecture; we would expect classes such as component, interface, computing node, etc. to appear in this domain. This model aligns with the ideas of the PIM in the OMG's MDA framework (www.omg.org/mda). In this particular system we have chosen to realize the architecture using Java's J2EE. This corresponds to the MDA's PSM.

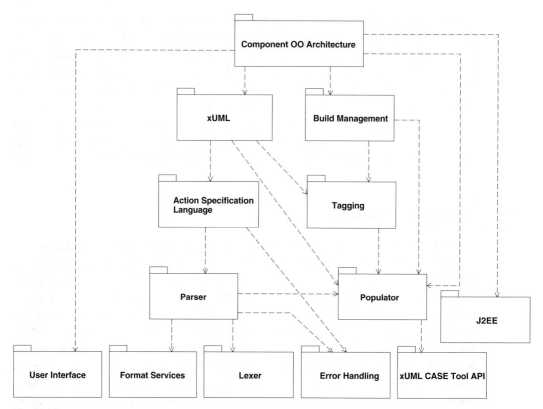

Fig. 13.12 Explicit design-centric code generation

The primary difference between this domain chart and the formalism-centric one is that the application domain uses the xUML metadomains to populate instances of design artefacts (such as component and interface) and then uses the PSM (J2EE in this case) to output the appropriate code.

We do not have the space here to examine this approach in any depth, but we can state some of the pros and cons of the formalism versus explicit design-centred approach.

The formalism-centric approach is most applicable when the design is 'regular', that is there are few exceptions to the way that we generate code for a particular model element type (such variances are driven by the use of a tagging model, see Section 13.8). If such regularity is present in our architecture, then the code generator can often be produced more quickly, since there is only one (rather than two) process steps in producing the code.

A disadvantage of the formalism-centric approach is the merging of the design and implementation, so that it is often problematic to change the generator to another language later on. The lack of an explicit design makes understanding and later modifying the code generator more difficult. If the design is highly variable (i.e. we have a number of ways to generate various xUML^2 constructs) instantiating an intermediate design model facilitates writing the code generator.

Having seemingly given a hint that the explicit design model approach has benefits that outweigh its costs, it must be stated that the vast majority of current code generators are based upon the formalism-centric approach. However, as architectures become more sophisticated and as a result often less regular, the design model approach is starting to be used. It is best not to view these methodologies as antagonistic, but rather as differing tactics to implement the same goal. It is conceivable that a mixed approach could be employed for an architecture, where the regular parts are directly generated from the metamodels, whilst the highly varying parts are design modelled explicitly. There is also no reason why the split between these approaches may not vary as the requirements on the architecture evolve over time.

13.5.3 Archetypes and mechanisms

The code that we have seen so far has been generated in a way that involves lexical substitution rules specified in ASL, forming the **archetypes** of the architecture. Another facet of the architecture are code elements which are not subject to translation rules at code generation time, these are **mechanisms**. As an example of these, let us turn our attention back to the archetype for the AIRCRAFT class and consider some of the ASL statements that are likely to apply to it. We shall, perhaps, want to `find` instances with certain attribute values, or perhaps perform a `countof` in order to find out how many aeroplanes we are controlling. This implies there should be a container for the instances of AIRCRAFT. In Java we could use a basic container type, such as a hash map. Since every JAVA class has a common ancestor (the 'Object') then we can contain our AIRCRAFT archetype instances in this generic container and cast them back to be AEROPLANE class types when we retrieve one with a `find-one` at run time. The 'UML' class (see the following code) is another example of a mechanism that allows us to capture the idea of a universal architectural identifier; every archetypal class we generate will inherit from this mechanism. This allows instances to be inserted, retrieved and deleted from a hash map container (see the following 'InstanceContainer' class). The Java example is a start at meeting these requirements using such mechanisms.

In any architectural design, there is a trade-off between the use of mechanisms and archetypes. Often archetypes display better type safety (this is less of an issue with a language like Java that safely checks casts at run time). The cost of using an archetype is increased translator complexity.

```
import java.util.*;

class InstanceContainer {
  private HashMap theInstances = new HashMap ();
  public void newInstance (UMLInstance theNewInstance) {
    theInstances.put (new Integer (theNewInstance.archIdentifier),
    theNewInstance);
  }
  public void deleteInstance (UMLInstance theInstance) {
```

```
      theInstances.remove(new Integer (theInstance.archIdentifier));
    }
    public UMLInstance findone (UMLInstance theInstance) {
      return (UMLInstance) theInstances.get(new Integer
        (theInstance.archIdentifier));
    }
    public int countof () {
      return theInstances.size();
    }
}

class UMLInstance {
  private static int noOfInstances = 0;
  public UMLInstance () {
    noOfInstances++;
    archIdentifier = noOfInstances;
  }
  public int archIdentifier;
  public int hashCode() {
    return archIdentifier;
  }
  public boolean equals (Object o) {
    return (o instanceof UMLInstance) &&
      (archIdentifier == ((UMLInstance)o).archIdentifier);
  }
}
```

13.6 Production of the code generator

We have already seen that the metamodels in the code generator may be considered in much the same way as any 'ordinary' xUML analysis models. With this in mind, the strategy that we use for production of the code generator itself will come as no surprise. We code generate it! Such a technique of 'bootstrapping' is common in compiler development of course.

In order to run the executable domains, bridges and mechanisms, comprising our complete code generator, we require an xUML virtual machine. As with any architecture we must assess its requirements. Fortunately, all we require is a minimal architecture: single task, non-persistent, non-distributed, with no real-time constraints. This will be adequate to meet our need (again we can draw an analogy here with the simple compilers that are used to bootstrap compiler developments). With this in mind, there are a number of commercial architectures for xUML, usually aimed at model simulation and debugging which will fulfil our requirement. In fact, some architecture tools, come with such a bootstrapping architecture, as well as all the basic architecture domains, services and population mechanisms discussed. There is an added bonus to using a simulation and debug architecture to produce our code generator: as with any xUML analysis project, we are unlikely to get it right first time, so we can use the facilities of the simulation architecture to debug our code generator. This is another example of the value of reuse of an architecture.

Once we have fully specified our code generation rules, built and tested our mechanisms, then we populate the code generator domains and run the models, using our bootstrapping architecture. The result of the run is our code generator, ready for test.

13.7 Testing the architecture

There are a number of areas that we must consider when testing our architecture. The good news is that because we have used the powerful idea of separation of concerns, we do not have to test everything when it comes to considering our architecture. In order to perform most of these tests, it will be necessary to instrument the architecture. Such instrumentation is code that is generated for trace or testing purposes. In most cases, we shall turn off its generation for the finished system. In some cases, however, it is desirable to leave trace code in place and have the ability to turn on the trace on the live system. This will give us the ability to provide an analysis level view of the state of the system leading up to some unforeseen calamity. Understanding the instances of the analysis classes, their attribute values and the events that have been consumed is much more informative than using a source level debugger on a core file!

13.7.1 Code generation testing

We must satisfy ourselves that the code that is being produced is correct and compliant with any external standards that we have to meet. Coding guidelines (as used in elaborative developments) are useful in specifying the architecture – in fact more useful in certain constrained ways. We can capture, and systematically apply good practices such as use of exception declarations, throwing and placing each block in a try catch construct (or similar). Unlike manual coding guidelines, the code generator will never forget to apply a rule.

Code layout is often considered important. We may put effort into producing code that is beautifully formatted. Whilst this is often considered a 'virtue' for handcrafted code, its merits are strictly limited for code that is produced by a code generator, bearing in mind that we shall always maintain the models and forward engineer the code (i.e. the code is not a maintained deliverable, **the model is the code**!). Therefore, it is rarely appropriate to spend much effort in a code generator getting the code layout 'just so!' (although post-processing it with some form of pretty printer may make initial review easier).

There are a number of tools that will help us in the task of assessing the quality of the generated code. The first, of course, is the compiler itself, since syntactically incorrect code will not pass compilation (or at least should not!). There are other errors, however, that may creep in that will not be spotted by the compiler. For example, in a C++ architecture, we may forget to call all the necessary destructors on deletion of objects, so that our architecture would exhibit memory leakage. There are static analysis tools that can help here, as well as the expedient of soak testing. We may also wish to use tools that assess code complexity through function point analysis or other techniques. Another sensible technique is to use code walkthroughs strictly limited in scope to the code produced for one, or a small number, of UML constructs. For instance, examine the code produced for one small typical analysis class.

13.7.2 xUML virtual machine compliance

We must ensure that our generated system works in accordance with the semantics specified in xUML. For instance, do signal events generated between two instances always arrive in the order in which they were sent? We do not need the analyst-derived models for the finished system in order to perform such tests. In fact, such models are likely to be poor material for this class of testing. We require abstract test suites (which are available from third parties) that exercise our implementation of the xUML virtual machine and check that no semantic violations occur. The ability to buy in, or reuse previously produced abstract test suites, will greatly speed this part of the architectural testing. Thus such testing, as well as the production of the architecture, is independent of the concurrent analysis effort that produces the xUML domain models.

13.7.3 Infrastructure testing

Once we are satisfied that the code generator is producing code that is correct and compliant with xUML semantics, we can test that we are exploiting any infrastructural elements correctly. For instance, we should test our use of CORBA and any services or facilities that we have specified. In this step we should also consider the effects of failures – what happens if a node goes down, etc?

13.7.4 Robustness, soak and stress testing

We should test our architecture to see whether it can meet both its performance and robustness requirements in the presence of typical and worst case loads. If possible, go further and attempt to induce failure, characterize the failure levels and ask whether the operational system will ever be likely to encounter these in the future. If it is a distributed architecture, we must consider situations where one or more nodes suddenly fail and assess how the architecture manages such partial failure.

13.7.5 Performance testing

It is not sufficient to ascertain that our architecture is functionally correct and stable under various loadings. We must also assess its performance in a wide variety of situations. If we fail some of the tests then changes to the code generation rules must be sought. In the last resort, increasing the architecture complexity by extending the tagging scheme may be necessary (see Section 13.8).

13.7.6 Host and target testing

The first stages of architecture testing (Sections 13.7.1, 13.7.2 and, to some extent, Section 13.7.3) may be performed on the host platform. If a performance model is available for the host then some performance estimates can also be performed (or even modelled through the use of real-time simulators). This means that the architecture can be progressed to a very large extent even if the target platform is not currently available.

13.8 Varying the architecture – tagging

Up until this point we have used a constant mapping from an analysis construct to its implementation, for example each instance has become a 'UMLInstance' in Java. Each, and every, class in our analysis models will follow this pattern. We have seen that we can conditionally generate code for model elements by walking the metamodel (the case we saw earlier was the situation that we only produce write accessors for attributes that are not part of a class identifier). Whilst this scheme certainly works and produces a straightforward regular mapping, it may not be the most efficient in all cases.

Tagging is a way for the analyst to supply extra information about model elements, which may be used by the translation system. The information is based upon facts from the viewpoint of the analysed domain, not of the architecture (i.e. tags are business rather than design-oriented).

Let us consider the Air Traffic Control (ATC) domain again, in a slightly different form, with the classes AIRCRAFT and RUNWAY, given in Figure 13.13. The abstraction of AIRCRAFT is a model of an aircraft that is currently under control of air traffic control within an air traffic control zone. The RUNWAY is an abstraction of a runway that the ATC, potentially, has available to it to assign aircraft movements. It is apparent from an understanding of the problem domain that the number of runways will not change over the lifetime of the running system (although the availability of a runway may change), whilst the population of aircraft will be highly dynamic, as aeroplanes fly in and out of the controlled

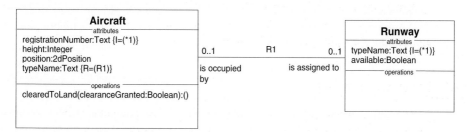

Fig. 13.13 A modified ATC class diagram

zone. The analyst may tag the RUNWAY as having a static population, whilst tagging the AIRCRAFT as having a dynamic population. These 'facts' are based upon the knowledge of the subject matter of air traffic control.

How can the architect exploit this knowledge if he or she chooses to? An obvious area is to modify the choice of container that is used for instances of a class within its *Class*. The dynamic Aircraft would continue to use a hash map container, whilst the static RUNWAY employs an array. If we now classify the xUML^2 *Class* into dynamic and static, using a tag, then the translator can choose the appropriate container for the *Class* currently being code generated.

13.8.1 General tagging schemes

It is the job of the architect to devise a tagging scheme that may be exploited by the code generator and make this available to the analyst teams. In general, a tagging scheme will comprise the following elements:
- Name for the tag (e.g. instance population); description from the viewpoint of the business requirements (e.g. describes whether instances will be created or destroyed at run time);
- Range the allowable values (enumerated values 'static', 'dynamic'; if 'static' then value for population must be supplied, e.g. 'static' 42; if 'dynamic' then upper and lower bounds may be specified, if not then these are assumed to be zero and infinity);
- Allowable xUML^2 model elements that may be tagged with the tag (*Class* in this case);
- A statement as to whether all such model elements must be tagged or whether, in the untagged case, a default tag value will be assumed (e.g. default = 'dynamic').

We give an example of a typical tagging scheme in Figure 13.14. It should be noted that the actual scheme used would vary depending upon the architecture being used, the characteristics the architecture must deliver and the nature of the business requirements. For instance, it is pointless identifying a dynamic instance population tag, if in an embedded application all instances are static.

If we later generate code for a different architecture that does not support some or all of the original tag scheme, then such tags will be ignored.

Consider Figure 13.13, as a simple example of an analysis model that may be tagged using part of this scheme. The analyst would have tagged the AIRCRAFT class with an Instance Population tag of 'dynamic', minimum value zero, maximum value 500 (in this example 500 is deemed suitable as a maximum figure since the business logic, derived from aviation authority requirements, will prevent more than 500 aircraft being in the airspace). If no upper limit can be ascribed, then the scheme must allow the value 'unlimited' to be set. The runway would be tagged with an Instance Population tag of 'static' and a constant value for the number of runways under control of the system (e.g. six). It can be seen that these tag values are driven from the business requirements, whilst the tag schema definition is driven from the technical features that may be exploited in the architecture.

The architect must also bear in mind that the more complicated the tagging scheme, the harder an analyst will find it to use. The judicial use of default tag values can greatly simplify the job of the analyst. The architect should consult with the analyst teams to find

System generation

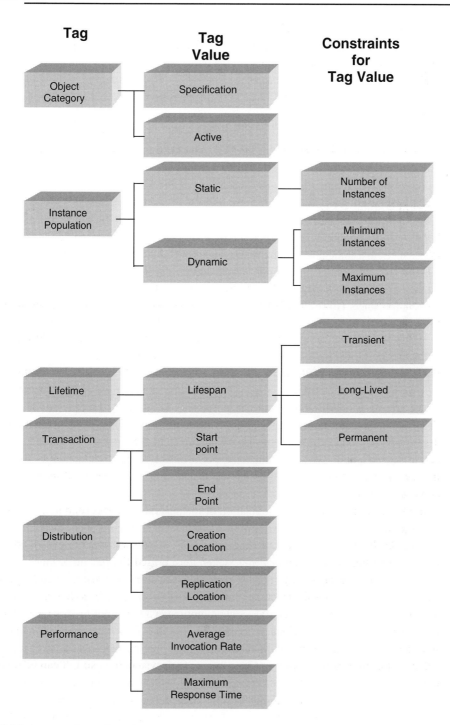

Fig. 13.14 An example tagging scheme

the most common value for default tags, although this should be tempered with the desire that the default value is 'safe'; for instance, with the population case it is always safe to make the population 'dynamic' in so much as a static instance population would have redundant services to create and delete instances.

13.8.2 Tagging versus process model exploration

In the simple example described, we relied upon the analysts to set a tag on classes, describing whether their instance populations are static or dynamic. In actual fact, this could be discovered from the ASL that has been written for the process models. The translator could look at all the ASL (barring initialisation segments) and see if a create or delete operation was ever called for a particular class; if it is not then it must have a static instance population. It could be imagined that the translator may also be able to derive the number of instances in a static population, by looking at the initialisation files that create the initial population (although this problem may not be generally tractable). It is thus conceivable to move some of the onus from the analyst ascribing tags to model elements to a process model exploration stage. However, this option will involve significantly increased complexity in the translator.

13.9 Optimizing the architecture

There are many optimizations that can potentially be applied to all classes of architecture, such as tuning mechanisms and applying more sophisticated code generation rules. Tools such as code profilers can aid the architect in this task. There are, however, a whole host of possibilities for enhancing performance of the architecture by examining or even mutating the source analysis models in a systematic, rule-driven way prior to code generation.

A simple form of such pre-optimization of the analysis model is to look for any model elements that are not used by the current process models. For instance, for each attribute of a class, we can see if it is both read and written to (this does not apply, of course, to attributes that form part of an identifier, which must only be written on instance creation). If an attribute is never written (outside the context of instance creations) then we would not choose to create a write accessor for it. If it is never read, then we could eliminate it altogether. There may also be cases where entire classes are not used, so again we can eliminate any code for them (and, optionally, provide warning to the analysts that redundant classes are present in the models). This may be intentional since in an iterative development approach an early iteration may seek to generate code when only part of the model is fully specified. Classes that are not used in one build may well be used in a future one. Therefore, the 'pre-optimization' would not seek to 'fix' the model but could gently inform the analyst of the situation.

Relationships are also fertile ground for exploring model optimizations. For instance, even though the xUML assumes that a relationship may be navigated in both directions, processing within the model may only navigate it in one direction, therefore we do not have to generate the code to hold the backward references for the relationship. If a relationship is never linked then we could eliminate it altogether (checking, of course, that it is not subsequently navigated – if it is then we can raise an error at this stage, alerting the analyst of this logical error that would have normally only been found at run time). In addition, if a relationship is only linked, and then subsequently never manipulated, then we could still eliminate it, although we would also have to remove the ASL process that caused the link. This is done only in the model as presented to the code generator. The link statement is not removed from the xUML model as perceived by the analyst. Such an optimization takes us into the realms of invasive model optimization (see Section 13.9.1).

In general, this form of optimization complements pre-checking of the models, allowing warnings of any potentially illogical or unused constructs to be passed back to the analysts.

13.9.1 Invasive model optimization

It is possible to write processing within the translator that mutates the analysis models prior to code generation. The notion initially strikes one as bizarre, since we have emphasized that the model is the code! However, this approach, used judiciously, can provide scope for considerable optimization. However, this is usually at the cost of greatly increased translator complexity. We shall just touch upon this subject to give a flavour of the possibilities that await the bold architect!

As a simple example of such a mutation-optimization, let us revisit the ATC example given in Figure 13.3. Recall that an aeroplane must have a specification. We could produce a denormalized form of this, where we embed the specification class in each AIRCRAFT class. This eliminates the association Rl and makes any code that requires information from the aircraft specification faster (at the cost, of course, of increased size due to the repetition of the specification facts in each AIRCRAFT instance). The optimizing translator must also change any ASL that refers to association Rl, in order to reflect that the specification class is now part of AIRCRAFT. For instance, if we have some processing defined within the AIRCRAFT class where one of its processes is to check the maximum height for this type of aeroplane, then we might have the following ASL:

```
theAircraftSpec = this -> Rl
theMaxHeight = theAircraftSpec.maxHeight
```

The optimizing translator would have to change this to:

```
theMaxHeight = this.maxHeight
```

This ASL realizes that the navigation of the Rl association is not now required and that the read accessor on the specification class has now become a read accessor on the AIRCRAFT class.

This form of optimization can become very complex if the subsumed class has associations other than Rl; in this case the translator must spot where other classes try to navigate to this, now, non-existent class.

Another form of this optimization may be employed when we are using technologies such as an RDBMS to store class instances. The RDBMS may offer referential integrity mechanisms, such as cascade deletes, that are indicated by a declarative trigger. In essence, the RDBMS takes care of deleting instances of classes that are related to a newly deleted instance via a mandatory association. If these are used then the code generator must look for the associated 'clean-up' processing on deletion of the object that causes the cascade delete and remove it (typically unlinks of relationships). If the clean-up processing does not accord with the RDBMS's view of the clean-up (e.g. where an analyst has forgotten to delete an instance on the end of a trail of association links) then an error must be raised and the model translation process halted.

Such mutation-optimization is a prime candidate for inclusion in later versions of the architecture. It should rarely be attempted in the first version where the architect should concentrate on correctness rather than on such complex model transformations.

It should be noted that the code generator should not change any of the analysis models, as they are presented to the analysts and held in the xUML tool model repository, since apparently redundant model elements are typically introduced in iterative development as placeholders for future work. Any changes to the source xUML models must be performed by the analysts, although these can be facilitated by suitable code generation informational and warning messages. Therefore, mutation-optimizations are performed solely at code generation.

13.10 The role of the design model

In an elaborative development the design model plays a crucial role. It acts as the 'glue' between the abstract, incomplete definition of the user requirements (the analysis) and the executable implementation. The design model itself is an entanglement of both analysis and implementation concerns. It provides us with a 'picture' of the code.

In a translation approach we define generic mappings from analysis constructs to their implementation based either on xUML metamodel elements or elements of an abstract design. The instantiated design view is therefore a derived product. In essence, we have captured the translation rules that would allow us to derive the instantiated design view automatically. So this begs the question, do we require an instantiated design model at all? The short answer is no, since the product is derived we certainly do not want to maintain it alongside our other models. However, this form of secondary view is useful in two ways.

First, when producing the initial architectural design, it is often convenient to work from a reference model, or analysis model 'snippets', and produce an instantiated design model. Working at the specific design level is often considered the most efficacious way to capture design principals. Once the initial design has been validated then it is possible to generalize the design as the set of maintained mapping rules for the translator. The specific design model can then be discarded.

Secondly, we can employ reverse engineering tools to produce a pictorial representation of the code produced by the translator. This is useful to validate the translation rules (as part of architecture testing) and to provide architectural training for new users. Such views should always be derived from the implementation, remembering that the implementation itself was directly derived from the analysis models. In this way we uphold the philosophy that the analysis models and the mapping rules are the maintained artefacts and all else is derived either from the application of the mapping rules or use of a reverse engineering tool. Reverse engineering tools fit with a process that focuses on maintaining the code. Reverse engineering allows the design documentation to be kept in step with the code. In a translational process the analysis model (PIM) and the mapping rules are maintained. The code (and the design model if required) are derived from these. The code is therefore not a maintained artefact. Changing the code should be actively discouraged since the direct link between analysis models and code has then been broken. If a developer stridently asserts that it is essential that some generated code be changed then find out why since the developer may have found a genuine problem or a necessary optimization. In this case, reflect the required change in the mapping rules and tagging model as appropriate. If the situation is so specialized that it would not be cost effective to update the translation mapping (e.g. because it would never be required anywhere else), then treat the software as manually-coded and adopt the same process as for legacy code or manually-produced code.

Reverse engineering of the code, outside the context of architectural testing or training, is commonly not a useful activity and adds needlessly to the documentation overhead of the project.

We can extend the 'matrix' view of analysis and design to include the design view as well as the implementation run-time view, see Figure 13.15. This full matrix view is useful to visualize that in a translational approach, once we have fully defined the mappings on the left of the matrix and performed the analysis indicated by the instantiation arrows on the top, the other products are derived by the architecture and run-time system.

13.11 The development life cycle for the translational approach

Production of the architecture is, in many ways, similar to any software project and must be managed with a risk-driven approach. There are many techniques that we can employ to reduce and manage risk throughout the life cycle of the architecture; its creation, test, deployment and maintenance.

One of the important differences of building an architecture for a translational approach, as opposed to the design step in traditional elaboration, is that the construction of the architecture can take place in parallel with the analysis activity. Figure 13.16 shows that since the foundation of the architecture is the xUML itself, we can construct the architecture in parallel with the analysis activity. In fact, if a suitable architecture already exists, or can be purchased, then the right-hand side of this process model disappears altogether, dramatically reducing the costs and risks of the software project.

As discussed in Section 13.7, the architecture can be tested independently of the business analysis models. This allows tuning and refining the architecture before any analyst models

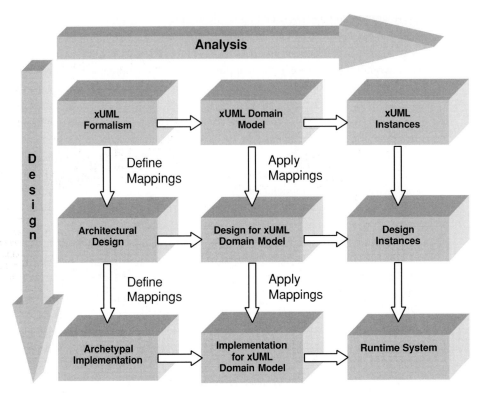

Fig. 13.15 The complete matrix view of analysis, design and run time

are available to the architecture team. Thus, we have broken away from the debilitating waterfall life cycle of the traditional elaborative approach (Figure 13.2).

The architecture definition should, of course, start with gathering pertinent requirements. A standard list of requirements to support xUML semantics may be obtained from third parties, rather than engineered from scratch using the xUML formal definition. The other constraints and performance requirements come from customer and marketing functions (among others). Once these have been sufficiently characterized, the architects should assess whether there is scope to reuse, adapt or purchase an existing xUML architecture. If this can be done then there are clear benefits in reducing risk and enhancing timescales.

The architecture team should not attempt to produce an intricate architecture as their first release. They may choose not to support some of the xUML or ASL requirements in the first release, for instance ASL set-theoretic operations, such as `union-of` or `intersection-of`, or creation and deletion of objects, if these are not required in an embedded application. In addition, requirements such as persistence, replication, security, distribution, etc. should be phased in over a number of releases. Emphasis must be placed on correctness before architecture and model optimizations are considered.

The architects and analysts must be prepared to have an ongoing dialogue. The analysts will refine requirements on performance, throughput, responsiveness and so on, whilst the

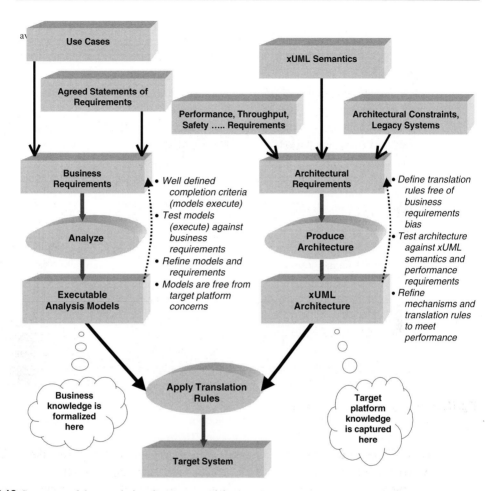

Fig. 13.16 Summary of the translational process model

architects will inform the analysts what xUML constructs are supported in a particular architectural release. Unsupported xUML features can be avoided or worked around by the analysts and any impact on the models isolated and minimized. For instance, a release of the architecture may lack persistence mechanisms, in this case the architects may supply a simple file I/O domain or suggest the use of ASL initialisation segments as a temporary workaround.

The architecture should therefore be produced as a series of iterations, adding extra capability on each release. The first iteration of the architecture is likely to be formalized in writing only (i.e. as a set of hand-coding guidelines). These are then applied to a small reference model (expressed in xUML) and the results should be critically appraised on the target platform. Once the hand-written guidelines produce code that is approaching the required production quality, then these should be formally captured by entering them in an automated code generator.

The makeup and choice of individuals for the architecture team will be critical for its success. In general, successful architecture teams are small, comprising two to four individuals. These people should be expert in xUML and also be very competent technologists in the given target platform(s). Communication is also a skill that will be required to elicit architectural requirements from customers and to brief and be briefed by the analyst teams (that are often much more numerous than the architects).

As we have already emphasized, in architectural modelling we still consider the domain as being the primary unit of reuse. Most of the domains that comprise a software architecture may already exist (they can be bought or reused from a previous project). This affects the fundamental cost structure of a software project using translation. Subsequent projects, requiring the same or similar architectural features, are considerably cheaper and less risky to produce. In order to achieve such reuse, the architects must act as publicists for the work that they have carried out and ensure that the architecture is properly documented and characterized. Coupled with reuse of analysis service domains, reuse of the architecture should greatly enhance the software teams' productivity over a number of years within an organization.

13.12 Defining the architecture – tool support

In order to effectively support the process described here, we require a toolset that allows us to populate the xUML metamodel from the 'ordinary' PIMs produced by the analysts. The next step is the ability to define xUML models of the translation rules, applied in either a formalism- or design-centric way (see Section 13.5) and finally we require a 'bootstrap' architecture to run our translation rules on. There are already such tools available, for instance, see (www.kc.com). The number will undoubtedly increase as MDA matures and grows.

13.13 Conclusion

This chapter has provided a flavour of the basis for exploiting xUML and the MDA process to produce an implementation by translating the analysis models. This has a profound effect on the development life cycle, freeing us from the many pitfalls of the standard waterfall approach of elaborative design (and its myriad of derivatives). We have not had the space to explore a full architecture but the reader should have an overall notion of how the xUML metamodel can be exploited to capture our design strategy and allow translation from PIM to PSM.

Formalizing the design in an executable fashion, by way of a code generator, means that extremely valuable intellectual property (IP) is unambiguously captured. We can exploit this IP by reusing the architecture in future developments. This has a profound positive affect on the economics and risk profile of future developments. There is an additional benefit in the way that it shields the organization from the risk that key development personnel leave with IP that has not been sufficiently characterized to allow transfer to other developers.

One of the key aspects of translational development is that the analysis models (or PIMs, to use the MDA term) are the maintained artefacts. This means that an application specified in an xUML model can be ported from one technology base to another without any change to the xUML model itself. Organizations have already achieved this using this approach. This makes it much easier for companies to adopt new technologies since they no longer have to rewrite their software but simply define how xUML maps to the new technologies. There is a second key bonus to the translational approach, since the mapping of any xUML model to a given architecture is a highly valuable and reusable item, particularly if fully automated in a code generator. This means that an organization, which has developed a code generator for a particular platform-specific architecture, can reuse it for a range of applications. This double bonus arises directly from the principle of separation of concerns, which has long been espoused in software engineering and is now finally able to deliver its full potential, since MDA, which is a translational process, maintains the separation of concerns throughout the development life cycle.

One conclusion that may be erroneously drawn from this discussion is that in the future far fewer developers will be needed on software projects. It is true that the architecture team is generally much smaller than the legions of programmers that are often required on an elaborative development; however, the emphasis is now placed on building executable models, so we require individuals that are equipped with good abstraction skills and are prepared to tease out requirements from customers and fulfil them in precise analysis models.

The radical approach espoused by translation should not deflect us from producing the architecture using best software development practices, focusing on a risk-driven iterative development. This means that we start small and increase complexity and sophistication as we go. We can also successfully exploit design patterns and language idioms for our archetypes and mechanisms within the architecture so there is no need to throw out all of those software engineering texts!

In summary, the production of rigorous and testable xUML analysis models, coupled with a translational approach to system design, allows the software development industry to move from a cottage-based craft to a well-founded engineering discipline.

14 Case study

14.1 Introduction

This book is accompanied by a CD that contains an xUML model captured within an iUML database[1]. Actually the CD contains a number of example databases – this case study refers to the model contained in the `GasStation.uml` database. The model is of a simplified Gas Station – intended to be easy to comprehend, but sophisticated enough to illustrate the use of most aspects of xUML within a MDA context.

The model consists of a number of domains, each containing a number of typical xUML concepts (classes, attributes, associations, states, signal events, etc.), together with illustrations of how to specify behaviour (methods, state entry actions, etc.), using the xUML action semantic compliant language called ASL. Additionally, there are examples of counterpart relationships and bridges.

The model also contains a set of use cases with accompanying sequence diagrams showing the interactions that occur between the domains.

14.2 Summary of system requirements

The following set of requirements statements is a summary of the requirements for the Gas Station system – it should not be considered a complete statement of requirements:
- The Gas Station is divided into two main parts – the forecourt consisting of a number of pumps where fuel of various grades may be dispensed, and an accompanying shop which sells basic auto-products (screenwash, bulbs, wipers, oil, etc.), and essential food items (milk, bread, etc.), confectionary, magazines, and so on;
- Each pump is fitted with a number of nozzles, each capable of delivering only one grade of fuel supplied from a designated tank;
- Each fuel grade has a specified price. This may be changed by the system administrator to support price fluctuations but once a fuel delivery has commenced any change in price made during that delivery should not be reflected in the final cost to the customer – the advertised price at the start of the delivery is the price **throughout** the delivery;

---

[1] iUML is the xUML modelling and simulation tool from Kennedy Carter Ltd.

- When a nozzle is removed from its holster the attendant is alerted to the fact that a customer is waiting and requires the respective pump to be enabled. Once enabled, if the customer depresses the nozzle trigger then fuel delivery will commence;
- Once fuel delivery has commenced, the customer may pause the delivery by releasing the trigger on the nozzle. Depressing the trigger again will restart it. Pausing and restarting may be done as many times as required, however there is a maximum amount of time that any one pause may last, after which the delivery is automatically suspended. Resumption of the delivery after this point requires that the attendant re-enables the pump;
- When the nozzle is returned to its holster the delivery is considered to have been completed;
- Concurrent use of multiple nozzles from the same pump shall be prevented;
- An archived summary of the details of fuel deliveries, transaction payments, etc. shall be maintained, which will be available for inspection in the long term – it is anticipated that this data might be used for trend analysis, tax assessment, etc.;
- Shop items are identified using a bar code;
- Shop purchases are made by scanning the item using a bar code reader;
- If a bar code is not recognised and the purchase needs to proceed, the checkout operator will need to provide an agreed price and the item is recorded as the purchase of an unknown item;
- The system administrator may add and remove types of item to the shop's inventory;
- The system administrator may change the price of an item type;
- At the checkout a customer may purchase as many items as required in a single transaction – these items may be fuel purchases and/or shop purchases;
- A checkout transaction may be paid for in a number of ways – cash, credit card, debit card and cheque;
- For card-based purchases, credit authorisation shall be sought from the appropriate bank.

14.3 Use cases

There are eight use cases specified within the model. A use case diagram containing each of the eight use cases defined within the model is shown in Figure 14.1.

The following are short explanations of each use case:
- **Make fuel delivery** – where a customer drives into the gas station and pulls up alongside one of the pumps. The customer then removes one of the nozzles from its pump holster and inserts it into the car's tank filler tube. Removal of the nozzle from the holster alerts the attendant to a customer waiting at a pump – the attendant enables the pump and the customer depresses the nozzle trigger to pump fuel into the car's tank. Once the required amount of fuel has been delivered the customer releases the trigger and pumping stops. The customer then replaces the nozzle into its pump holster, at which point the fuel delivery is deemed to have been completed.
- **Go to checkout** – where a customer identifies the items he or she wishes to purchase and pays for them using one of a number of payment methods. A purchased item can be

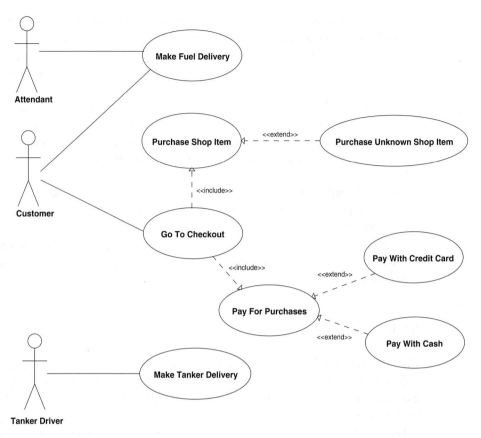

Fig. 14.1 Gas Station System use case diagram

either a fuel delivery or a shop item. In the case of a fuel delivery, the customer declares the pump number from which the delivery was made and the attendant requests a list of all the currently unpaid deliveries from that pump – this will typically be a short list (often with only a single entry). Between them, the attendant and customer identify the delivery that the customer wishes to pay for and it is added to the checkout transaction. The customer may have also elected to purchase one or more items from the shop – this aspect of the system was considered to be a discrete part of the overall system's behaviour and as such this use case **includes** the following use case:

- **Purchase shop item** – where items are scanned using a bar code reader to identify the product and its cost which is added to the checkout transaction. Occasionally the item may not be recognised (maybe due to a read failure or because the item on the shelf is not listed in the shop's inventory). This specialized extension to the use case has been modelled by **extending** it with the following use case:
- **Purchase unknown shop item** – where, if the attendant is authorised to do so, he or she can set a price and, if agreed with the customer, the item is added to the checkout transaction and entered into the system as the purchase of an unknown item.

- **Pay for purchases** – where once all the intended purchases have been added to the checkout transaction and the total cost calculated, the customer pays using one of a number of payment methods (only credit card and cash payments have been modelled). Each of these payment methods is fundamentally different and, as such, have been modelled with a separate use case which **extends** this payment use case:
  - **Pay with credit card** – where the customer elects to pay for the checkout transaction using a credit card that is swiped to read the card details. This triggers an automatic credit validation check with the appropriate bank using a modem connection. If the credit check is passed, the payment is made and the checkout transaction deemed paid for.
  - **Pay with cash** – where the customer elects to pay for the checkout transaction with cash and any change due is returned to them. Obviously this type of payment method is much simpler than paying with a credit card.
- **Make tanker delivery** – where a fuel tanker arrives at the forecourt and fills a fuel tank. A number of checks are performed to ensure that only fuel of the appropriate grade is put into the tank.

14.4 The domains

The use cases, together with the system requirement statements were used to identify the major subject matters (domains) of concern within the Gas Station system. These domains, and their mutual dependencies, have been captured in the 'Domain Model' of the Gas Station System.

The domain model is replicated in Figure 14.2 for ease of reference. A brief summary of the responsibilities of each of the domains is outlined as follows.

14.4.1 Fuel sales

This domain is a model of the requirements for managing the delivery of fuel using a fuel nozzle located at one of a number of self-service pumps situated on the forecourt of the gas station.

The domain is responsible for ensuring that only one fuel delivery at a time may be made from a given pump and for defining the behaviour of the pump when the trigger is depressed and released on a nozzle.

The domain also monitors the amount of fuel available from each of the tanks and takes all nozzles that are supplied by it out of service if the tank level falls below a specified threshold level.

Responsibility for controlling the pump hardware is delegated to another domain – hardware interfacing is not the same subject matter as fuel deliveries and thus if the pump hardware is upgraded at some point it has no effect on this domain.

Additionally, payment for a fuel delivery is treated in the same way as the purchase of items from the station shop. As such, payment of a fuel delivery is not supported within

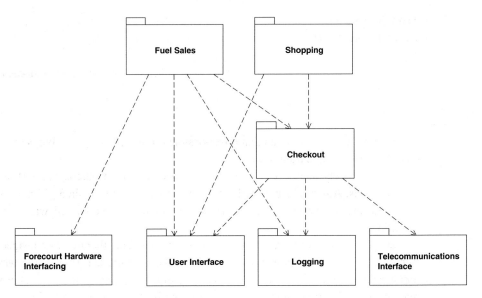

Fig. 14.2 Domain model for gas station system

this domain, rather it is delegated to a generic service domain (in the case study this is the Checkout domain).

The key class abstractions within this domain are:
- PUMP;
- NOZZLE;
- DELIVERY;
- TANK.

You will find descriptions for each of the classes, attributes, etc. within the model itself.

14.4.2 Shopping

This domain is a simplified stock control system. All items within the shop are identified using a unique bar code. When an item is purchased the stock level is decremented by one. New deliveries of items increase the stock level.

Purchases made within one shopping transaction are tracked such that purchasing trends may be subsequently analysed.

It is anticipated that bar codes may occasionally be misread or are unreadable, or that an item may be on display within the shop for which there is no record within the stock control system – as such, a single special item number is reserved for 'Unknown Items'. If an attempt to purchase such an item is made, the checkout operator must supply an agreed price for the item.

Payment for shopping transactions are delegated to a generic service domain (again, like for the fuel sales domain, this is the checkout domain).

The key class abstractions within this domain are:

- ITEM TYPE;
- PURCHASED ITEM;
- CUSTOMER.

You will find descriptions for each of the classes, attributes, etc. within the model itself.

14.4.3 Checkout

This is a generic service domain that is responsible for managing the purchases of a number of items within a single transaction.

In the case study there are two types of item that may be purchased; fuel deliveries and shop items. Any number of such items may be purchased within a given transaction, so multiple fuel deliveries and multiple shop items may be purchased within the same transaction.

Given that the two types of item that may be purchased are themselves managed within other domains (fuel deliveries within the fuel sales domain, and shop items within the shopping domain), each checkout transaction item has a counterpart relationship with its respective specialization using a counterpart generalization (see Figure 14.3).

Payment for a transaction may be made using a number of different payment methods (e.g. cash, credit card, debit card, etc.). For card-based payments, the appropriate bank authority must be sought. The mechanism for interfacing electronically to the bank is delegated to an external domain – communications interfacing is not the same subject matter as that of checkout, and thus if the mechanism for electronically communicating with the bank is changed it has no effect on this domain. In the case study, it is assumed that the means of communicating with a bank is achieved using a telephone link, but this could easily be upgraded by replacing the 'Telecommunications Interface' package and modifying the affected bridges.

The key classes within this domain are:
- TRANSACTION;
- TRANSACTION ITEM;
- OPERATOR;
- CHECKOUT DESK.

You will find descriptions for each of the classes, attributes, etc. within the model itself.

14.4.4 Telecommunications interface

This generic domain has not been provided within the case study but it should be assumed to provide a means for two computers to communicate electronically using a well-defined communications protocol.

14.4.5 Forecourt hardware interfacing

Again, this domain has not been provided within the case study model but it should be assumed that it is responsible for controlling hardware devices (pumps, actuators, switches,

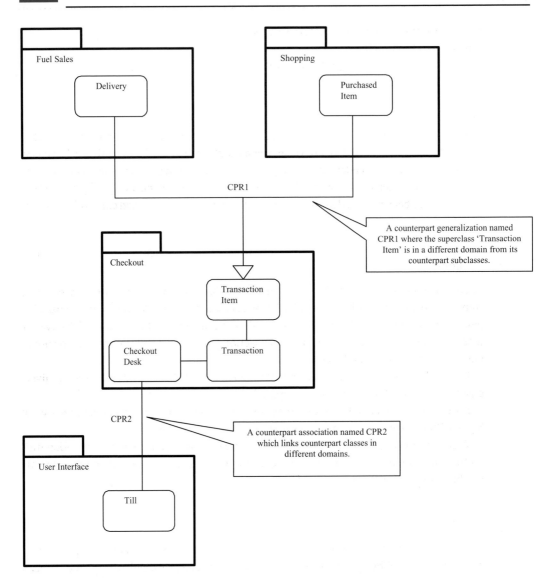

Fig. 14.3 Counterpart relationships in the gas station system

etc.), and for monitoring the state of hardware inputs (micro-switches on nozzle triggers and pump holsters, flowmeters, etc.). Changes to such inputs may be alerted to interested 'client' domains (e.g. if a customer removes a nozzle from its holster, then the respective nozzle in the fuel sales domain changes state as a result of being 'alerted' to this fact by the arrival of a signal event).

The key class abstractions in this domain might be:
- INPUT DEVICE;
- OUTPUT DEVICE;
- BINARY INPUT DEVICE;

- ANALOGUE-TO-DIGITAL CONVERTER;
- BINARY OUTPUT DEVICE;
- DIGITAL-TO-ANALOGUE CONVERTER.

14.4.6 User interface

In this case study it has been assumed that a suitable user interface can be purchased as a commercial off the shelf (COTS) item consisting of the necessary cash till hardware, keyboard(s), bar code reader, and displays. As such, it has not been analysed using xUML.

However, in order to support simulation of the gas station system, a very simple prototype test model has been provided in order to show the intended behaviour and expose the required interfaces.

14.4.7 Logging

There are a number of events that occur within the gas station system that need to be summarised and archived – things like historical records of fuel deliveries and transaction payments (where, for fuel deliveries, we might need to record details of amount delivered, date and time, type of fuel, etc. and for transaction payments we typically need a record of amount, date and time, and payment method).

Rather than build a logging mechanism for each type of thing that needs to be archived, a generic logging domain has been proposed.

It would support a number of different types of log, each capable of maintaining a set of log entries for which any number of characteristics may be stored.

This domain has not been provided within the case study model, but has been identified on the domain model to illustrate the proposed ideas – a simple example logging domain can be found in Chapter 11.

14.4.8 General

Now that you have an appreciation of the purpose of each domain it may be worth exploring some of their detail by opening up the gas station system database and having a look at each of the domains. Start with the class diagram, then perhaps the CCD, and finally some of the operations and state models. This should give you an appreciation of the structure and detail of the models – further details can be investigated if you so desire.

14.5 Characteristics of the model

This section highlights some important and desirable characteristics of an xUML model – characteristics that you might like to embed within your own models. Such characteristics are illustrated using the gas station model, so you might like to open the iUML database at this point and browse the respective parts of the model that illustrate the following points.

Let's start by focusing on the fuel sales domain – in particular the class diagram, observing the following:

- **No Controller or Manager classes** – Each class within the domain is an abstraction of part of the problem space – there are no classes that have been introduced purely to 'manage' or 'control' the behaviour of the domain. Such classes represent an undesirable characteristic which is most readily observed in models that contain a small number of classes that interact with most of the other classes, where each interaction is primarily via the invocation of 'get' and 'set' operations provided by classes that for all intents and purposes simply represent a data structure. This is hardly object-oriented. All of the 'intelligence' in such models is thus contained in these manager/controller classes rather than being apportioned to the classes with the appropriate responsibility. Additionally, the use of manager/controller classes often gives rise to bottlenecks since all threads of control have to pass through them.
- **No Factory classes** – Each class is responsible for providing one or more operations that create a new instance of that class – there are no 'factory' objects that take responsibility for creating the instances of the other (problem-oriented) classes within the model. Factory classes are platform-oriented solutions that have no place in a PIM. They can of course be justifiably employed in the PSI.
- **Association roles** – Association roles have been used at both ends of each association to describe the meaning of the relationship. You will notice that all the roles have been specified – none are assumed – and they are not merely the names of the classes involved in the association (as often illustrated and which, of course, adds no value to the model at all).

  A little time spent thinking about the roles assigned to associations can pay enormous dividends with respect to readability and understanding of the meaning of the association. The role naming style employed within the model is one where you can make meaningful sentences out of a combination of the two related class names, the destination role phrase and the destination multiplicity. For example, let's look at the association R5:

```
Nozzle isLocatedAt 1 Pump
```

...and in the other direction:

```
Pump isLocationFor 1 or more Nozzles
```

This style has been used by many modellers to clarify the semantics of the association and to make the intended abstraction more understandable. Weaker styles often fail to consider some of the subtleties of the association semantics, resulting in different interpretations in the intended abstraction, incorrect multiplicities being specified and even failure to recognise the existence of multiple associations between the same pair of classes.

- **Use of associations** – It is worth reminding ourselves of the definition of an association:

> **Method definition**
>
> An **association** is the abstraction of a relationship that holds systematically between objects.

The fact that one object may 'communicate' with another by sending it a signal event is not a valid reason to abstract an association between their respective classes. Associations are **not** there to support message passing per se, rather they are there to support the querying required within the action specifications. Of course you may need to navigate an association in order to determine which object(s) to invoke an operation on or to generate a signal event to, but such 'messages' do not **flow** along the association.

If an association is not navigated at least once within the processing contained within the model then its presence must be questioned.

- **Avoidance of domain pollution** – If we consider the classes within the fuel sales domain, the viewpoint for the abstractions underlying every class is that of Fuel Sales, therefore we get classes such as DELIVERY, FUELGRADE, PUMP etc. There is no mention of the way we control the hardware necessary to deliver fuel (that is the subject matter of the forecourt hardware interfacing domain) or the way that payments are received (the subject matter of the checkout). Each domain deals exclusively in the currency of its stated subject matter.

Now let's take a look at the way the classes and objects communicate with one another in order to achieve the domain's purpose. To do this take a look at the CCD for the fuel sales domain, and observe the following:

- **Hierarchy of control** – If the responsibilities of the domain have been distributed amongst the classes within the domain in an object-oriented fashion then a natural hierarchy of control should emerge. Some classes will have a **strategic** purpose within the domain – they will typically carry out their responsibilities by delegation (for example by invoking operations on other objects). In contrast there will be classes in the domain that have an **obedient** nature, consistent with their limited awareness of the domain responsibilities as a whole.

    Within a large domain one sees this pattern of communication successively repeated, where classes with much purpose and intelligence delegate work to less intelligent classes and often coordinate their life cycles as the work is carried out (Shlaer and Mellor, 1992).

    You can observe whether such a hierarchy of control exists by examining the communication paths defined on the CCD and checking that communication is, in general, only between classes in adjacent layers of the class hierarchy.

    A side effect of organising behaviour within the domain in this fashion is that you will minimise the coupling between the classes (see the next item) – a highly desirable characteristic for maintenance purposes.

- **Minimal coupling** – The CCD summarises the communication amongst the classes within a domain. As such, it succinctly exposes the degree of coupling[2] between those classes. Our aim is to have minimal coupling – this greatly improves the maintainability of the domain since the ripple effect of changes is correspondingly constrained.

    If, upon examining a CCD for a domain, it is found that, in general, classes interact with many other classes then it is worth reassessing the class responsibilities and checking whether an effective hierarchy of control (see the preceding point) has been established.

Now let's take a look at some of the characteristics of the state models.

---

[2] 'Coupling' here is used to refer to the degree of interactions between objects, that is the number of operation invocations and transmission of signal events.

- **Cyclic versus born-and-die** – The classic forms of state machine can be observed in the model. The NOZZLE class, from the fuel sales domain, is an example of a cyclic life cycle, typical of a model of a piece of equipment that is instantiated before the model is executed (in an initialisation segment). The DELIVERY class is an example of a hybrid between cyclic and born-and-die. It is 'born' in the `Initialised` state and can either 'die' due to the customer returning the nozzle to its holster before delivering any fuel, or simply completes (as a result of completing the delivery), and remains as an instance in the model.
- **Statechart versus state table** – Statecharts are the usual way for modellers to specify and visualize state-dependent behaviour. They are a great tool to reason about required responses to signal events. However, they have a weakness: they are poor tools when considering the completeness of the state model. If we consider the NOZZLE class – it has 14 states and responds to 14 different signal events. That means to complete the model we have to consider 14^2 different effects, not a trivial task in a graphical representation of the state model! This is where the state table comes into its own. It forms an agenda to allow us systematically to consider the effect of receiving any signal in every state. Failure to do this often results in neglecting to deal with critical state/signal combinations that ultimately result in undesirable behaviour in the implementation. Analysis of the state behaviour can only be said to be complete when the modeller has specified the desired effect for **all** state/signal combinations.

Finally, let's take a look at how the domains are integrated together using **bridges** that provide the following beneficial characteristics to the model:

- **Anonymous coupling of domains** – When one domain makes use of services provided by another domain we say that a service dependency exists between the two domains. The domain requiring the service should not know (or care) which domain provides the service or how it is implemented. Conversely, the domain providing the service should not be aware of its clients.

    By implication the client service invocation should be couched in terms of the subject matter of the client domain, and the provided service expressed in terms of the subject matter of the providing domain. For example, when in the fuel sales domain it is determined that fuel should be pumped from a specific NOZZLE (see the NOZZLE state model) and a terminator operation called `startPumping` is invoked. So far so good – everything is in the context of fuel sales.

    In the gas station system, the pumps are software-controlled through the use of input/output devices that are modelled in the forecourt hardware interfacing domain. Let's say that the impeller for each nozzle is controlled by a binary output device where, for example, setting it to '1' turns the impeller on, and resetting it to '0' turns it off. Furthermore, let's say that there are two operations provided within this domain called `setBinaryOutputDevice` and `resetBinaryOutputDevice`. Again, this is reasonable since the provided operations are couched in terms of the subject matter of hardware interfacing.

    So, we need to map the invocation of the `startPumping` operation in the fuel sales domain to the `setBinaryOutputDevice` operation provided within the forecourt hardware interfacing domain. This mapping is implemented using a bridge operation in xUML and would be specified in a build set for the gas station system (see Chapter 12 for more details).

The benefits of using bridges to specify such mappings are:

- The client and server domains are completely unaware of each other, that is they are anonymously coupled;
- The required and provided operations within each domain are defined in terms of the subject matter of the respective domain, that is the domains themselves are unpolluted by knowledge of how a required operation is actually implemented;
- The side effects of replacing a domain with an alternative are limited to changes in the bridges. For example, if the way we pump fuel is changed to some other technology, we would want to replace the forecourt hardware interfacing domain with one that deals with the subject matter of this new technology – which undoubtedly will have a different set of provided operations. Using bridges limits the effect of such a change to redefining the mapping in the respective bridges – no change will be required in the client domain.
- **Contained pollution** – As explained in Section 10.6.3.3, when integrating systems you cannot avoid polluting parts of the system with knowledge of other parts – however, using bridges you can at least contain the pollution to be within the bridges.

Bridges thus make reuse of domains much more realistic since the domains really can remain unpolluted by knowledge of other parts of the overall system.

14.6 The build set

It is recognised that system developers often build systems in an incremental fashion, adding more and more functionality until the system can be released. Additionally, it is likely that the system will be made available to customers as a series of ever-improving releases. To support this, iUML has the concept of a **build set**.

A build set is defined for each release of the system, each defining the following aspects of that release:

- **The versions of each domain included in the release** – The domain model defines which domains are required in the **completed** system – the build set defines which of those domains (and their versions) are included in the respective **release** of the system – note that not all domains are necessarily included in all releases;
- **The bridge implementations included in the release** – Each included domain typically has a number of services that it delegates to other domains – these 'required' services are modelled within that domain as terminator operations which may be implemented as a bridge method. The bridge method typically maps the required service onto one or more services (operations) provided by other domains.

Those bridges that are required to be implemented within a particular release are defined within the build set. Note that not all bridges need to be implemented in any given release – they can be stubbed out.

- **The counterpart relationships supported in the release** – Where counterpart relationships[3] are used within a system consisting of multiple domains, it is the build set that defines which counterpart associations exist within any specific release. Note that not all the intended counterpart relationships of the final system need to be defined (after all, not all the domains may be available for integration within any given release of the system). The counterpart relationships used within the gas station system are graphically represented in Figure 14.3.

You may recall that the checkout domain is a generic domain that manages the purchase of a number of transaction items within a single transaction. Each transaction item may be either a fuel delivery or an item purchased from the shop. Thus, a TRANSACTIONITEM is a generic abstraction of one of a number of specializations. In the gas station system these specializations are DELIVERY and PURCHASEDITEM, which exist in other domains.

Additionally, each checkout has an abstraction in both the user interface domain and the checkout domain, that is there are counterpart abstractions of the same real-world thing in those two domains. Linking these abstractions using a counterpart association allows behaviour defined in one abstraction to be mapped to its counterpart in the other domain. For example, let's say that when the checkout operator completes a transaction at their checkout interface, that the total value of the transaction is displayed to that operator. Using a counterpart association this is simple – you simply navigate across the counterpart association 'CPR2' from the respective instance of TILL to the counterpart instance of CHECKOUT DESK. From there it is possible to get the current transaction, and the total value is determined by summing the individual values of each of the items in that transaction. Without the counterpart association, there would need to be an alternative way of determining the counterpart instance of CHECKOUT DESK.

One could imagine that there would be several other counterpart relationships in the final version of the gas station system. For example, each Nozzle (in the fuel sales domain) might have a number of counterpart input/output devices (in the forecourt hardware interfacing domain), associated with it – a binary input device which indicates whether the nozzle trigger is depressed or not, another which indicates whether the nozzle is in its holster or not, and perhaps a binary output device for turning the associated pump on and off.

14.7 Viewing the case study models

The CD that accompanies this book includes a copy of iUML*ite* – a free version of the iUML Modeller and Simulator product available from Kennedy Carter Ltd (www.kc.com). You can use it to view and execute the case study models included on the CD. You will also find a number of additional databases supplied on the CD – see the CD Installation Instructions at the back of the book for further information.

iUML*ite* includes a set of on-line manuals that describe how to use the comprehensive set of modelling and simulation features.

---

[3] A counterpart relationship is a relationship that exists between classes in different domains.

14.7.1 Opening an iUML database

The iUML databases may be opened for viewing or editing using the supplied iUML*ite* program. To achieve this, do the following:
1. Run 'Modeller Lite' and open the required database using either the **Open Repository** icon in the toolbar, or the **File → Open Repository...** menu option, and then;
2. Select the respective database in the **Open UML Repository** selection dialog.

You can then use the model browsing features of iUML*ite* to explore the models. If you are unfamiliar with iUML*ite* it is worth spending some time exploring the basic features of the toolset by studying the iUML Tutorial (one of the manuals supplied in the iUML*ite* installation).

14.8 Executing the case study models

The following outlines the steps that need to be carried out to execute the case study models. Step 1 needs to be done only once unless you change the models and want to see the effect of your changes – in which case you will need to rebuild those parts that you have changed.

14.8.1 Step 1 Building the simulation executables

In order to execute the models contained in the Gas_Station.uml database you must first generate the simulation executables – this is known as the 'build process'. The build process involves the following:
1. Writing each domain version to the build area (the build area is simply a folder used to generate the simulation executables);
2. Building the domain version executables;
3. Writing the Build Set (contained in the 'Gas Station System' project) to the build area;
4. Building the Timer domain;
5. Building the Build Set.

Details of how to perform these operations may be found in the **Building Models for Simulation** section of the Simulator User Guide.

14.8.2 Step 2 Starting the simulation

Once the build process has been completed you will be ready to execute the case study models.

As explained earlier in this chapter, the Gas Station system consists of multiple domains integrated together using bridges. To execute the models you will need to simulate the Build Set – details of how to do this can be found in the **Starting the Simulation** section of the Simulator User Guide.

14.8.3 Step 3 Initializing the simulation

The simulation will need to be initialized before execution can begin. Initialization simply creates the initial instances of each of the classes, associations, etc. in each of the domains. So, in the case of the Fuel Sales domain, for example, after initialization there will be a number of Fuel Grade objects, Tank objects, etc.

At startup the simulator expects to run the initialization segments before it does anything else. You can execute these by using one of the control buttons in the simulator toolbar. For details on the meaning and use of each of these buttons see the **Initial Window Displayed by the iUML Graphical Simulator** section of the Simulator User Guide.

14.8.4 Step 4 Executing the models

The scenario to be executed is as follows: a customer arrives at the forecourt of the gas station and makes a fuel delivery using one of the nozzles at a particular pump. The customer then enters the shop, selects an item of confectionary and approaches the attendant in order to pay for both items.

If you have had a chance to review the models prior to executing them you will have realized that they are stimulated by the actions of the users of the system – for example, when a customer removes a nozzle from the pump holster and uses it to make a fuel delivery, or when the attendant scans an item that is to be purchased from the shop. In the absence of the real pump hardware, tills, etc. the executing models need to be stimulated in some other way. This is achieved using a set of 'test methods'. A number of these have been written already and are available for use in the simulation. Consult the **Test Methods and Timers** section of the Simulator User Guide for details on how to invoke the supplied test methods.

For the scenario outlined above the provided test methods should be executed in the order specified in Table 14.1. Whilst the model is executing you can browse the state of the executing model by examining the objects in each domain and navigating the associations to find related objects. For details on how to use these and other features of the iUML Simulator see **Features of the Graphical Interface** in the Simulator User Guide.

You might like to experiment with the invocation of further test methods – for example, the purchase of a shop item that is not known to the system.

Additionally, you can see what happens when alternative but predicted scenarios are executed. For example, during a fuel delivery the customer releases the trigger that halts the delivery of fuel, then depresses it again to resume the fuel delivery.

You might then go on to observe how the model behaves when you execute the test methods in ways in which the system is not expected to work, and to think about whether this scenario could actually occur in practice. If so, you might like to consider how the models would need to be changed in order to exhibit the desired behaviour.

Finally, you might want to enhance the models in some way – for example, by providing the functionality to support the payment of a transaction with a credit card, or providing alternative initial conditions, or alternative ways of stimulating the models.

Case study

Table 14.1 Test methods for the Gas Station system

Domain	Test Method Name	Description
Fuel sales (FUEL)	CUSTOMER Removes Nozzle 2 From Pump Holster	When a nozzle is removed from the pump holster the attendant is alerted to the fact that a customer is waiting to proceed with a fuel delivery.
Fuel sales (FUEL)	ATTENDANT Enables Pump For Nozzle 2	The attendant enables the pump thus making it possible for the fuel delivery to begin.
Fuel sales (FUEL)	CUSTOMER Delivers 20 units of fuel from Nozzle 2	The customer depresses the nozzle trigger and delivers 2.2 litres of fuel. As the units are successively delivered you should observe that the details of the fuel delivery appear on the simulated attendant interface (the Application Simulation window).
Fuel sales (FUEL)	CUSTOMER Delivers Fuel Unit for Nozzle 2	
Fuel sales (FUEL)	CUSTOMER Delivers Fuel Unit For Nozzle 2	
Fuel sales (FUEL)	CUSTOMER Releases Trigger on Nozzle 2	The customer stops the fuel delivery by releasing the trigger on the nozzle.
Fuel sales (FUEL)	CUSTOMER Replaces Nozzle 2 in Pump Holster	By placing the nozzle back in the holster, the customer completes the fuel delivery.
User interface (UI)	ATTENDANT Starts A Transaction	The customer approaches the attendant in order to pay for a fuel delivery, so the attendant starts a new transaction.
User interface (UI)	ATTENDANT Adds A Fuel Sale To A Transaction At Till 1	The attendant asks the number of the pump from which the delivery was made and gets a list of unpaid deliveries made from that pump (see the simulated attendant interface again). The attendant confirms which one the customer is paying for (if there is more than one unpaid delivery from that pump), and adds it to the transaction. At this point you will need to type the delivery number at the 'Select a delivery' prompt in the simulated attendant interface.
User interface (UI)	ATTENDANT Scans A Mars Bar At Till 1	The customer also purchases a Mars bar.

CD installation procedure

This section explains how to install:
- the iUML*Lite* software, documentation and tutorials;
- the example databases.

If you experience any problems with the installation process please contact technical support on +44 1483 226180 or by email at support@kc.com.

Step 1 Installing the iUML*Lite* software and example databases

Administrator privileges are required during the first section of the installation to allow installation of the required up-to-date Microsoft Foundation Classes in the WINNT folder. Follow the procedure as outlined below:

1. Log in as administrator;
2. Insert the distribution CD in the drive;
3. Invoke the program 'setup.exe' from the CD (if the installation procedure does not begin automatically);
4. Follow the instructions on screen.

The remaining parts of the installation procedure should be undertaken as if logged in in your usual way (this may require you logging out as administrator and logging back in again in your usual way).

In all subsequent instructions the folder in which you installed the software and example databases is referred to as the <**installation folder**>.

Step 2 Making a user copy of the example databases

In order to avoid any potential file permission problems that may result from installation of the example databases on some versions of the Windows operating system, users are encouraged to make a user copy of the example databases.

The example databases can be found in the <installation folder>\Databases folder. You should copy this folder (and all its contents) to a user-accessible part of your file system.

There are three example databases supplied:
- **Gas_Station.uml** – this iUML database contains the models referred to in the Case study described in Chapter 14. It can be viewed and edited using iUML*Lite*.

- **Simple_Gas_Station.uml** – this iUML database contains a simpler version of the Gas Station problem referred to in the iUML Tutorial that accompanies iUML*Lite*. You can use this tutorial to familiarize yourself with the features of iUML*Lite*.
- **Traffic_Lights** – this folder contains an iUML database (TrafficLights.uml), and accompanying documentation relating to a simple traffic light problem that can be used to explore the basic features of an xUML model.

If you intend to make changes to the example databases you should always use your copy rather than those installed in <installation folder>\Databases.

References

Albrecht, A. J. (1979). *Measuring Application Development Productivity*, IBM
Alexander, C. (1964). *Notes on the Synthesis of Form*, Harvard University Press
Alexander, C., M. Silverstein, S. Angel, S. Ishikawa, & D. Abrams (1975). *The Oregon Experiment*, Oxford University Press
Alexander, C., S. Ishikawa, & M. Silverstein (1977). *A Pattern Language*, Oxford University Press
Allen, R., & D. Garlan (1994). 'Formalizing architectural connection' (in the Proceedings of the 16th International Conference on Software Engineering, Sorrento, Italy, May 1994, 71–80)
Beck, K. (1996). *Smalltalk Patterns: Best Practices*, Prentice-Hall
Boehm, B. (1981). *Software Engineering Economics*, Prentice-Hall
  (1988). 'A spiral model of software development and enhancement', *Computer*, **21(5)**, 61–72
Booch, G. (1986). *Software Engineering with Ada* (2nd edn), Benjamin Cummings
  (1987). *Software components with Ada*, Benjamin Cummings
  (1991). *Object-oriented Design with Applications*, Benjamin Cummings
  (1993). *Object-oriented Analysis and Design with Applications*, Benjamin Cummings
Booch, G., I. Jacobson, & J. Rumbaugh (1998). *The Unified Modeling Language User Guide*, Addison-Wesley
British Standards Institute (2002). www.bsi.org.uk
Brooks, F. P. (1975). *The Mythical Man-month*, Addison-Wesley
Buhr, R. (1984). *System design with Ada*, Prentice-Hall
Buschmann, F., R. Meunier, & H. Rohnert (1996). *Pattern-oriented Software Architecture*, Wiley
Coad, P. (1997). *Object Models: Strategies, Patterns and Applications*, Yourdon Press Computing Series, Prentice-Hall
Cockburn, A. (2001). *Writing Effective Use Cases*, Addison-Wesley
Cook, S., & J. Daniels (1994). *Designing Object Systems: Object-oriented Modelling with Syntropy*, Prentice-Hall
Cusamano, M. A., & R. W. Selby (1997). *How Microsoft Builds Software*, Communications of the ACM 40, June 1997, 53–61
DeMarco, T. (1978). *Structured Analysis and System Specification*, Prentice-Hall
Finch, J. K. (1951). *Engineering and Western Civilization*, McGraw-Hill
Fowler, M. (1997). *Analysis Pattern: Reusable Object Models*, Addison-Wesley
Gamma, E., R. Helm, R. Johnson, & J. Vlissides (1994). *Design Patterns: Elements of Reuseable Object-oriented Software*, Addison-Wesley
Gilb, T. (1988). *Principles of Software Engineering Management*, Addison-Wesley
Gram, C., & G. Cockton (eds.) (1996). *Design Principles for Interactive Systems*, Chapman and Hall
Harel, D., & M. Politi (1998). *Modeling Reactive Systems with Statecharts: The Statemate Approach*, McGraw-Hill

Hatley, D. J., & I. A. Pirbhai (1988). *Strategies for Real-time System Specification*, Dorset House

Hopcroft, J. E., & J. D. Ullman (1979). *Introduction to Automata Theory, Languages and Computation*, Addison-Wesley

Jackson, M. (1983). *System Development*, Prentice-Hall

Jacobson, I. (1992). *Object-oriented Software Engineering: A Use Case Driven Approach*, Addison-Wesley

(1994). 'Basic use case modeling', *ROAD*, **1(2)**

Jacobson, I., G. Booch, & J. Rumbaugh (1999). *The Unified Software Development Process*, Addison-Wesley Object Technology Series, Addison-Wesley

Jones, C. (1991). *Systematic Software Development Using VDM*, Prentice-Hall

Krutchen, P. (1999). *The Rational Unified Process: An Introduction*, Addison-Wesley

McCall, J., P. Richards, & G. Walters (1977). 'Factors in software quality', NTIS, AD-A049-014, 015, 055

McMenamin, S., J. Palmer (1984). *Essential Systems Analysis*, Prentice-Hall

Martin, J., & J. Odell (1992). *Object-oriented Analysis and Design*, Prentice-Hall

(1997). *Object-oriented Methods: A Foundation*, Pearson Education

May, E., & B. Zimmer (1996). *The Evolutionary Development Model for Software, Hewlett-Packard Journal*, **47(4)**, 39–45

Mellor, S. J., & M. J. Balcer (2002). *Executable UML: A Foundation for Model-driven Architecture*, Addison-Wesley

Meyer, B. (1988). *Object-oriented Software Construction*, Prentice-Hall

Mowbray, T. J. (1997). 'The seven deadly sins of object-oriented architecture', *Object*, March 1997, 22–4

Naur, P., & B. Randell (eds.) (1969). Software engineering (report on a conference sponsored by NATO Science Committee, Garmisch, Germany), NATO

Nielsen, K., & K. Shumate (1988). *Designing Large Real-time Systems with Ada*, McGraw-Hill

Object Management Group (2001). *OMG Unified Modeling Language, v1. 4*, OMG

(2002). *UML 1.4 with Action Semantics, Final Adopted Specification*, OMG

Page-Jones, M. (1988). *Practical Guide to Structured Systems Design*, Yourdon Press Computing Series, Prentice-Hall

Potter, B., J. Sinclair, & D. Till (1996). *Introduction to Formal Specification and Z* (2nd edn), Prentice-Hall

Royce, W. W. (1970). *Managing the Development of Large Software Systems*, IEEE Wescon

Rumbaugh, J. (1999). (Private communication to one of the authors)

Rumbaugh, J., M. Blaha, W. Premerlani, F. Eddy, & W. Lorenson (1991). *Object-oriented Modeling and Design*, Prentice-Hall

Rumbaugh, J., I. Jacobson, & G. Booch (1998). *The Unified Modeling Language Reference Manual*, Addison-Wesley Object Technology Series, Addison-Wesley

Schneider, G., & J. Winters (1998). *Applying Use Cases: A Practical Guide*, Addison-Wesley

Selic, B., G. Gullekson, P. T. Ward, & J. McGee (1994). *Real-time Object-oriented Modeling*, Wiley

Shlaer, S., & S. Mellor (1988). *Object-oriented Systems Analysis: Modeling the World in Data*, Prentice-Hall

(1992). *Object Life Cycles: Modeling the World in States*, Prentice-Hall

Soley, R. (2000). *Model-driven Architecture*, Object Management Group (OMG)

Sommerville, I. (2001). *Software Engineering* (6th edn), Addison-Wesley

Sun Microsystems Inc. (2000). *Scaling the N-tier Architecture*, Sun (www.sun.com)

References

Symons, C. R. (1991). *Software Sizing and Estimating: MkII Function Point*, Wiley

Tyrrell, S. (2001). *The Many Dimensions of the Software process*, ACM Crossroads

Ward, P. T., & S. J. Mellor (1985). *Structured Development for Real-time Systems: Essential Modeling*, Prentice-Hall

Warmer, J., & A. Kleppe (1999). *The Object Constraint Language: Precise Modeling with UML*, Addison-Wesley

Wilkie, I. T., & S. J. Mellor (1999). *A Mapping from Shlaer–Mellor to UML*, Kennedy Carter Ltd/Project Technology Inc

Wilkie, I. T., A. King, M. Clarke, C. Weaver, C. Raistrick, & P. Francis (2002). *UML ASL Reference Guide, ASL Language Level 2.5 Revision D*, Kennedy Carter Ltd

Wirfs-Brock, R. (1990). *Design Object-oriented Software*, Prentice-Hall

Yourdon, E. (1978). *Modern Structured Analysis*, Prentice-Hall

Yourdon, E., & L. L. Constantine (1978). *Structured Design: Fundamentals of a Discipline of Computer Program and Systems Design*, Prentice-Hall

Index

acceptance tests, 79–80
actions *see* entry actions
action languages, 222, 225–226, 240–244
Action Semantics, 5, 32
Action Specific Language (ASL), 15, 16–17, 49, 71–73
   and associations, 218–219
   and instance handles, 216–217
   and platform independence, 221
   class manipulation, 217–218
   coding hints, 244–247
   definition, 214
   example, 220–221
   features of, 215–220
   history of, 4
   object manipulation, 217–218
   operations, 219–220
   overview, 212–215
   test methods, 239, 240
   usage hints, 247–248
active classes, 68, 151, 189
activity diagrams, 13, 32
actor generalization/specialization, 92, 93
actor inheritance, 92, 93
actors, xiii, 10, 83–84, 85, 171
actual patterns, 251
Ada language, 10–11
advanced bridges, 310–311, 315–318, 325
air traffic control case study, 101, 102, 104, 109
alarm domains, 106
analysis phase, 13–14, 27–30, 39
   artefacts, 56–59
anonymous coupling, 9, 296, 314
   Gas Station case study, 374–375
anti-patterns, 287–291, 331
application domains, 105, 109
application-level patterns, 292
archetypes, xiii, 11, 333, 349–350
architectural frameworks, 103
architecture *see* software architecture

architecture domains, 106, 109
ASL^2, 345–346
assigner patterns, 274–275
association classes, xiii, 132–133
association role phrase tense patterns, 256–257
≪association terminator≫ classes, 313–315, 319
association timeframe pattern, 256–257
associations, xiii, 121–122, 127–146
   and referential attributes, 147–148
   ASL manipulations, 218–219
   conditional, 129–130
   definition of, 128
   descriptions, 131–132
   Gas Station case study, 372–373
   multiplicity, 129, 130
   navigation, 246
   role phases, 128–129
asynchronous interactions, 167
asynchronous stimuli *see* signals
ATM case study, 83, 85, 87–88
   with an extend, 89–90
   with an include, 91, 245, 121, 124–125
attributes, xiii
   data types, 125
   in class graphics, 126
   optimizing, 356
   redundant, 147–148
   referential, 147–148
   visibility of, 153–154
audit patterns, 283

behaviour modelling, 16, 162–163
beta releases, 79
blocking contracts, 295–296, 298
Boehm model of software development, 22
Booch method, 10, 11
'born-and-die' state machines, 193–194, 374
bridge operations, xiii, 184–185, 300–311, 315–317
   invoking, 316–317
   summary, 318–319

bridges, xiii, 74–75, 297, 300, 304–311, 315–316, 318–319
   in build sets, 322, 375
build-set specific bridge implementations, 322, 375
build sets, 65, 75, 78, 321–322, 324
   Gas Station case study, 375–376

cannot happen effects, 198
class-based operations, 164, 180, 183
class collaboration diagrams, xiii, 68–71, 168–169
   and classes, 49
   and sequence diagrams, 173–174
   banking domain, 301
   construction, 173–178
   in Gas Station case study, 373
   layers, 174–178
   patient admin domain, 49
class collaboration models, xiii, 170–178
   domain interfaces, 170–172
   in train management domain, 236
class count costing method, 58
class diagrams, xiii, 15, 16, 60, 165
   for banking domain, 300
   for patient admin domain, 48
   for train management model, 227–229
   life cycle, 122
   overview, 120–122
class models, 47, 48, 109
class-scoped operations, 164
classes
   active, 151, 189
   and CCDs, 49
   and domains, 101
   and multiple associations, 133, 134
   and state charts, 48
   as xUML model layer, 34
   association classes, 132–133
   attributes, 121, 124–125
   definition of, xiii, 123–124
   descriptions, 123–124
   dynamic, 151
   graphic for, 121, 125–126, 127
   identifiers, 146–147
   interactions, 168
   operations, 180, 183, 185
   optimizing, 356
   passive, 68, 151
   pollution, 222–223, 224
   potato diagrams for, 127
   responsibilities, 175–177
   static, 151
   tabular representation, 126
client domains, 103, 104, 300–302
closed contracts, 295–296, 297–298, 311
closure notification, 295, 298

COCOMO costing method, 58
code generation, 17, 5–52, 75–77
   executable modelling, 76
   multiple platforms, 244, 245
code generators, 5, 342–347
   commercial, 62
   production of, 350–351
   testing, 351–352
cohesion, 9, 309
collaboration diagrams, xiii, 15, 166–169
   see also class collaboration diagrams
Collin's patterns, 254
comments, in ASL, 221
commercial code generators, 62
compatibility patterns, 260–265
completion transition, 194
complexity
   of models, 13
   of state machines, 206–209
component diagrams, 12, 32, 60
composite patterns, 275
conditional associations, 129–130
configuration management case study, 92–93
constraints on development, 339
construction phase, 78–79
contention resolver patterns, 274
contract specifications, 45, 46
contracts, 293
   types of, 294–296, 297–298, 310
CORBA standard, 8, 352
costing methods, 58
counterpart associations, 101, 157, 159, 313–317, 370
counterpart classes, 45–46, 311–313
counterpart generalizations, 313, 317–318, 369, 370
counterpart relationships, 312, 313, 370, 376
counterparting, 311, 341, 343
coupling, 9, 169, 186, 296, 309, 325
   in Gas Station case study, 373
cyclic state machines, 192, 374

data flow diagrams, 10
data structures, choice of, 242, 243
data types, 125
delayed signals, 195–196
deletion of objects, 194–195
deletion states, 199
deliverables, 60–61
dependencies, 43–44, 100, 104
deployment diagrams, 12, 32, 60
design, changes to, 80–81
design documentation, 37
design model, 358–359
design patterns catalogues, 330–331
design process, 338–340
design requirements, 338–339

Index

design testing, 51
developer allocation, 65–66, 75
development cycle, translational approach, 359–362
documentation, 37
   associations, 131–132
   classes, 123–124
   design, 37
   domains, 117–118
   use cases, 85–87
domain-based operations, 164, 180, 184, 185, 307
domain charts, xiii, 100, 101, 103–104, 108, 109, 117
domain level sequence diagrams, 15, 44–45, 57, 65, 111–114, 173–174
domain-scoped operations, 164
domains
   analogous, 108, 110
   and «terminator» classes, 299
   and CCDs, 49
   and class collaboration model, 170–172
   and class diagrams, 48, 66–67
   and terminators, 171–172
   as xUML model layer, 14, 34
   boundaries, 157–159
   client, 103, 104
   common behaviour in, 114
   contracts, 45, 46
   dependencies, 43–44, 104
   documenting, 117–118
   Gas Station case study, 367–371
   generic, 110–111, 112
   identifying, 108–115
   in train management model, 227–229
   initial, 57
   integration of, 74–75, 319, 321–325
   interactions between, 111–114
   interfaces, 61, 171, 172, 293
   logging, 105, 107
   low risk, 74, 78
   mission statements, 103, 117
   overview, xiii, 31–32
   partitioning, 14, 100–103, 118–119, 222
   pollution of, 222, 373
   purchased, 61
   reuse of, 115, 241
   server, 103, 104
   service, 114–115
   specifying, 43
   testing, 50
   types of, 104–108
   version roles, 117–118
   versions, 65, 321, 323, 375
dynamic classes, 151
dynamic classification patterns, 268–270
dynamic classification of classes, 138–140, 144, 145
dynamic modelling, 172–178, 188–211

effectiveness, of software engineering, 21
effects, in state transition table, 196–198, 199–200
Eiffel language, 10
elaboration phase, 59–78
   completion criteria, 78
   deliverables, 60–61
   iteration of, 64
elaborative development, 35–37, 76, 327–330
encapsulation, 170
$ENDUSE operation, 305, 316, 318
entity relationship diagrams, 10
entry actions, xiv, 188
essential model, 9
events, xiv
event-responses, 10
evolutionary development model, 22
executable descriptions, 60
executable modelling
   code generation, 76–77
   for prototyping, 76
   history of, 11–12
   PIM verification, 76
executable UML *see* xUML
expertise
   acquisition of, 39
   modelling, 6
explicit design model approach, 341, 347–349
extend relationship, use cases, 88–90, 93
external events, as actors, 83–84

feasible combinations patterns, 261
final state vertex, xiv, 195
Finch schema of engineering disciplines, 20
find operation, 246–247, 307
finite state automatons, 211
foreign classes, 323, 325
forking bridges, 308, 309
formalism centric approach, 341–347, 348–349
$FORMAT operation, 344
function point analysis costing method, 58
functional partitioning, 99–100

Gas Station case study, 17, 364–379
   anonymous coupling, 374–375
   associations, 372–373
   'born-and-die' state machines, 374
   build set, 375–376
   class collaboration diagram, 373
   counterpart associations, 370
   counterpart generalizations, 369, 370
   counterpart relationships, 370
   coupling, 373
   cyclic state machines, 374
   domain model, 368
   domain pollution, 373

Gas Station (*cont.*)
  domains, 367–371
  executing, 377–379
  hardware interfaces, 369–371
  hierarchy of control, 373
  'manager' classes, 372
  pollution, 375
  service domains, 369
  simulation, 377, 379
  state tables, 374
  statecharts, 374
  system requirements, 364–365
  telecommunications interface, 369
  test methods, 378, 379
  use case diagram, 366
  use cases, 365–367
  user interface, 370
generalization–specialization
  and inheritance, 143–144
  hierarchies, 133–140
  in xUML, 141–143
generalizing classes *see* superclasses
generic required operations, 318
goal levels, in use cases, 95–96
GUI domains, 73–74

hardware interface, Gas Station case study, 369–371
Herel state charts, 211
heuristic methods, 20
Hierarchical State Models, 10
hierarchy of control, Gas Station case study, 373
hierarchy patterns, 275–277
host platforms, testing on, 353

iCCG, 5
identifiers, xiv
  classes of, 146–147
ignore effects, 197–198
implementation domains, 51, 107–108, 109
implementation-free analysis, 28
implementation model, 9
implementation strategies, 62
improvement, as engineering goal, 21
inception phase of project, 54, 56–59
include relationships, 91, 93
infrastructure testing, 352
inheritance, 143–144, 145
inheritance-oriented relations, 317
initial costings, 58
initial risk assessment, 58–59
initial state vertex, xiv, 194
initial domain model, 57
initialisation segments, ASL, 239–240
instance creation patterns, 279–280
instance deletion patterns, 277–279

instance handles, 216–217, 218, 247
instantiations, 335–338
integration
  of domains, 74–75, 293–325
  of PIMs, 17, 45–46
Intelligent Object Oriented Analysis (I-OOA), 4
interaction diagrams, 60
inter-domain invocations, 235–238
interface classes, 49
invasive model optimization, 357–358
iterative development, 62–63
iUML, 364, 376–377, 380–381
iUML*Lite* software, 380

Kennedy Carter, 17
key letters, 183, 184
key use case set, 56, 65

layers
  in class collaboration diagrams, 174–178
  in models, 14, 34
legacy system development, 327–330
legal combinations patterns, 261
life cycles
  and MDA, 81
  of class objects, 47
links, xiv
  operations, 314, 316
lists patterns, 270
local variables, naming, 246
logging domains, 105, 107
logging patterns, 283–286
low risk domains, 74, 78

maintainability, models, 80
maintainability
  as system goal, 21
  of PIMs, 62
managed risk list, 61
'manager' classes, Gas Station case study, 372
manual implementation, 62
many-to-many-to-many association patterns, 258
mappings
  between domains, 46
  from PIMs to PSMs, 24–25, 41–42, 50–52
matrix view, 359, 360
McCall's quality factors, 21
Mealy state machines, 33, 211
meaningless effects, 199–200
mechanisms, xiv, 349
message sequence diagrams, 169
metamodels, xiv, 41–42, 326, 334–338
middleware, 22
mission statement, in domains, 103, 117
model-driven architecture (MDA)

advantages of, 2–3
definition of, 5–7
in xUML, 15
life cycles, 81
overview, 22–23
summary of, 12–13, 27, 42–43
model layers, xUML, 14, 34
model view, xUML, 13
modelling patterns, 17, 249–292
modelling, dynamic, 188–211
modelling expertise, 6
modelling, operations, 179–187
models, xiv
    as prime definitions, 39–40
    in MDA, 23
    maintainability of, 80
Moore machines, 33, 68, 210
Mowbray patterns model, 249–250
multiple associations, classes, 133, 134
multiple case use diagrams, 91
multiple classification patterns, 265–268
multiple classification, 135, 136, 140, 144
multiple platforms, code generation, 244, 245
multiplicity, xiv, 129, 130
multivalued associative patterns, 258–260
mutation, 356, 357–358

names
    in class graphics, 126
    of local variables, 246
navigating associations, 213, 246, 316
non-blocking contracts, 295, 297–298, 311
non-existent state, 200
normalization, 149–151, 152
N-tier architectural frameworks, 103

object blitzes, 57, 108, 122, 154–161
object level collaboration diagrams, 166–168
object level sequence diagrams, 57, 166–167, 169
Object Management Architecture, 8
Object Management Technique, 11
`object` operation, 307
object tables, 126, 144–146
object-based operations, 164
object-level sequence diagram, 46
object-oriented methods, history, 4, 10
objects
    coupling, 169
    creating, 194, 244–245
    deleting, 194–195, 246
    interactions between, 166–168, 169
    manipulation, in ASL, 217–218
    operations on, 181–183, 185
object-scoped operations, 164
OCL language, 10

OMG
    history, 7–8
    Object Management Architecture, 8
    Precise Action Semantics, 49
    Request for Proposal on Action Semantics for the UML, 225
    website, 17
Object Modelling Technique, 4
open contracts, 294, 295, 297
operations, xiv, 14, 16, 34, 162, 164–166, 179–187
    compartment, 126, 165
    describing, 71–73
    in ASL, 219–220
    in train management model, 234–235
    names, 183, 184
    numbers, 183, 184
    types of, 179–181
    visibility of, 186, 187
optimization
    of relationships, 356–357
    of system, 51, 242–244
ordered items patterns, 270–272
orthogonal properties patterns, 265
outside-in modelling, 10

package diagrams, 15, 321
packages, 15, 100
paradigms in design, 340
partitioning
    of systems, 99–119
    of state machines, 206–209
    of subject matter, 1q00–103
    procedural hints, 160–161
    scalable, 30–32
passive classes, 68, 151
patterns, xiv, 10–11, 249–292
    application-level, 292
    design patterns catalogues, 330–331
    genesis of, 291
    levels, 249–250
    for plant control, 286–287
    usage hints, 287–291
performance requirements, 338–339
    and use cases, 96–97
    testing, 352
personnel roles, in object blitzes, 159–160
pervasive services, 103
pipe and filter architectural frameworks, 103
plant control patterns, 286–287
Platform-Specific Implementation, 24
Process Specification Language (PSL), 4
platform-specific models (PSMs), 24–25
platform-independent models (PIMs), 23–25, 60, 120
    building, 46–50, 151–152
    integration, 17, 45–46

platform-independent (*cont.*)
  maintenance, 62
  mappings from, 17
  non-redundant, 149–151
  testing, 50, 61
  verification, 76
platforms
  definition of, xiv, 23
  independence, 6–7, 221
  see also host platforms *and* target platforms
pollution
  control of, 221–224
  example, 222–223, 224
  in Gas Station case study, 375
polymorphic behaviour, 165
polymorphic operations, 185–186
polymorphic signals, xiv, 185, 204–206, 210
potato diagrams, 127, 132
predictability, 21
presentational patterns, 251
primary actors, 83
primary scenarios, xiv, 64–65, 86–87
process activation tables, 10
process input/output domains, 105, 106
processes, 19
programming by contract, 10
project life cycles, 16
project vision document, 54, 55, 56
prototyping, in executable modelling, 76
provided interfaces, xiv, 171, 299, 308–309
provided services, 293, 298–299
purchased domains, 61

quality, 21
quantitative models, 61

rapid prototyping, 22
Rational Unified Process (RUP), 16, 53–54
real world domains, 101, 105
real-time systems, 10, 12, 68, 287
recursive design approach, 12
redundant attributes, 147–148
referential attributes, 147–148
reflexive transition, 197
regression testing, 75
repeatability, 21
Request for Proposal on Action Semantics for the UML, 225
required interfaces, xiv, 172, 297, 299, 308–309
required services, 293, 296–298
requirements
  definition of, 56
  of design, 338–339
  changes to, 80

natural language specification for, 97
  see also use cases
requirements authority, 56
resource allocation domains, 105
resource allocator patterns, 274
resource requester patterns, 272–4
responsibilities, 170, 175–177
reuse
  of domains, 40–41, 43, 61, 115, 241
  of expertise, 39
  of mappings, 25, 326–327
reverse engineering, 37, 329–330, 358–359
risk assessment, initial, 58–59
robustness requirements, 339
role migration, 139–140
role phases, 128–129
role subclasses, 139, 206–209
rolenames, 128
ROOM method, 12
round-trip engineering, 80
Royce formalization, 21–22

scalable partitioning, 30–32
scenario specification versions, 322
scenarios, xiv, 86–87
  see also primary scenarios *and* secondary scenarios
secondary actors, 83
secondary scenarios, 65, 78, 86–87
self-directed signals, 202–203
semantic patterns, 251
semantic shifts, 157
sequence diagrams, xiv, 57
  see also domain level sequence diagrams *and* object level sequence diagrams
server domains, 103, 194, 300–302
service domains, 105–106, 109, 114–115, 369
Shlaer–Mellor models, 4
signals, 47, 164–165
  definition, xiv, 188
  delayed, 195–196
  in class graphic, 165
  self-directed, 202–203
  transmission rules, 203–204, 206
simulation, 238, 377, 379
software architecture, 38–39
  and code generation, 75–77
  and xUML^2, 342
  history of, 8–12
  optimizing, 356–358
  testing, 351–353
software development, 20–22, 26–27
software engineering, 20–22
≪specialization terminator≫ classes, 318, 319
specification classes, 141–142
specification patterns, 251–254

specific–generic relationships, 319
spider state models, 290–291
spiral model of software development, 22
staff skill analogy costing method, 58
stakeholders list, 55, 56
state actions, 71–73, 231–234
state-dependent behaviour, 16, 162–163
state diagrams, 10
state-independent behaviour, 16, 162–163
state machines, 33, 189
   'born-and-die', 193–194, 374
   complexity of, 206–209
   cyclic, 192, 374
   design hints, 209–210
   execution, 200–201
   Mealy machines, 33, 211
state modelling, 10, 16, 188
state models, 47, 162–163, 165–166
state-tables, 47
state transition diagrams, 13
state transition tables, 13, 15, 196–200, 374
statecharts, xiv, 13, 15, 33, 47, 60, 69
   design hints, 191–192
   examples, 48, 69, 232, 374
   overview, 189–191, 192
   use of, 68
states, xv, 14, 34, 188
static classes, 151
stereotypes, xv
Structure Charts, 9, 10
Structured Analysis, 3–4, 9–10
Structured Design, 9
Structured Programming, 9
stubbed bridge implementations, 322
subclasses, 135–140, 141, 142
subject matter partitioning, 100–103
subtype-migration patterns, 268
success guarantee, 86
superclasses, 140–141
   and state machines, 206–207
   operations, 185–186
synchonize and stabilize development model, 22
synchronous interactions, 47, 164, 167, 185
Syntropy method, 11
system generation, 326–363
system level class diagrams, 156–159
system level object blitz, 108, 154–161
system optimization, 242–244
system partitioning, 16
system requirements, Gas Station case study, 364–365
system versions, 321

tabular representation, classes, 126
tagging, 346, 353–356

target language pollution, 223
target platforms, testing on, 353
telecommunications interface, Gas Station case study, 369
≪terminator≫ classes, 166, 186, 297, 299, 319
   and bridge operations, 185
   in banking domain, 301–302
   see also terminators
terminators, xv, 49, 171–172
   see also ≪terminator≫ classes
test methods, ASL, 239, 240
   Gas Station case study, 378, 379
testing, 351–353
   acceptance, 79–80
   in legacy development, 329
   of construction phase, 79
   PIMs, 50, 61
   system, 51–52
textual requirements statements, 57
thread count costing method, 58
three-tier architectural frameworks, 103
timing, 195–196
'to do' list patterns, 280
top-down modelling, 10
tracking, 21
Traffic Lights case study, 381
train management model, 226–238
   class collaboration model, 236
   class diagram, 227–229
   domain chart, 227
   domains, 175–178, 227–229
   operations, 234–235
   state actions, 231–234
   statechart, 232
transition effects, 197
transition phase, 79–80
transitions, xv, 188, 197
   definition of, xv, 188
   reflexive, 197
translation-driven development, 331–338
   benefits, 37–40
   development cycle, 359–362
   example, 331–334
tree patterns, 275
type patterns, 251
types of values, 125, 247
Tyrrell criteria for engineering goals, 21

UML
   as basis for xUML, 12–15
   Virtual Machine, 6–7, 340, 350, 352
unlink operation, 314, 316
unmodelled domains, 73–74
unordered operations patterns, 280–283

usage hints
  PIMs, 151–152
  use cases, 94
use case diagrams, 84–85, 89, 91
use case-based costings, 58
use case realizations, 66–67
use case scenarios, 61
  see also primary scenarios *and* secondary scenarios
use cases, xv, 13, 15, 16, 82–83
  abstract, 94
  and domains, 46–47
  and project requirements, 56, 57, 94
  ATM example, 87–88, 89–90
  concrete, 94
  configuration management example, 92–93
  descriptions, 66, 86–87
  function of, 82–83
  goal levels, 95–96
  history of, 4, 11
  identifying, 44–45, 83–84
  in Gas Station case study, 365–367
  modelling, 82–98
  testing, 50
  usage hints, 94
  with extend relationship, 88–90, 93
  with include relationship, 91, 93
$USE operation, 305, 316, 318

user defined types, 125
user interface, Gas Station case study, 370

VDM language, 10
verification, 238
viewpoint authorities, 56
viewpoints, 56
visibility, xv
  of attributes, 153–154
  of operations, 186, 187

waterfall development, 21–22, 63, 328
websites, 17
weighting costing method, 58

xUML
  and elaborative development, 330
  and system generation, 50–52, 326, 331–338, 340–363
  and UML, 12–15
  definition of, 19, 25–26
  model layers, 14, 34
  notations used, 15, 32–34
xUML^2, 335, 340–346

Yourdon models, 9

Z language, 10